Microbiology: Yeast and Fungi

Microbiology: Yeast and Fungi

Edited by **Dean Watson**

SYRAWOOD
PUBLISHING HOUSE

New York

Published by Syrawood Publishing House,
750 Third Avenue, 9th Floor,
New York, NY 10017, USA
www.syrawoodpublishinghouse.com

Microbiology: Yeast and Fungi
Edited by Dean Watson

International Standard Book Number: 978-1-68286-148-6 (Hardback)

Printed in the United States of America.

Contents

Preface

The world is advancing at a fast pace like never before. Therefore, the need is to keep up with the latest developments. This book was an idea that came to fruition when the specialists in the area realized the need to coordinate together and document essential themes in the subject. That's when I was requested to be the editor. Editing this book has been an honour as it brings together diverse authors researching on different streams of the field. The book collates essential materials contributed by veterans in the area which can be utilized by students and researchers alike.

This book contains some path-breaking studies in the field of microbiology with particular emphasis on yeast and fungi. It traces the progress of this field and highlights some of its key concepts and applications. The chapters included in this book focus on structures and classification of yeast and fungi, their physiology, applications of yeast and fungi in different fields that of utmost significance and bound to provide incredible insights to readers. Some of the diverse topics covered in this book address the varied branches that fall under this category. The book aims to provide a comprehensive overview of the discipline as well as the recent developments and researches in the field of yeast and fungal science.

Each chapter is a sole-standing publication that reflects each author's interpretation. Thus, the book displays a multi-facetted picture of our current understanding of applications and diverse aspects of the field. I would like to thank the contributors of this book and my family for their endless support.

Editor

Molecular cloning and characterization of *STL1* gene of *Debaryomyces hansenii*

Juan Carlos González-Hernández

Laboratorio de Bioquímica del Departamento de Ingeniería Bioquímica, Instituto Tecnológico de Morelia; Avenida Tecnológico 1500. C. P. 58120. Morelia, Michoacán, México. E-mail: jcgh1974@yahoo.com.

Debaryomyces hansenii is often found in salty environments. This yeast species is not only halotolerant, but also halophilic. Its genome sequence is known completely, but the mechanisms behind its halotolerance are poorly understood. It was compared to the STL1 protein sequence of *Saccharomyces cerevisiae* against the translated sequence from the *D. hansenii* genome sequence database released by Génolevure. An ORF (DEHA0E01122g) was found with 54% homology and 39% identity with Stl1p from *S. cerevisiae*. *DhSTL1* was heterologously expressed successfully in a *S. cerevisiae* (BY4741) wild type and in another strain lacking its own system for the glycerol transport (*STL1*) gene. The *DhSTL1* gene in transformed *S. cerevisiae* strains showed a slight but significant difference in the doubling times in growth curves obtained in liquid YNB-ura medium, with glycerol as carbon source. *DhSTL1* gene in transformed *stl1* yeast strain showed phenotype growth at pH 7.5 under salt stress conditions (glucose as carbon source). The kinetic parameters of transport and glycerol accumulation conferred by *DhSTL1* in the *S. cerevisiae* transformant strains did not show significant differences. An increase in the transcript level of *DhSTL1* gene in the presence of saline stress at pH 5.6; whereas, at 7.5 pH, it was expressed in all evaluated conditions.

Key words: *Debaryomyces hansenii*, glycerol, transport, salt tolerance.

INTRODUCTION

In yeasts, defense responses to salt stress are based on osmotic adjustment by osmolyte synthesis and cation transport systems for sodium exclusion. Polyols, and especially glycerol, are the major osmolytes produced by yeasts (Blomberg and Adler, 1992). In *Saccharomyces cerevisiae*, physiological and molecular studies have previously shown the presence of an active uptake system driven by proton motive force (Holst et al., 2000), only operative when glycerol is the carbon and energy source, this was the first report on a gene product involved in active transport of glycerol in yeasts. According to previous studies in other yeast species, more halotolerant than *S. cerevisiae*, like *Pichia sorbitophila* (Lages and Lucas, 1995), *Zygosaccharomyces rouxi* (Zyl et al., 1990) or *Debaryomyces hansenii* (Lucas et al., 1990), glycerol was found to be actively transported, establishing and

maintaining a glycerol gradient in the presence of high salt concentrations and counterbalancing glycerol's natural leakage.

Ferreira et al. (2005) identified genes involved in active glycerol uptake in *S. cerevisiae* by screening a deletion mutant collection comprising 321 strains. They found that deletion of *STL1*, which encodes a member of the sugar transporter family, eliminates active glycerol transport. Stl1p is present in the plasma membrane of *S. cerevisiae* during conditions in which glycerol symport is functional. Both the Stl1p and the active glycerol transport are subject to glucose-induced inactivation. These last authors concluded that, the glycerol proton symporter in *S. cerevisiae* is encoded by *STL1*.

When *D. hansenii* was subjected to increased NaCl stress, there was a decrease of intracellular K^+ and an increase of intracellular Na^+ (Norkrans, 1968; Norkrans and Kylin, 1969). However, the total salt level in the cells was not sufficient to balance the water potential of the medium. For this reason, additional osmotically active solutes must be accumulated (Chen and Wadso, 1982). Many eukaryotic microorganisms accumulate polyols when exposed to osmotic stress (Adler et al., 1985; Brown, 1978), and, in *D. hansenii*, a positive correlation was demonstrated between the intracellular glycerol level and the salinity of the surrounding medium (Adler et al., 1985; Andre et al., 1988; González-Hernández et al., 2005).

In addition, tolerance to a sudden osmotic dehydration shock is higher in cells carrying an increased amount of intracellular polyols (Adler and Gustafsson, 1980). The two polyols produced and accumulated by *D. hansenii* under saline stress are glycerol, which is the major internal solute in exponentially growing cells and arabinitol, which predominates in stationary-phase cells (Adler and Gustafsson, 1980; González-Hernández et al., 2005).

In *D. hansenii* a striking relationship between the intracellular glycerol concentration and the metabolic activity, measured as heat production rate, has been reported when cells were grown in presence of 2.7 M NaCl (Gustaffson, 1979), suggesting a specific role for glycerol when *D. hansenii* is exposed to these conditions. Determinations of intracellular glycerol levels with respect to the osmotic volumes revealed that increases in intracellular glycerol may counterbalance up to 95% of the external osmotic pressure due to added NaCl (Reed, 1987).

Lucas et al. (1990) reported that, NaCl concentration and glycerol accumulation are linked through a putative sodium-glycerol symport that uses the sodium gradient as a driving force for maintaining the glycerol gradient.

This transporter would also accept potassium as co-substrate. Furthermore, it has been observed that, the glycerol uptake is accompanied by proton uptake when extracellular NaCl is present and that the protonophore (CCCP) induces collapse of the glycerol gradient, supporting earlier proposals that the intracellular Na^+ concentration is kept lower by an active Na^+-H^+ exchange mechanism (Luyten et al., 1995).

It is also interesting that, this halotolerant-halophilic yeast has the natural capacity to adapt to high salt concentrations, Na^+ being accumulated without producing any apparent toxicity. On the contrary, cells show some better functional characteristics in the presence of NaCl (González-Hernández et al., 2004).

Yeasts have revealed along the years that, the information contained in their genes may be successfully transferred to other microorganisms, plants and other higher eukaryotes, resulting in the improvement of resistance capabilities (Domínguez, 1998). Thus, heterologous expression of genes associated with salt stress resistance remains one of the most important goals in salt stress research. However, the mechanisms underlying long-term resistance to high salt concentrations are yet poorly understood. Considering the important role of glycerol in osmoregulation, described for *D. hansenii*, it is of great importance to characterize the glycerol transport from both molecular and physiological points of view, in order to understand the mechanisms underlying salt resistance in this yeast. Here, describing with the available tools, the molecular and physiological characterization of a glycerol transporter from this halophilic and alkaline tolerant yeast.

MATERIALS AND METHODS

Yeasts strains and bacteria

The type strains *D. hansenii* PYC2968 (CBS767) and *S. cerevisiae* BY4741 (MAT a; his3Δ1; leu2Δ0; met15Δ0; ura3Δ0) were used as genetic background strains; *S. cerevisiae* Y05831 (BY4741Mat a; his3Δ1; leu2Δ0; met15Δ0; ura3Δ0; YDR536w::kanMX4) was obtained from EUROSCARF consortium (Table 1). Competent cells of *Escherichia coli* XL1-Blue (Invitrogen) were also used for plasmid (Yep352) selection and propagation (Hanahan, 1985). JCGHZERO is the wild type with the empty plasmid, JCGHSTL1 is the wild type with the plasmid plus *DhSTL1*, JCGH*zero* is the *STL1* mutant with the empty plasmid, and JCGH*stl1* is also the mutant with the plasmid plus the *DhSTL1* gene.

Growth media

D. hansenii strain was routinely maintained in YPD medium (10 g yeast extract, 10 g peptone, 20 g glucose and 20 g agar per liter). The transformant strains were routinely maintained on solid YNB-ura (w/o amino acids) medium with 2% glucose, supplemented, when required, with the adequate requirements for prototrophic growth (Pronk, 2002). *E. coli* XL1-Blue strain was routinely maintained in Luria-Bertani medium (LB) at 37 °C; ampicillin (100 µg ml⁻¹) and 5-bromo-4-chloro-3-indolyl-ß-D-galactopyranoside (X-Gal, 4 µg ml-1) were used as supplements (Sambrook, 1989) when required.

Cloning strategy and DNA manipulations

Sequence data for *D. hansenii* were obtained from Génolevures Consortium website (Génolevures, 2001), by performing tblastn search with the Stl1p sequence from *S. cerevisiae* against the *D. hansenii* genomic sequence. Using the BLASTP 2.2.14 program (NCBI, Bethesda, MD, USA) (Altschul et al., 1997), ORF revealing homology to *S. cerevisiae* Stl1p protein were identified. Based on the nucleotide sequences of this ORF together with the contiguous upstream and downstream regions, primers were designed to amplify a region comprised between approximately -1000 bp from the ATG start codon and +200 bp after the TAA stop codon of the gene. Specific primers designed and modified to incorporate a restriction site (underlined) for *Bam* HI forward primer (5'-CGGGATCCCGTTTTGTCTTTGCTGACTCCC-3'), and for reverse primer, *Pst* 1; (5'-AACTGCAGAACCAATGCATTGGTCACGGTTCAAGTGTCTTAAA-3', to amplify STL1 ORF and the flanking regions (-1000 bp, +200 bp). PCR amplification was carried out in an Eppendorf thermocycler with DNA polymerase from BIOTOOLS, for 30 cycles, at 64 °C (annealing temperature chosen according to the primer characteristics). Using this approach, one fragment of 2.6 kb was

Table 1. List of bacteria, yeasts strains, and plasmids used in the present work

Plasmids and strains	Characteristics	Source
Bacteria		
Escherichia coli DH5α		(Hanahan, 1985)
Strains		
D. hansenii (CBS767)	Type strain	Portuguese yeast culture collection
S. cerevisiae (BY4741) (wild type)	MAT a; his3Δ1; leu2Δ0; met15Δ0; ura3Δ0	EUROSCARF
S. cerevisiae (BY4741) (*stl1*)	BY4741Mat a; his3Δ1; leu2Δ0; met15Δ0; ura3Δ0; YDR536w:: kanMX4	EUROSCARF
JCGHZERO	BY4741 (YEp352 empty)	This work
JCGHSTL1	BY4741 (JCGHpSTL1)	This work
JCGH*zero*	BY4741 *stl1* (Yep352 empty)	This work
JCGH*stl1*	BY4741 *stl1* (JCGHpSTL1)	This work
Plasmid		
YEp352	Yeast episomal vector, 2 µm, URA3 yeast marker and Amp[R]	(Hill et al., 1986)
JCGHpSTL1	YEp352 derivative containing *DhSTL1* gene	This work

obtained, using *D. hansenii* CBS 767 genomic DNA as a template. The amplified products were digested with *ECoR* I and *Hind* III, purified using the purification kit "GFX PCR DNA and Gel Band Purification" (GE Healthcare). The fragment was cloned into the XL1-Blue *E. coli* strain (YEp352, see Table 1), characterized at the molecular level, and used to transform a wild type *S. cerevisiae* strain and *stl1 S. cerevisiae* strain by the lithium acetate method (Geitz and Schiestl, 1995). Current plasmid isolation was performed by alkaline extraction as described in Birnboim and Doly (1979), modified as in Sambrook et al. (1989). For plasmid isolation from yeasts, the procedure described by Hoffman and Winston (1987) was followed. Agarose gel electrophoresis and restriction site mapping were performed according to standard methods (Sambrook, 1989). Yeast genomic DNA from *D. hansenii* was isolated (Cryer, 1975), after a previous treatment with liticase. Yeast genomic pDNA was isolated using QIAprep Spin Miniprep Kit Protocol according to manufacturer's directions. Constructs were named as: JCGHZERO for plasmids without *DhSTL1*; JCGHSTL1, for plasmids containing the *DhSTL1* gene (Table 1).

Transformants were selected on minimal medium with methionine, leucine, and histidine; 40 transformants (JCGHZERO), 147 transformants (JCGHSTL1), 48 transformants (JCGH*zero*), and 119 transformants (JCGH*stl1*) were obtained. One representative clone from each transformant was used for heterologous expression studies.

Isolation of RNA and northern blot analysis

Total RNA was extracted from exponential phase yeast cells after 6 h of incubation under saline stress conditions. Cells were then collected by centrifugation, frozen using liquid nitrogen and kept at -80℃ until RNA extraction was performed. Total RNA was extracted by the hot phenol extraction protocol (Schmitt et al., 1990), modified by Daniela Castro (personal communication), as described below. Frozen cells were re-suspended in 470 µl of 100 mM sodium acetate, pH 5.0, 5 mM $MgCl_2$ plus 1/10 volume of 10% SDS (w/v), 5 µl DEPC, and 500 µl glass beads, and vortexed for 1 min. After vortexing, cells were subjected to three hot phenol extractions

(5 min at 65℃) with 1 volume of phenol:chloroform:isoamyl alcohol (25:24:1), pH 5.0, and one extraction with 1 volume of chloroform : isoamyl alcohol at room temperature. RNA precipitation was performed as described by Schmitt et al. (1990). Total RNA was fractionated through formaldehyde-agarose gels and transferred to N[+]-Hybond membranes (GE Healthcare). Hybridization was performed with a digoxygenin (DIG)-labelled probe prepared from an internal fragment of *DhSTL1* labelled using the DIG system (Roche) by random priming, according to manufacturer's instructions. Hybridizations were performed in DIG Easy Hyb (Roche), at 50℃ (16 h). Membranes were then washed under high-stringency conditions and exposed to X-ray films for a maximum of 48 h.

Growth assays

Specific growth rates were determined in liquid YNB-ura medium (2% (w/v) of glucose or glycerol) starting from a 50 ml pre-inoculum of cells grown in the same medium. The pre-inoculum was grown at 28℃ for 24 h in an orbital shaker at 180 rpm and used to inoculate 250 ml flasks containing 100 ml of liquid medium at 0.2 of the initial optical density. Growth was followed by measuring the absorbance at 640 nm, with a Shimadzu spectrophotometer, model UV-160A. The ability of transformant strains to grow in the presence of different NaCl and KCl concentrations (0.6, 1.5 and 3.0 M) at two different pH (5.6 and 7.5, adjusted with Trizma base), and two carbon sources (2% (w/v) of glucose or glycerol) was assessed on solid YNB-ura media with KCl and NaCl to the desired final concentrations. Transformants were grown for 24 h in 25 ml of YNB-ura liquid medium (2% glucose) to a final density of approximately 3×10^7 ml[-1]. Plates were inoculated with serial 10-fold dilutions of these cultures and incubated at 28℃. Growth was recorded after 1 or 2 weeks (in the case of the 3 M NaCl or KCl medium).

Glycerol transport measurements

Initial uptake rates and accumulation ratios of radiolabeled glycerol were measured using the method described by Lucas et al. (1990) with modifications. Cells were grown to exponential phase (OD_{640} nm 0.4-1.0), harvested by centrifugation, washed twice, and resuspended in ice cold water to a final concentration of 0.25 g/ml and kept on ice. The cell suspension (25 µl) was mixed with 100 mM MES or MOPS buffer (pH 5.6 or 7.5, 50 µl; adjusted with Trizma base) in 10 ml conical centrifuge tubes. After 2 min at 28°C in a water bath, the reaction was started by adding 25 mM aqueous solution of D-[U-^{14}C]-glycerol (specific activity of ≈ 700 cpm/nmol) at appropriate concentrations. After 10 s, the incorporation was stopped by diluting with 5 ml of ice cold water. Cells were immediately collected on Whatman GF/C filters (Whatman, Maidstone, England) at reduced pressure, washed with 5 ml cold de-mineralized water, and immersed in vials containing 5 ml of scintillation fluid (Optiphase `Hisafe´ 2, Perkin Elmer; Life sciences manufactured by Fisher Chemical Products, England). The radiolabeled glycerol taken up by cells was measured in a scintillation counter (Beckman LS3801). All determinations were performed in triplicate and referenced to a blank made by inverting the sequence of addition of glycerol and water.

The initial uptake rates were also measured in the presence of 0.6 M of NaCl or KCl. In order to determine the transport-driven in/out accumulation ratios of ^{14}C-glycerol, 25 µl of cell suspension (0.25 g/ml) was mixed with 5 µl glucose [1 M] and 30 µl of 100 mM MES or MOPS (adjusted at pH 5.6 or 7.5 with TRIZMA base) in 10 ml conical centrifuge tubes and incubated at 28°C. After 2 min of incubation, the reaction was started by adding 40 µl (25 mM) of ^{14}C-glycerol (specific activity ≈ 700 cpm/nmol). At appropriate time intervals, 10 µl aliquots were taken and filtered through Whatman GF/C filters at reduced pressure, washed twice with 5 ml ice-cold water, and transferred to vials containing 5 ml scintillation fluid. The accumulation ratio assays were also performed in the presence of 0.6 M of NaCl or KCl. Efflux of the radiolabel upon the addition of nonradioactive glycerol at a final concentration of 1 M or of the protonophore carbonyl cyanide m-chlorophenylhydrazone (CCCP) at a final concentration of 50 µM was tested. The *D. hansenii* and *S. cerevisiae* intracellular volumes, used to calculate the intracellular glycerol concentrations were determined previously (González-Hernández et al., 2004).

Reproducibility

All assays were repeated at least three times, and the data reported are means or representative values.

RESULTS

The release of the complete genome sequence of *D. hansenii* by the Génolevure consortium, led me to search for *D. hansenii* sequences with homology to genes involved in glycerol transport in order to clone and characterize the putative transporter (s). By performing a tBlastn search, It was compared the STL1 protein sequence of *S. cerevisiae* against the translated sequence from the *D. hansenii* genome sequence database released by Génolevure. An ORF (DEHA0E01122g) was found with 54% homology and 39% identity with Stl1p from *S. cerevisiae*; located in chromosome E (anti-sense strand) and previously annotated as a gene of the sugar permease family STL1 uptake transporter protein of 510 amino acids (aa). The chromosomal regions corresponding to *DhSTL1* were amplified together with it own putative promoter and

termination regions. The amplified fragments were cloned and used to transform a wild type and a *S. cerevisiae* mutant (*stl1*). Table 1 lists plasmids and strains used for cloning this transporter.

In order to assess the possible contribution of the DhStl1p protein under study to the glycerol transport, growth of *D. hansenii* and the *S. cerevisiae* transformants was evaluated in YNB-ura liquid mineral medium without stress conditions with glucose or glycerol as carbon sources. Figure 1 illustrates the growth results obtained with the transformants. It is very clear that the presence of *STL1* gene disturbs growth on glucose in both types of transformants (Figure 1A) whereas growth on glycerol is improved (Figure 1B). Although Figure 1 only represents one set of results, they were consistent in three sets of experiments. This effect was more evident in the transformant obtained with the *stl1* mutant of *S. cerevisiae*. In this case the doubling time decreased from 55 to 35 h. Table 2 shows the doubling times of the growth curves of Figure 1; *S. cerevisiae* transformant cell grown in glucose showed similar or lower doubling time values; in the glycerol curves of *S. cerevisiae* transformants, the doubling times showed clearly the contribution of *DhSTL1* in the *S. cerevisiae* (*stl1*) strain.

Figure 2 shows the ability of *D. hansenii* and transformants strains to grow on solid YNB-ura media without salt, and the presence of NaCl and KCl (0.6 M, of each salt), at pH 5.6 (A), and 7.5 (B), and glucose as carbon source. The phenotype conferred by *DhSTL1* was evaluated, inserted into the YEp352 plasmid, and cloned into a wild type *S. cerevisiae* strain and into another lacking the *STL1* gene, obtaining JCGHZERO, JCGHSTL1, JCGH*zero*, and JCGH*stl1*. Growth of these transformants was assessed in plates with 0.6 M, 1.5 M, and 3.0 M of NaCl or KCl. The *S. cerevisiae* transformant strains and *D. hansenii* grown in presence of glucose without salt were able to grow in plates in presence of the salts (0.6 and 1.5 M). It is worth mentioning that the phenotype of Na$^+$ or K$^+$ tolerance in the *S. cerevisiae* (*stl1*) transformant was conferred by *DhSTL1* gene; slightly at pH 5.6, and clearly at pH 7.5, growth was recorded after 1 week. The plates growth in the presence of 1.5 and 3.0 M of the salts (data not shown); in the presence of 1.5 M of salts the results were similar to those obtained in the plates with 0.6 M of salts; the plates with 3.0 M of salts did not show growth of *S. cerevisiae* transformants, only *D. hansenii* had the capacity to grow after 2 weeks in presence of KCl or NaCl (3 M). In the presence of glycerol as carbon source, *D. hansenii* grew in all experimental conditions evaluated in the presence of 3.0 M of the salts, showing a slight growth after two weeks (data not shown); the *S. cerevisiae* transformant cells showed growth more clearly in YNB-ura medium without salt at both pHs; the phenotype conferred by *DhSTL1* in JCGHSTL1 transformant strain was observed in the presence of 0.6 M of salts; the other *S. cerevisiae* transformant cells did not show a clear phenotype; at 1.5 M of salt a slight growth was observed, whereas at 3.0 M

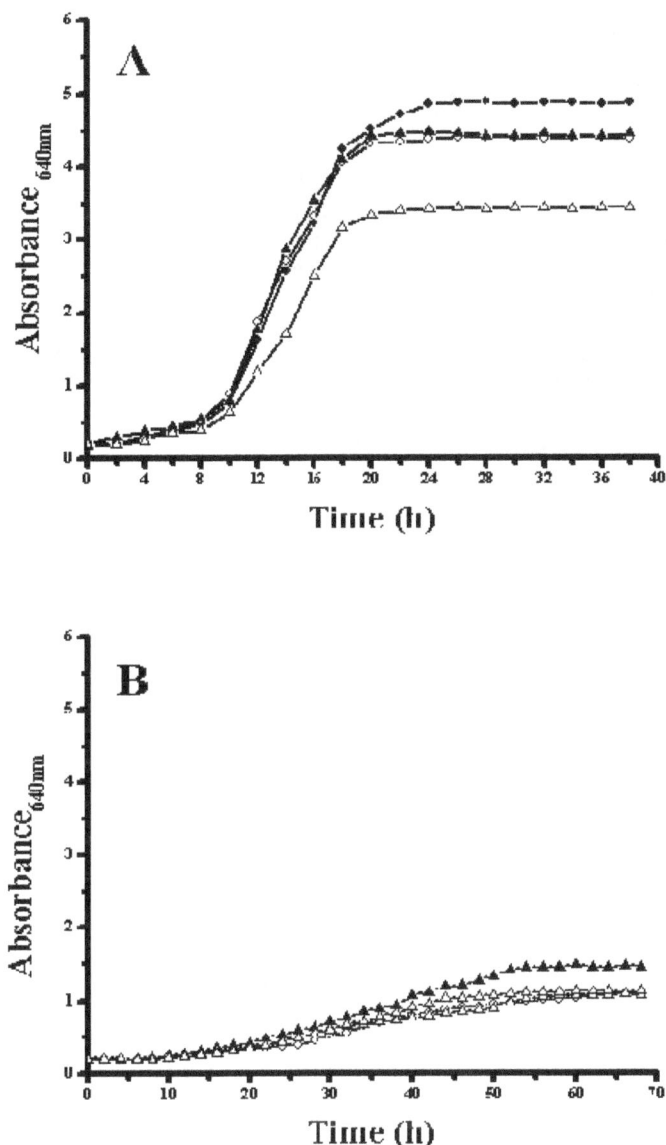

Figure 1. (Gonzalez-Hernández JC) growth of *S. cerevisiae* transformants cultivated in YNB-ura medium (2% (w/v) glucose [A]

Table 2. Doubling time of D. hansenii and S. cerevisiae transformants cultivated in YNB-ura medium (2% (w/v) glucose and 2% (w/v) glycerol).

	Doubling time (h)	
	Glucose	Glycerol
D. hansenii	9.41	9.38
JCGHZERO	2.52	32.3
JCGHSTL1	2.55	31.35
JCGH *zero*	2.72	55.42
JCGH *stl1*	3.13	35.51

Doubling time was determined from the exponential phase of the curves. Cells were grown as described under Materials and Methods. The results are the means of three experiments.

no growth was observed (data not shown).

The kinetic properties of the glycerol transport system in *D. hansenii* and *S. cerevisiae* transformants is shown in Table 3, at pH 5.6 and 7.5. The results of glycerol kinetic parameters can be summarized as follows: a) in *D. hansenii* (pH 5.6) the Vmax values are similar, the presence of NaCl or KCl increased the affinity for ^{14}C-glycerol uptake; b) the Vmax values for all *S. cerevisiae* transformants (pH 5.6) are around 6-fold lower as compared with *D. hansenii* strain; c) the Km values (pH 5.6) for *S. cerevisiae* transformant yeasts did not show significant differences between the cells incubated with or without salt; d) in *D. hansenii* (pH 7.5), the Vmax values are similar too, but the presence of NaCl or KCl decreased the affinity for ^{14}C-glycerol uptake; e) the Vmax values for all *S. cerevisiae* yeast strains (pH 7.5) are around 6 or 10-fold lower as compared with *D. hansenii* strain; f) the Km values (pH 7.5) for *S. cerevisiae* transformants decreased the affinity for ^{14}C-glycerol uptake in the cells incubated with salt; g) at both pHs, in the JCGH*zero* strain, a simple diffusion (D) of ^{14}C-glycerol transport was observed; h) JCGH*stl1* conferred glycerol transport phenotype at both pHs, it was saturable and adjustable to one component in *stl1* strain.

The influence of extracellular pH on maximum glycerol accumulation ratio in *D. hansenii* was studied at 5.6 and 7.5 pH values Cells grown in YNB-ura medium with glycerol as carbon source showed accumulation of labeled glycerol (Figure 3) against gradient. CCCP prevented slightly accumulation and elicited a significant efflux of labeled glycerol when added after 20 min of incubation. A similar behavior caused the addition, at the same time of 1 M of cold glycerol after 20 min incubation time. Figure 3A shows that, the presence of KCl increased the in/out accumulation rate more than NaCl at pH 5.6; whereas, at pH 7.5, the salts decreased the in/out accumulation rate. In/out ratios presented significant variations in comparison with the value previously obtained at pH 5.6, and CCCP prevented glycerol accumulation at the pH value tested. In Figure 3B, incubation in the presence of a salt diminished the accumulation rate of labeled glycerol, the results obtained in the Figure 3A showed that maximum accumulation ratio did not exceed 20 times in *D. hansenii* without salt, whereas the accumulation rate of *D. hansenii* without salt at pH 7.5 increased 1-fold as compared with the experiments at pH 5.6.

The accumulation rates of labeled glycerol was evaluated in the *S. cerevisiae* transformant strains (Figure 4), these were also incubated in the presence of CCCP, or cold glycerol (data not shown); in all conditions, the incubation decreased the glycerol accumulation rate, the CCCP elicited efflux of labeled glycerol when it was added at the start of the incubation reaction or after 20 min of incubation. Another point to describe about this experiment are the very low ratios of accumulated exposed to high NaCl concentrations (Adler and Gustafsson, 1980). Some authors (Serrano, 1996) have considered the

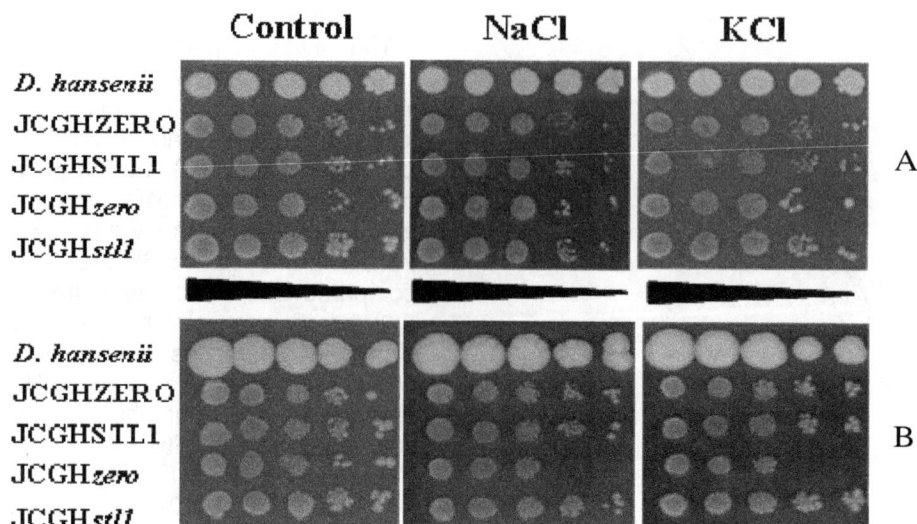

Figure 2. (Gonzalez-Hernández JC) Growth of *D. hansenii* and transformants of *S. cerevisiae* containing YEp352 with or without fragment of *STL1* genomic DNA from *D. hansenii*. Cells were grown for 24 h in 5 ml of YNB-ura liquid medium up to final density approximately 3 × 10^7 ml^{-1}. Plates were inoculated with serial 10-fold dilutions (the black arrow indicates the dilution direction) of these cultures onto YNB-ura medium with glucose as carbon source (pH 5.6 [A], and pH 7.5 [B]), and different growth conditions (YNB-ura medium, 0.6 M of NaCl, and KCl, respectively), and incubated at 28ºC. *D. hansenii*, and the *S. cerevisiae* transformants (JCGHZERO, JCGHSTL1, JCGH*zero*, JCGH*stl1*). Growth was recorded after 1 week. Data are representative of three experiments.

Table 3. Kinetic parameters of ^{14}C-glycerol transport (pH 5.6 and 7.5) of *D. hansenii* and *S. cerevisiae* transformants cultivated in YNB-ura medium, and incubated in the absence or the presence of 0.6 M NaCl or KCl.

	pH 5.6		pH 7.5	
	Vmax	Km	Vmax	Km
D. hansenii	0.77	0.68	0.56	0.30
0.6 M KCl	0.78	0.49	0.52	0.60
0.6 M NaCl	0.79	0.49	0.40	0.43
JCGHZERO	0.13	0.90	0.051	0.38
0.6 M KCl	0.19	1.05	0.08	0.38
0.6 M NaCl	0.12	0.77	0.15	0.73
JCGHSTL1	0.086	0.69	0.034	0.10
0.6 M KCl	0.13	0.48	0.024	0.13
0.6 M NaCl	0.12	0.60	0.034	0.12
JCGH *stl1*	0.14	0.56	0.14	0.71
0.6 M KCl	0.12	1.12	0.17	1.12
0.6 M NaCl	0.20	0.92	0.11	0.57

	D = mmol g^{-1} d. wt.	
	pH 5.6	pH 7.5
JCGH *zero*	0.14	0.56
0.6 M KCl	0.12	1.12
0.6 M NaCl	0.20	0.92

The transport of labeled cations was measured as described under Methods. (Vmax = mmol h^{-1} g^{-1} d. wt; Km = [mM]; D = diffusion coefficient). Results represent the means ± SEM (n = 3).

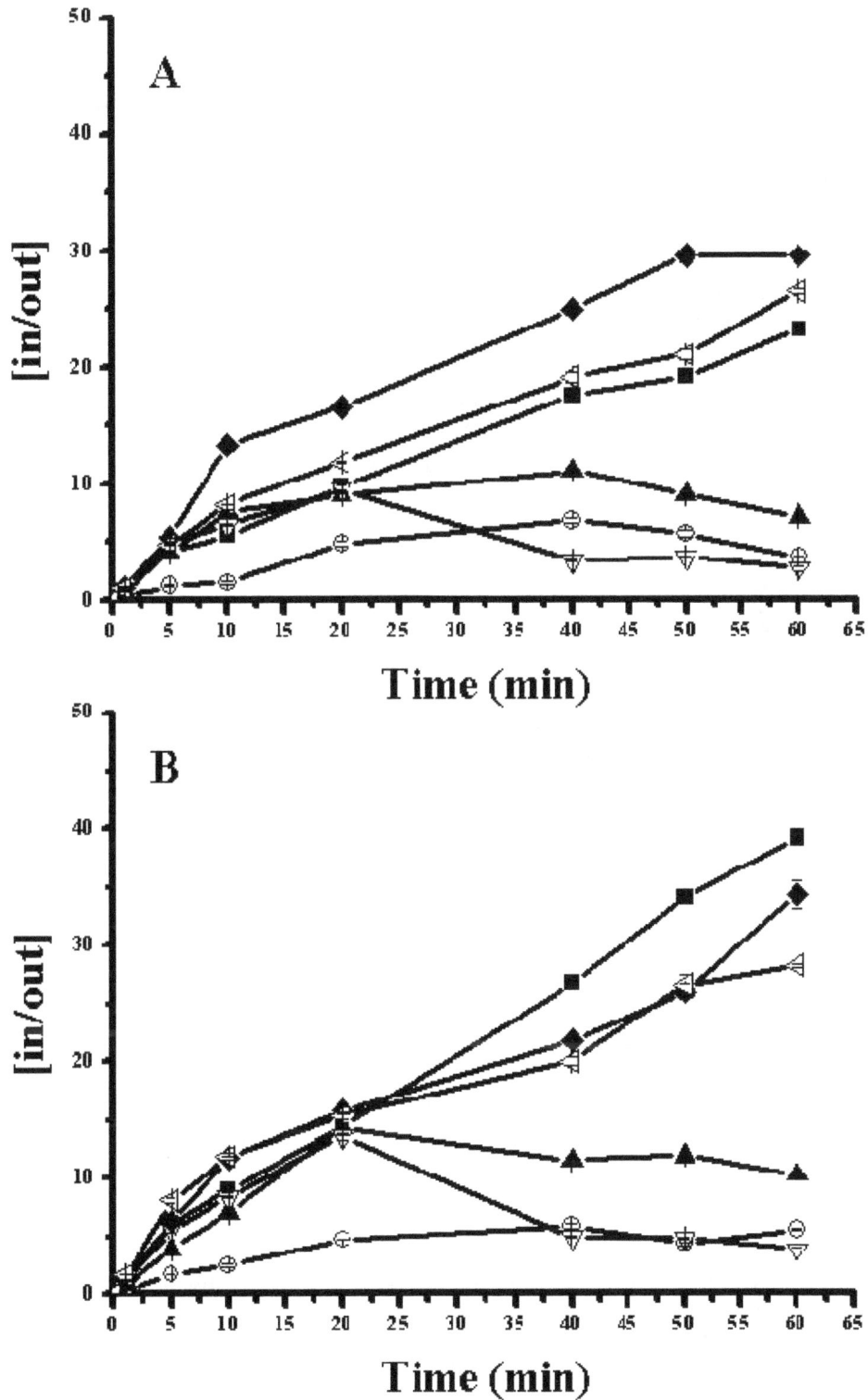

Figure 3. Accumulation of 14C- glycerol at pH 5.6 (A) or 7.5 (B) by D. hansenii grown in YNB-ura medium (2% glycerol), 25 μl cell suspension (0.25 g/ml), 5 μl glucose [1 M], and 30 μl 100 mM MES or MOPS (accumulation transport was also assessed in the presence of 0.6 M of NaCl or KCl). The experiment was started by adding 14C-glycerol. Aliquots of 10 μl cell suspension were taken at the indicated times and handled as described under Methods. Results represent the means ± SEM (n = 3). *D. hansenii* (■), Accumulation of radiolabel was prevented by adding 50 μM of CCCP (○), Efflux of radiolabel after the addition of 50 μM of CCCP at 20 min of incubation (▲), Efflux of radiolabel after the addition of nonradioactive glycerol at 20 min of incubation (▽, D. hansenii incubated with 0.6 M of KCl (◆), D. hansenii incubated with 0.6 M of NaCl (◁).

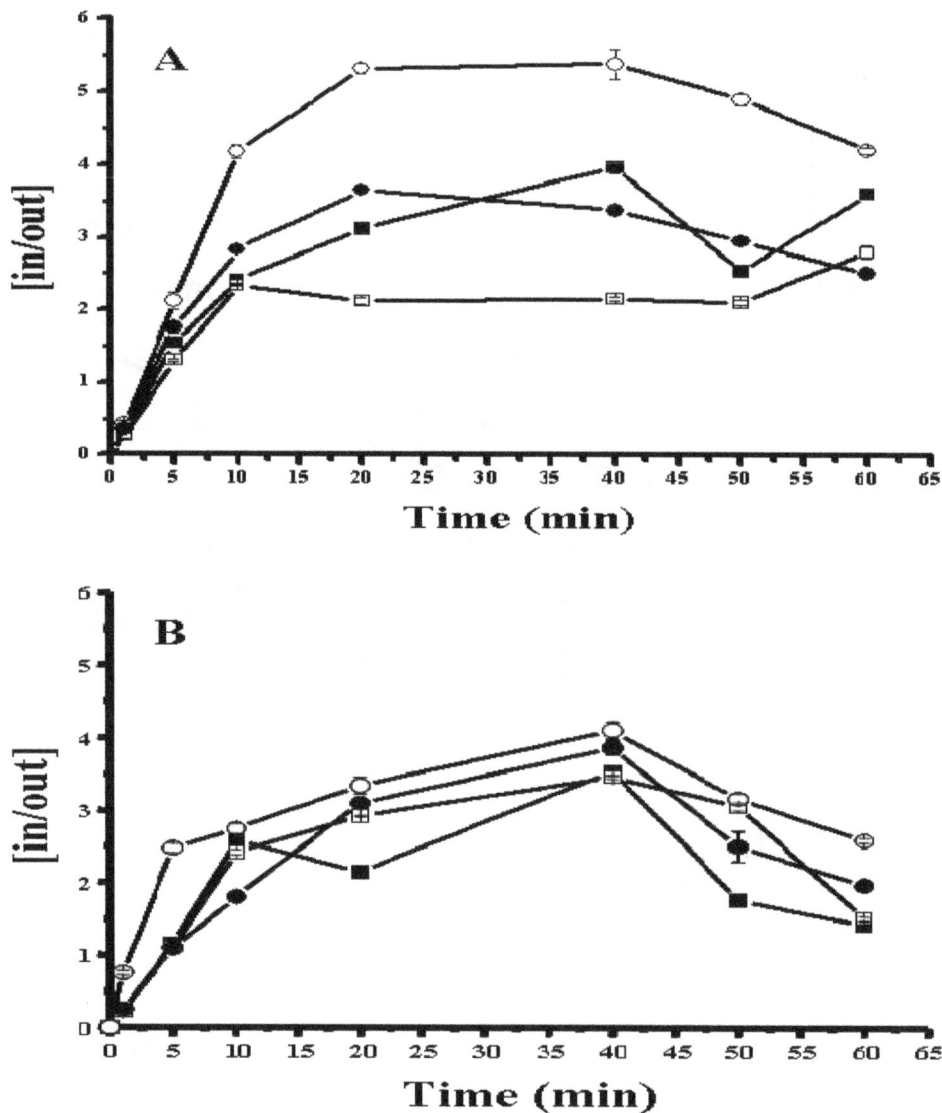

Figure 4. Accumulation of ^{14}C-glycerol at pH 5.6 (A) or 7.5 (B) by *S. cerevisiae* transformants grown in YNB-ura medium (2% glycerol), 25 μl of cell suspension (0.25 g/ml), 5 μl glucose 1 M, and 30 μl 100 mM MES or MOPS. The experiment was started by adding ^{14}C-glycerol. Aliquots of 10 μl cell suspension were taken at the indicated times and handled as described under Methods. Results represent the means ± SEM (*n* = 3). JCGHZERO (■), JCGHSTL1 (□), JCGH*zero* (●), JCGH*stl1* (○).

possibility of obtaining genes from *S. cerevisiae* involved in salt resistance, with the objective of eventually transferring these genes to plants. It would be worthwhile to consider at least the identification of one or several key genes from yeasts, such as *D. hansenii*, glycerol, which were almost 10-fold lower as compared with the ratios of accumulated glycerol in *D. hansenii*. Figure 4 shows the in/out accumulation rate obtained in the *S. cerevisiae* transformants strains grown in YBN-ura medium without salt in the presence of glycerol as carbon source at pH 5.6 or 7.5; at both pHs. The phenotype conferred by *DhSTL1* to the JCGHSTL1 transformant was observed,

which increased the in/out accumulation rate of labeled glycerol, more accumulated glycerol at pH 5.6 than 7.5 was observed.

The northern blot analysis was performed for *DhSTL1* transcripts in *D. hansenii* using total RNA prepared from *D. hansenii* cells that had been grown in YPD medium (Figure 5); cells were grown to exponential phase on YPD glucose medium and shifted to the same medium with or without 0.6 M of KCl or NaCl. *DhSTL1* gene expression was observed slightly in absence of salts (pH 5.6), whereas, in the presence of salts, it was observed an increase in the transcript level of *DhSTL1* at

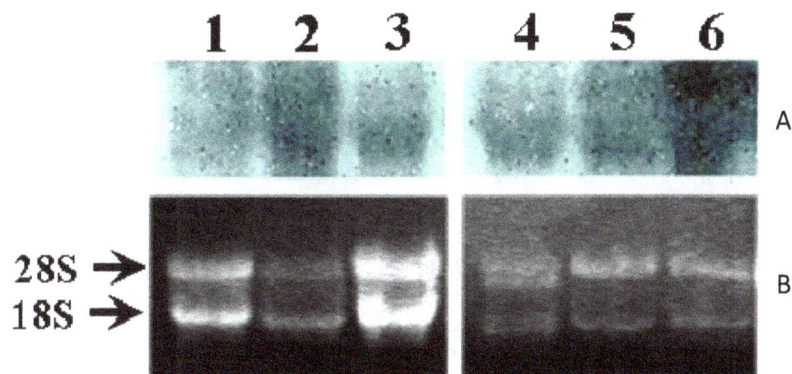

Figure 5. (Gonzalez-Hernández JC) Northern blot analyses. Total RNA (30 µg) was extracted and loaded onto 1.0 % agarose gel, transferred onto a membrane, and hybridized with DIG-labeled *STL1* Easy Hyb (Roche). Total RNA extracted from cells grown in YNB medium and incubated for 6 h in YPD medium without salt (lanes 1,4); with 0.6 M of KCl (lanes 2, 5) or 0.6 M of NaCl (lanes 3, 6). For lanes 1-3, the cells were incubated at pH 5.6; and for lanes 4-6, at pH 7.5 (A). Ethidium bromide-stained (0.01 µg ml^{-1}) pattern (B). Arrowheads indicate the positions of ribosomal RNA. Data are representative of three experiments.

pH 5.6; at 7.5 pH (Figure 5).

DISCUSSION

High NaCl concentrations cause loss of cellular water, which leads to cell shrinkage and growth arrest (Blomberg and Adler, 1992). As part of the adaptation to high NaCl concentrations, compatible solutes are produced and high levels of these solutes are obtained inside the cell, resulting in re-entry of lost water, regaining cell volume (Brown, 1978; Yancey, 1982). Both *D. hansenii* and *S. cerevisiae*, produce high intracellular levels of glycerol as the main compatible solute when which are naturally halotolerant or halophilic to the same purpose. *STL1* was identified during a screen for genes encoding membrane proteins involved in glycerol utilization, and it has been previously assigned as an active glycerol/H$^+$ symporter in *S. cerevisiae* (Ferreira et al., 2005).

The molecular biology of *D. hansenii* is poorly established, and recently the genome has been explored in the Génolevures project (Génolevures, 2001). In this project, the genomes of *S. cerevisiae* and other yeast species of the Hemiascomycetes class are compared. In that sense, the available genetic tools were used (*D. hansenii* genome database) to search for sequences of genes involved in glycerol transport in yeasts. This procedure allow to detect sequences and design primers to amplify and clone successfully the complete Stl1 protein sequence, as well as the upstream and downstream regions eventually involved in their regulation.

One of the phenotypes of the *S. cerevisiae*

transformant strains is a slow, and low growth in YNB-ura medium when glycerol is the carbon and energy source, this experiments are similar to those reported by Lages and Lucas (1995). Ferreira et al. (2005), reported that deletion of the *STL1* gene in both BY4742 and W303-1A genetic backgrounds resulted in cells that grew poorly on glycerol, the present experiments were made on BY4741 genetic background, obtained from EUROSCARF consortium. The genomic clones carrying the *DhSTL1* gene (from which the probe was derived) includes upstream and downstream non-coding sequences, therefore the expression of the *DhSTL1* gene was under the control of its own promoter.

To identify if Stl1 protein is involved in active uptake of glycerol in *D. hansenii*, the ability of *S. cerevisiae* transformant strains to complement *D. hansenii* growth phenotypes were evaluated (Figure 2); this was verified by cloning *DhSTL1* gene and inserting in YEp352 plasmid. The JCGH*stl1* transformant strain showed a slight but significant difference in the doubling times in growth curves made in liquid YNB-ura medium, with glycerol as carbon source. In an experiment performed in solid YNB-ura medium, In *STL1* gene-deleted transformant strains, the, *DhSTL1* depicts a slow growth on glycerol as a sole carbon source in the absence or presence of salts, but a phenotype was observed in the experiments; whereas, in glucose-grown cells, the phenotype of *DhSTL1* in *stl1* transformant strain is clearer.

S. cerevisiae has earlier been found to possess the ability to take up glycerol from the surroundings against a concentration gradient (Lages and Lucas, 1997). The uptake was shown to be driven by electrogenic proton symport. Recently, it has been reported (Ferreira et al.,

2005), that the cloning glycerol proton symporter in *S. cerevisiae* is encoded by the *STL1* gene. The successful of the *DhSTL1* gene in *S. cerevisiae* wild type, and in the *STL1* gene-deleted strain to characterize and identify if this protein is responsible for the glycerol transport in *D. hansenii*. All the transport and accumulation assays were evaluated in cells grown in YNB-ura medium using glycerol as carbon source, considering that in the *S. cerevisiae* transport system was under glucose repression and inactivation, glucose-grown cells presented, instead, a lower affinity permease for glycerol, probably a facilitated diffusion (Lages and Lucas, 1997).

Earlier reports, in which glycerol transport was evaluated in cells grown in mineral medium with vitamins and 2% of glucose as carbon source at 25ºC, confirmed that *D. hansenii* showed a constitutive active glycerol transport system that is not subject to glucose repression and mediates glycerol accumulation as a function of extracellular NaCl concentration (Lucas et al., 1990). The results were evaluated too at pH 7.5, because *D. hansenii* is considered an alkaline-tolerant yeast. In this sense, the performed experiments showing that glycerol transport was saturable and adjustable to one component at both pHs, 5.6 and 7.5. The kinetic parameter (pH 5.0) of glycerol transport reported in *S. cerevisiae* show an affinity uptake system with Km of 1.7 mM and Vmax of 441 μmol h^{-1} g^{-1} dry weight, the values obtained in the present experiments and the values reported in the presence of glucose as carbon source (Lucas et al., 1990) for *D. hansenii* showed a higher affinity uptake and velocity rate compared with the values reported for *S. cerevisiae* (Lages and Lucas, 1997). The *S. cerevisiae* transformant strains used in this work revealed that the *DhSTL1* gene conferred phenotype for JCGH*stl1* transformant strain.

As has been published for *S. cerevisiae* (Lages and Lucas, 1997), most of the strains able to grow on glycerol as the single carbon and energy source show evidence of inductive active transport. The existence of an H$^+$-glycerol symport induced by growth on glycerol is consistent with a role of such a carrier in glycerol catabolism. In contrast, a constitutive transporter can more easily be associated with both the salt stress response and glycerol assimilation.

Another way to analyze the components of glycerol uptake is to measure glycerol accumulation against a concentration gradient. The accumulation of compatible solutes, such as glycerol, in the yeast *S. cerevisiae*, is a ubiquitous mechanism in cellular osmoregulation. The rate of glycerol uptake is strongly reduced during growth at high osmolarity, indicating that yeast cells possess mechanisms that control the transport rate (Luyten, 1995; Sutherland, 1997). The action of a protonophore, eliminating ΔpH and lowering intracellular pH (Serrano, 1991), can affect active uptake and prevent the consequent accumulation, but the possibility cannot be disregarded that it also affects enzymes from the first

steps of catabolism, creating artifacts.

Rep et al. (2000) reported that expression of this *STL1* gene is undetectable under normal conditions but strongly induced after osmotic shock (in presence of NaCl and/or sorbitol), as confirmed by Northern blot analysis. The slight expression observed of the *DhSTL1* gene without salt (pH 5.6), but, in the presence of salts, It was observe an increase resulting from Stl1p expression (Figure 5). The obtained results at more alkaline pH showed similar Stl1p expression pattern in absence or presence of salts. Importantly, the levels of Stl1p and glycerol accumulation activity were directly correlated, when the experiments were evaluated at pH 5.6. The increased levels of Stl1p appearing after shifting to KCl or NaCl salts, and the accumulation rate in *D. hansenii* wild type strain was higher (pH 5.6) in the presence of salts (Figure 5). The slight expression of this transport and a higher level of salt-stress resistance suggest that this could be an evolutionary advantage for growth under such conditions. Correspondingly, glycerol symport has previously been shown to be repressed by glucose, induced by growth on nonfermentable carbon sources, and transiently detectable during diauxic shift upon growth on glucose (Lages and Lucas, 1997). *STL1* gene is also highly and transiently induced by osmotic shock during exponential growth on glucose-based media (Posas, 2000; Rep, 2000). The rapid appearance of Stl1p under these conditions suggests a role for the glycerol symporter during the immediate response to osmotic shock. This might be important in nature, considering the extreme, diverse, and rapid changes in environmental conditions yeasts may experience. Because yeast cells leak a substantial amount of the produced glycerol into the medium (Shen, 1999), this induction of Stl1p is not surprising.

D. hansenii yeast appears to respond to salinity stress in a similar manner to *S. cerevisiae* during early exponential phase, increasing the production of glycerol in response to salinity stress, whereas cells in late exponential phase show conservation of glycerol intracellularly, as in *Z. rouxii* (Zyl, 1990); the molecular basis for these differences in glycerol permeability is unknown.

None of the upstream regulatory proteins involved in glycerol transport had been studied and identified in *D. hansenii*. This paper allows suggesting that probably *DhSTL1* is not the principal gene involved in the glycerol transport in *D. hansenii*. Further investigation is being developed to know the actual role of *DhGUP1* and *DhAQPY1* in this solute transport; thus, the information is still scarce to present any regulatory mechanism implicated in the glycerol response functioning in *D. hansenii*. Due to the main role in osmoregulation played by glycerol, understanding its regulation in *D. hansenii* may allow us to comprehend the basis for the different halotolerance that characterizes these two yeasts. *D. hansenii* is not a friendly microorganism to work with.

The development of molecular tools for the manipulation of *D. hansenii* genes is a necessity an urgent task; the laboratory and others involved in this task have recently initiated the study of auxotrophic mutants or resistance molecular markers for this yeast, without any success.

ACKNOWLEDGMENTS

This work was supported by grant SFRH/BPD/19913/2004 to Juan Carlos González-Hernández, granted by FCT, Portugal.

REFERENCES

Adler L, Blomberg A, Nilsson A (1985). Glycerol metabolism and osmoregulation in the salt-tolerant yeast *Debaryomyces hansenii*. J. Bacteriol. 162: 300-306.

Adler L, Gustafsson L (1980). Polyhydric alcohol production and intracellular amino acid pool in relation to halotolerance of the yeast *Debaryomyces hansenii*. Arch. Microbiol. 124: 123-130.

Andre L, Nilsson A, Adler L (1988). The role of glycerol in osmotolerance of the yeast *Debaryomyces hansenii*. J. Gen. Microbiol. 134: 669-677.

Altschul SF, Madden TL, Schaffer AA, Zhang J, Zhang Z, Miller W, Lipman DJ (1997). Gapped BLAST and PSI-BLAST: a new generation of protein database search programs. Nucleic Acids Res. 25: 3389-3402.

Birnboim HC, Doly J (1979). A rapid alkaline extraction procedure for screening recombinant plasmid DNA. Nucleic Acids Res. 7: 1513-1523.

Blomberg A, Adler L (1992). Physiology of osmotolerance in fungi. Adv. Microb. Physiol. 33: 145-212.

Brown AD (1978). Compatible solutes and extreme water stress in eukaryotic micro-organisms. Adv. Microb. Physiol. 17: 181-242.

Chen A, Wadso I (1982). A test and calibration process for microcalorimeters used as thermal power meters. J. Biochem. Biophys. Methods. 6: 297-306.

Cryer DR, Ecclesmall R, Marmur J (1975). Academic Press, New York. Isolation of yeast DNA. In: Methods in Cell Biology (Prescott, D.M., Ed.) 12: 39.

Domínguez JM (1998). Xylitol production by free and immobilized *Debaryomyces hansenii*. Biotechnol. Lett. 20: 53-56.

Ferreira C, van Voorst F, Martins A, Neves L, Oliveira R, Kielland-Brandt MC, Lucas C, Brandt A (2005). A member of the sugar transporter family, Stl1p is the glycerol/H+ symporter in *Saccharomyces cerevisiae*. Mol. Biol. Cell. 16: 2068-2076.

Geitz RD, Schiestl RH (1995). Transforming yeast with DNA. Methods Mol. Cell. Biol. 5: 255-269.

Génolevures (2001). Genomic Exploration of the Hemiascomycete yeasts. http://cbi.labri.fr/Genolevures/. Center for Bioinformatics, Bordeaux.

González-Hernández JC, Cárdenas-Monroy CA, Peña A (2004). Sodium and potassium transport in the halophilic yeast *Debaryomyces hansenii*. Yeast 21: 403-412.

González-Hernández JC, Jiménez-Estrada M, Peña A (2005). Comparative analysis of trehalose production by *Debaryomyces hansenii* and *Saccharomyces cerevisiae* under saline stress. Extremophiles 9: 7-16.

Gustaffson L (1979). The ATP pool in relation to the production of glycerol and heat during growth of the halotolerant yeast *Debaryomyces hansenii*. Arch. Microbiol. 120: 15-23.

Hanahan D (1985). Techniques for transformation of *E. coli*. In: DNA Cloning: A Practical Approach (Glover, D.M., Ed.) 109-135.

Hill JE, Myers AM, Koerner TJ, Tzagoloff A (1986). Yeast/*E. coli* shuttle vectors with multiple unique restriction sites. Yeast 2: 163-167.

Hoffman CS, Winston F (1987). A ten-minute DNA preparation from yeast efficiently releases autonomous plasmids for transformation of *E. coli*. Gene. 57: 267-272.

Holst B, Lunde C, Lages F, Oliveira R, Lucas C, Kielland-Brandt MC (2000). *GUP1* and its close homologue *GUP2*, encoding multimembrane-spanning proteins involved in active glycerol uptake in *Saccharomyces cerevisiae*. Mol. Microbio. 37: 108-124.

Lages F, Lucas C (1995). Characterization of a glycerol/H+ symport in the halotolerant yeast *Pichia sorbitophila*. Yeast 11: 111-119.

Lages F, Lucas C (1997). Contribution to the physiological characterization of glycerol active uptake in *Saccharomyces cerevisiae*. Biochim. Biophys. Acta. 1322: 8-18.

Lucas C, Da Costa M, Van Uden N (1990). Osmoregulatory active sodium-glycerol co-transport in the halotolerant yeast *Debaryomyces hansenii*. Yeast 6: 187-191.

Luyten K, Albertyn J, Skibbe WF, Prior BA, Ramos J, Thevelein JM, Hohmann S (1995). Fps1, a yeast member of the MIP family of channel proteins, is a facilitator for glycerol uptake and efflux and is inactive under osmotic stress. EMBO J. 14: 1360-1371.

Norkrans B (1968). Studies on marine ocurring yeasts: respiration, fermentation and salt tolerance. Archiv. Fur. Mikrobiologie 62: 358-372.

Norkrans B, Kylin A (1969). Regulation of the potassium to sodium ratio and of the osmotic potential in relation to salt tolerance in yeasts. J. Bacteriol. 100: 836-845.

Posas F, Chambers JR, Heyman JA, Hoeffler JP, de Nadal E, Arino J (2000). The transcriptional response of yeast to saline stress. J. Biol. Chem. 275: 17249-17255.

Pronk JT (2002). Auxotrophic yeast strains in fundamental and applied research. Appl. Environ. Microbiol. 68: 2095-2100.

Reed RH, Chudek JA, Foster R, Gadd GM (1987). Osmotic significance of glycerol accumulation in exponentially growing yeasts. Appl. Environ. Microbiol. 53: 2119-2123.

Rep M, Krantz M, Thevelein JM, Hohmann S (2000). The transcriptional response of *Saccharomyces cerevisiae* to osmotic shock. Hot1p and Msn2p/Msn4p are required for the induction of subsets of high osmolarity glycerol pathway-dependent genes. J. Biol. Chem. 275: 8290-8300.

Sambrook J, Fritsch EF, Maniatis T (1989). Molecular Cloning: A Laboratory Manual, 2nd edn. Cold Spring Harbor Laboratory Press, Cold Spring Harbor, NY.

Schmitt ME, Brown TA, Trumpower BL (1990). A rapid and simple method for preparation of RNA from *Saccharomyces cerevisiae*. Nucleic Acids Res. 18: 3091-3092.

Serrano R (1991). Transport across yeast vacuolar and plasma membranes. In The Molecular and Cell Biology of the Yeast *Saccharomyces cerevisiae*. Genome Dynamics, Protein Synthesis pp. 523-585.

Serrano R (1996). Salt tolerance in plants and microorganisms: toxicity targets and defense responses. Int. Rev. Cytol. 165: 1-52.

Shen B, Hohmann S, Jensen RG, Bohnert H (1999). Roles of sugar alcohols in osmotic stress adaptation. Replacement of glycerol by mannitol and sorbitol in yeast. Plant Physiol. 121: 45-52.

Sutherland FC, Lages F, Lucas C, Luyten K, Albertyn J, Hohmann S, Prior BA, Kilian SG (1997). Characteristics of Fps1-dependent and-independent glycerol transport in *Saccharomyces cerevisiae*. J. Bacteriol. 179: 7790-7795.

Yancey PH, Clark ME, Hand SC, Bowlus RD, Somero GN (1982). Living with water stress: evolution of osmolyte systems. Science 217: 1214-1222.

Zyl PJ, Kilian SG, Prior BA (1990). The role of an active transport mechanism in glycerol accumulation during osmoregulation by *Zygosaccharomyces rouxii*. Appl. Microbiol. Biotechnol. 34: 231-235.

Interactions between multiple fungi isolated from two bark beetles, *Dendroctonus brevicomis* and *Dendroctonus frontalis* (Coleoptera: Curculionidae)

Thomas S. Davis[1], Richard W. Hofstetter[1], Kier D. Klepzig[2], Jeffrey T. Foster[3] and Paul Keim[3]

[1]Northern Arizona University, School of Forestry, P. O. BOX 15018, Flagstaff, Arizona, USA, 86011.
[2]USDA Forest Service, Southern Research Station, Pineville, Louisiana, USA.
[3]Center for Microbial Genetics and Genomics, Northern Arizona University, Flagstaff, Arizona, USA.

Antagonism between the fungal symbionts of bark beetles may represent a biologically significant interaction when multiple beetle species co-occur in a host tree. Since high density bark beetle populations rapidly and dramatically shift forest characteristics, patterns of competition between the obligate fungal associates of sympatric bark beetle species may have broad ecological effects. Primary and competitive resource acquisition between allopatric and sympatric isolates of mutualist fungi associated with the bark beetles *Dendroctonus frontalis* and *Dendroctonus brevicomis* were investigated. Growth assays at multiple temperatures suggest that primary resource acquisition by fungi growing in the absence of competitors varies regionally, and that optimal growth rate is likely to correspond to average summertime maximum temperatures. In competition assays, interactions were asymmetric between fungi isolated from sympatric beetle populations and fungi isolated from allopatric beetle populations: sympatric isolates out-competed allopatric isolates. However, competition between fungi from beetle populations in sympatry was found to be equal. These studies are the first to investigate interactions between the mycangial fungi of multiple *Dendroctonus* species, and the results suggest that competition is likely to occur when the mycangial fungi of multiple beetle species occur together.

Key words: Allopatric, competition, coexistence, mutualism, mutualist, mycangial fungi, sympatric.

INTRODUCTION

In many ecosystems, symbiotic associations are ubiquitous (Bronstein, 1994). For species that co-evolve with symbionts, interaction with symbionts is strongly correlated with population performance (e.g. fig/fig wasp, yucca/ yucca moth, ants/myrmecophytes, and beetle/fungus mutualisms; Bronstein, 1992; Huth and Pellmyr, 1997; Klepzig et al., 2001; Palmer et al., 2003); however, little research has investigated competitive symmetry among species with multiple symbionts (Palmer et al., 2003). *Dendroctonus* beetles (Coleoptera: Curculionidae) represent a useful system for studying interactions between symbiotic species (Six and Klepzig, 2004). *Dendroctonus* beetles associate with an extensive community of microorganisms, including mites, nematodes, fungi, yeasts, and bacteria (Whitney, 1982; Klepzig et al., 2001; Kenis et al., 2004; Kirisits, 2004; Scott et al., 2008). The composition and abundance of these microbial communities considerably impact beetle population dynamics (Bridges, 1983; Paine et al., 1997; Hofstetter et al., 2006), and previous studies have cited a need for investigating interactions among the microbial associates of *Dendroctonus* beetles (Klepzig and Wilkens, 1997; Harrington, 2005).

Dendroctonus species construct tunnels in the vascular tissue of host conifers in order to lay eggs (Wood, 1982). During excavation, tunnels are inoculated with fungal symbionts that grow throughout host tissues and deve-

*Corresponding author. E-mail: tsd3@nau.edu.

loping beetle larvae feed upon them (Harrington, 2005). In most cases *Dendroctonus* species are allopatric or colonize different tree species (Wood, 1982; Lieutier et al., 2004), so the opportunity for antagonism among fungal symbionts are avoided during larval development (Schlyter and Anderbrandt, 1993). However, in the ponderosa pine (*Pinus ponderosa* var *brachyptera*) forests of Southwestern North America, *Dendroctonus brevicomis* LeConte and *Dendroctonus frontalis* Zimmerman have been reported to co-colonize tissues of ponderosa pine (*Pinus ponderosa*) and with no apparent negative impacts on the fitness or fecundity of either beetle species (Davis and Hofstetter, 2009).

For *D. brevicomis* and *D. frontalis*, fungal symbionts in the genera *Ceratocystiopsis* and *Entomocorticium* confer important benefits to developing larvae (Klepzig and Wilkens, 1997; Hsiau and Harrington, 1997). For example, the presence of these filamentous fungi in feeding chambers is correlated with adult beetle size, fecundity, and nitrogen content (Bridges, 1983; Coppedge et al., 1995; Ayres et al., 2000). Also, beetles have evolved glandular structures (termed 'mycangia') that are used for transporting these fungi between host trees (Barras and Perry, 1971; Hsiau and Harrington, 1997; Yuceer et al., 2010). Due to the prevalence of association and interdependence of mycangial fungi and multiple *Dendroctonus* species, the beetle-fungal relationship is often considered a mutualism (Six, 2003). However, the sign (+, –) of interaction between multiple fungal symbionts of sympatric *Dendroctonus* species is unknown. But, interactions between fungal species that co-inhabit a niche have been shown to be antagonistic in many natural systems (Klepzig and Wilkens, 1997; Yuen et al., 1999; Murphy and Mitchell, 2001; Klepzig, 2006; Boddy, 2007; Licyayo et al., 2007). Uncolonized pine vascular tissue is a limiting resource for mycangial fungi; this should create an interface for antagonism between fungal mutualists when multiple beetle species co-colonize a host. Among sympatric *Dendroctonus* species that co-inhabit host tissue, competition between fungal symbionts of beetles may represent an interaction that limits beetle fitness or population growth.

Here, the authors report on patterns of primary and competitive resource acquisition by mycangial fungi of sympatric and allopatric *Dendroctonus* beetles. Throughout this report, "primary resource acquisition" is defined as the rate at which fungal isolates acquired resource area in the absence of competitors. "Competitive resource acquisition" is defined as the average proportion of trials in which individual fungal isolates acquired resource space that was occupied by another fungal isolate. The authors experimentally investigated the effects of temperature and biotic interactions on growth patterns of 33 isolates of mutualistic mycangial fungi associated with *D. brevicomis* and *D. frontalis*, from both sympatric and allopatric beetle populations. We ask two questions: (1) Does primary resource acquisition by mycangial isolates vary by fungal species, beetle species, or beetle populations? (2) Does the symmetry of competitive interactions by mycangial isolates vary with sympatry or allopatry of beetle populations?

MATERIALS AND METHODS

Beetle collection and acquisition of fungal isolates

Bark beetles used for fungal isolation were collected from three locations: Coconino National Forest in Arizona, U.S.A., Homochitto Ranger District National Forests in Mississippi, U.S.A., and Plumas National Forest in California, U.S.A. (Figure 1). In Arizona, populations of *D. frontalis* and *D. brevicomis* occur in sympatry and co-colonize ponderosa pine (*Pinus ponderosa* var *brachyptera*). In Mississippi only *D. frontalis* occurs and colonizes multiple pine species. In California only *D. brevicomis* occurs and colonizes *P. ponderosa* var *benthamiana*. There is moderate climatic variability between the three forests in terms of both annual mean precipitation and maximum temperature. On the Coconino National Forest annual precipitation averages 44.7 cm per year and mean summer maximum temperature is 25 °C (Hereford, 2007), on the Homochitto National Forest annual precipitation averages 162.5 cm per year and mean summer maximum temperature is 30 °C (Southern Regional Climate Center), and on the Plumas National Forest annual precipitation averages 38.8 cm per year and mean summer maximum temperature is 29.6 °C (Western Regional Climate Center).

To obtain mycangial fungi, live beetles were trapped in the field using Lindgren funnel traps baited with pine beetle pheromone lures containing frontalin, *exo*-brevicomin, and α-pinene (Synergy Semiochemicals Corp, Lot No. WPP10416). Beetles were placed individually into clear, size 0 gelatin capsules (Torpac, Lot No. 1100049271), and stored in the lab in dark environmental chambers at 5 °C until used for the isolation of fungi. All insects specimens used for microbial isolations mentioned in this study were collected between May 18, 2007 and July 18, 2007. Healthy female beetles were dissected and the thorax removed. Each thorax was surface sterilized using $HgCl_2$ and de-ionized water described by (Kopper et al., 2004) and then split dorsoventrally and placed in 2% malt extract media (Malt extract – MP Biomedicals LLC, Lot No. 6753J; Agar – BioServ, Lot No. 1740.01). The pH of the media was 4.7 ± 0.2 according to manufacturer specifications. Malt extract media (2%) and 95 × 15 mm Petri dishes (Fisherbrand) were used for all isolations and assays. Dishes containing isolates were sealed using Parafilm and incubated in dark environmental chambers at 15 °C until used in assays.

Fungal identification

Fungal colonies were determined to be one of two fungal genera, *Entomocorticium* and *Ceratocystiopsis*, and were putatively identified based on microscopic observations of hyphal morphology and degree of melanization (Klepzig et al., 2004). Twenty strains of mutualist mycangial fungi from *D. brevicomis* and *D. frontalis* in sympatry (9 *D. frontalis* and 11 *D. brevicomis*), and thirteen fungal strains from allopatric beetle populations (6 from *D. frontalis* in Mississippi and 7 from *D. brevicomis* in California) were isolated. Fungi from *D. brevicomis* and *D. frontalis* are *Ceratocystiopsis brevicomi* and *Entomocorticium* sp. B, and *Ceratocystiopsis ranaculosus* (J.R. Bridges and T.J. Perry) Hausner and *Entomocorticium* sp. A, respectively (Hsiau and Harrington, 1997).

The identity of fungal strains were confirmed by sequencing of internal transcribed spacer (ITS) regions 1 and 2 between the ribosomal RNA *genes* 18S and 28S. DNA was extracted using a Qiagen DNeasy plant kit (Valencia, CA) with the modified protocol

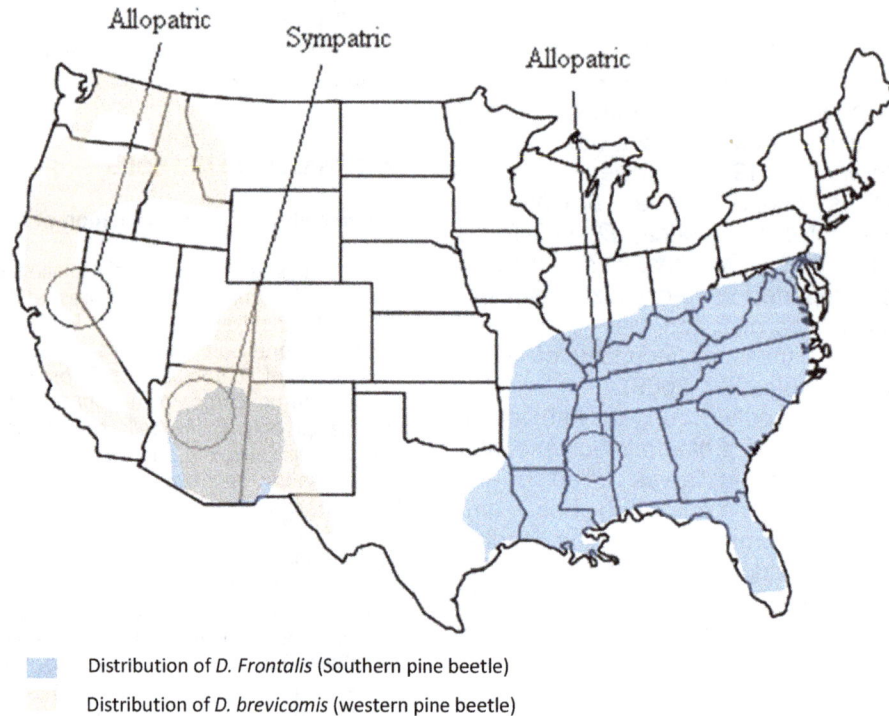

Figure 1. The locations of beetle populations where fungal isolates were collected. *D. brevicomis* and *D. frontalis* occur in sympatry in northern Arizona (Coconino National Forest), *D. brevicomis* in allopatry in California (Plumas National Forest), and *D. frontalis* in allopatry in Mississippi (Homochitto National Forest).

for yeasts, which includes a sorbitol buffer and lyticase enzyme to digest cell walls. The author used primers 5.8SF (5'-CGCTGCGTTCTTCATCG-3') and 5.8SR (5'-TCGATGAAGAACGCAGCG-3') from White et al. (1990) and paired them with newly developed primers ITS-18S (5'-CTTSAACGAGGAATNCCTAGTA-3') and ITS-28S (CATWCCCAAACWACYCGACTC) for ITS regions 1 and 2, respectively (Cindy Liu et al. unpublished manuscript).

The authors used the following parameters for a touchdown PCR: hot start 95°C for 4 min; then 20 cycles at 95°C for 30 s, 60°C for 1 min decreasing 0.5°C each subsequent cycle, 72°C for 1 min; 12 cycles at 95°C for 30 s, 45°C for 30 s, 72°C for 30 s; finishing with 72°C for 7 min. PCR reagents were used in the following final concentrations: Invitrogen PCR buffer 1 x (Carlsbad, CA), primers 0.2 uM each, MgCl$_2$ 2.5 mM, dNTPs 0.8 uM, and Invitrogen Platinum taq polymerase 1.4 U. PCR amplicons were cleaned up using ExoSAP-IT (USB, Cleveland, OH), cycle sequenced with ABI PRISM BigDye Terminator 3.1 (Applied Biosystems, Foster City, CA), and run on an ABI 3130 x l Genetic Analyzer. Sequences for both reads were edited and compiled in Sequencher 4.9 (Gene Codes, Ann Arbor, MI) and BLASTed against all GenBank accessions. Identifications were based on the highest identity value (complete match) and read length. Representative voucher specimens of fungi were preserved in 80% glycerol/20% malt extract broth (MEB; Difco; Lot No. 7306921) and placed in storage freezers held at -80°C in the Microbial Genetics and Genomics Center in Flagstaff, Arizona, U.S.A.

Primary resource acquisition by mycangial fungi

Radial growth is a primary mode of resource acquisition for fila-

mentous fungi, and primary resource acquisition was defined as the rate at which fungal colonies occupied media area in the absence of competitors. All fungal strains were incubated in dark environmental chambers at six temperatures of 5, 10, 15, 20, 25 and 28°C. Fungi were transferred from original isolates to sterile 2% malt extract agar by extracting a 1 x 1 mm section of growth media from hyphal tips of isolations during the linear growth phase using a flame-sterilized spatula. Hyphal growth was traced every 48 h beginning at day zero (initial transfer of colony) for 15 d. The growth rate for each fungal colony at each temperature was determined by dividing the distance between tracings by the number of days of growth. This study was replicated twice for each strain and growth rate was quantified in mm growth/day to 1 x 10^{-1} mm.

Competitive resource acquisition by mycangial fungi

Combative interaction between organisms is a secondary means of resource acquisition when limited resources are occupied by competitors (Tilman, 1982). The authors divided competitive resource acquisition into two parts: (1) Resource acquisition/capture: the mean frequency with which a fungal isolate colonized media resources occupied by a competing fungal isolate, and (2) Resource defense: the mean frequency with which a fungal strain resisted colonization by a competing fungal colony. Competitive interactions between mycangial fungi were tested using a pairwise approach. All fungi were paired in Petri dishes by placing 1 x 1 mm media sections containing fungal hyphae at opposing ends of the dish, and fungal strains were transferred from original isolate colonies as described above. Each isolate was also tested against itself. This full factorial design (33 x 33) was replicated twice (n = 2178 assays).

Variation in the growth patterns of these fungi on 2% malt extract

Table 1. Fungal growth rates (mm/day^{-1}) at multiple temperatures and passive / competitive resource acquisition patterns by fungal species and beetle host isolated from sympatric beetle populations. ANOVA results also shown. Bold values indicate significance differences in means at α = 0.05.

| | Beetle species | | | | | Fungal genera | | | | |
| | D. brevicomis | D. frontalis | | | | Ceratocystiopsis | Entomocorticium | | | |
Variable	Mean ± SE	Mean ± SE	F	df	P	Mean ± SE	Mean ± SE	F	df	P
5 °C	0.091 ± 0.033	0.040 ± 0.029	1.312	1.16	0.268	0.074 ± 0.027	0.057 ± 0.036	0.141	1.16	0.711
10 °C	0.425 ± 0.117	0.368 ± 0.103	0.133	1.16	0.720	0.446 ± 0.094	0.347 ± 0.127	0.372	1.16	0.550
15 °C	0.704 ± 0.150	0.665 ± 0.131	0.040	1.16	0.843	0.592 ± 0.121	0.777 ± 0.162	0.808	1.16	0.381
20 °C	1.086 ± 0.254	1.361 ± 0.222	0.660	1.16	0.428	1.016 ± 0.205	1.432 ± 0.276	1.410	1.16	0.252
25 °C	1.518 ± 0.299	1.999 ± 0.262	1.452	1.16	0.245	1.584 ± 0.242	1.933 ± 0.325	0.715	1.16	0.409
28 °C	1.250 ± 0.228	0.95 ± 0.152	1.193	1.16	0.306	1.320 ± 0.216 a	0.880 ± 0.125 b	3.100	1.16	0.116
Resource defense	0.488 ± 0.047	0.497 ± 0.041	0.016	1.16	0.900	0.594 ± 0.038 a	0.391 ± 0.051 b	**9.707**	1.16	**0.006**
Resource capture	0.386 ± 0.076	0.425 ± 0.067	0.141	1.16	0.712	0.445 ± 0.061	0.366 ± 0.183	0.571	1.16	0.460

*Letters indicate differences in means (Tukey's HSD test) by row.

media is consistent with that found in tree phloem (Rayner and Webber, 1984; Klepzig and Wilkens, 1997; Hofstetter et al., 2005). Following establishment of strains on dishes (1 - 2 d); hyphal growth was traced every 48 h for 30 d. Assays were done at 25 °C in the dark. The outcomes of paired competition assays are reported in terms of the mean frequency of resource acquisition and the mean frequency of resource defense by each fungal strain. These two observational metrics yielded four basic outcome categories for each colony in each pairing: (1) Fungal isolate A grew over media colonized by its paired competitor fungal isolate B; (2) Media colonized by fungal isolate A was grown over by its paired competitor fungal isolate B, (3) Both fungal isolates A and B successfully grew into others colonized area, or (4) Both fungal isolates A and B resisted overgrowth or formed a partition (Tuininga, 2005). Thus, each isolate in each trial received both a resource capture score and a defense score. Scoring for every pairing was verified microscopically (10 - 100 x magnification) by examining hyphal interactions.

The outcomes (mean frequency of resource capture and resource defense) of paired competition assays were converted "a posteriori" to binary values (0 = failure to defend/capture; 1 = successful defense/capture) for each isolate and averaged over all assays to yield an index of each fungal isolates' competitive performance on a continuous scale. Thus, each isolate received a relative frequency score (ranging from 0 - 1) that described the proportion of competitive resource acquisitions and defensive responses. For example, an isolate with a resource defense score of 0.87 indicates that the fungi successfully resisted colonization in 87% of competition trials.

Statistical analyses

All statistics were computed using JMP 7.0 software (SAS Institute). Statistical tests were prefaced by checking statistical assumptions. Assumptions of normality were verified using a Shapiro Wilk Test, and no transformations were required. In comparisons among fungi in sympatry (n = 20 sympatric fungal isolates), a two-way ANOVA was performed to analyze the fixed effects of beetle species, fungal species, and beetle species x fungal species interaction on response variables of radial growth rate, mean resource capture and mean resource defense. The beetle species x fungal species interaction effect did not contribute significant variation to the statistical model, so the interaction effect is not reported in the ANOVA summary for ease of display.

In comparisons between isolates across beetle populations (n = 33 fungal strains [20 sympatric/13 allopatric]), a one-way ANOVA was performed to analyze the fixed effect of sympatry or allopatry on response variables of radial growth rate, mean resource acquisition and mean resource defense. For this analysis sympatric isolates were considered as a single (Arizonan) population. The effects of temperature on fungal growth were analyzed as a fixed effect nested by location and differences between means were tested using contrasts. Statistical significance was established at α = 0.05 for statistical tests and ANOVA models for growth and competition assays were analyzed using F-tests to establish the significance of effects. Where differences in mean growth and competitive responses were detected in ANOVA models, directionality was established using Tukey's HSD Test.

RESULTS

Primary resource acquisition by mycangial fungi

Sympatric fungi: Growth rates of fungi from Arizona did not vary by beetle species, fungal species, or beetle species x fungal species interaction at 5, 10 °C, 15, 20, 25 or 28 °C (Table 1, Figure 2). Thus, fungi from sympatric beetle populations behaved statistically identically in terms of growth rates across temperatures by both fungal species and beetle species. However, fungal growth rates did consistently increase as temperature increased then declined once ambient temperature surpassed 28 °C. Entomocorticium species exhibited greater variability in growth rates across temperatures than Ceratocystiopsis species.

Sympatric and allopatric fungi: Growth rates did not vary among the three populations at 5 or 10 °C (Table 2, Figure 2). Primary resource acquisition from the sympatric fungi was significantly higher than the rate of primary resource acquisition by fungi from D. brevicomis in California at 15 and 20 °C, and fungi from D. frontalis in Mississippi were intermediate. There was no difference in primary resource acquisition by population at 25 °C, but at 28 °C primary resource acquisition by sympatric isolates

Table 2. Fungal growth rates (mm/day^{-1}) at multiple temperatures and passive / competitive resource acquisition patterns by fungal species and beetle host isolated from sympatric and allopatric beetle populations. ANOVA results also shown. Bold values indicate significance differences in means at α = 0.05.

Variable	Population			F	df	P
	Sympatric (*D. brevicomis* and *D. frontalis*) Mean ± SE	Allopatric (*D. brevicomis*) Mean ± SE	Allopatric (*D. frontalis*) Mean ± SE			
5°C	0.064 ± 0.034	0.034 ± 0.054	0.187 ± 0.060	2.199	2.30	0.137
10°C	0.408 ± 0.065	0.237 ± 0.079	0.356 ± 0.086	1.367	2.30	0.277
15°C	0.799 ± 0.084 a	0.428 ± 0.100 b	0.588 ± 0.108 ab	4.093	2.30	**0.032**
20°C	1.341 ± 0.180 a	0.577 ± 0.216 b	1.136 ± 0.233 ab	3.752	2.30	**0.041**
25°C	2.016 ± 0.174 a	0.841 ± 0.259 b	1.810 ± 0.280 a	6.386	2.30	**0.007**
28°C	1.100 ± 0.201 b	2.362 ± 0.241 a	2.987 ± 0.260 a	18.233	2.30	**<0.001**
Resource defense	0.800 ± 0.031 a	0.707 ± 0.038 b	0.616 ± 0.041 b	3.593	2.30	**0.046**
Resource capture	0.496 ± 0.078 a	0.185 ± 0.101 b	0.210 ± 0.109 b	3.533	2.30	**0.050**

*Letters indicate differences in means (Tukey's HSD test) by row.

was significantly lower than for allopatric isolates (Figure 2). Growth rates consistently increased with temperature for all fungal isolates until 25°C, where sympatric isolates showed a substantial decline but allopatric fungi achieved optimal growth.

Competitive resource acquisition by mycangial fungi

Sympatric fungi: Patterns of competitive resource acquisition varied significantly with fungal species but were not variable with beetle species (Table 1). Specifically, *Ceratocystiopsis* species had higher mean frequencies of resource defense than *Entomocorticium* species (Table 1). Thus, neither beetle species was associated with a consistently more competitive fungal symbiont. Sympatric and allopatric fungi; Competitive resource acquisition by fungal isolates varied significantly with sympatry and allopatry of source beetle populations (Table 2). Isolates from the

sympatric beetle populations exhibited significantly higher mean frequencies of competitive resource acquisition and resource defense of growth media than fungi isolated from either allopatric beetle population (Table 2). Fungi from opposing allopatric populations were not significantly different from each other.

DISCUSSION

Primary resource acquisition

The radial growth rates of fungi were found to vary by population at multiple temperatures (Table 2). This is in agreement with the findings of Six and Bentz (2007), which showed that ambient temperature was a mediator of fungal abundances, and that this variation was related to both site and seasonality in a *Dendroctonus ponderosae* system. Here, they show that the growth rates of mycangial fungi vary across beetle populations, which in the current study are separated by large

geographic regions. However, growth rates did not vary among fungal isolates for fungal species or beetle species within sympatric populations. The studies did not sample fungi from multiple sites within sympatric populations, so they might have detected greater variation in growth rates by assessing multiple sites within each region.

In contrast to the present study, Hofstetter et al. (2007) showed that *Entomocorticium* sp. A, exhibited optimal growth at a lower ambient temperature than *C. ranaculosus* in a *D. frontalis* system in Mississippi. In the present study, no differences in response to temperature were detected for growth by fungal species from sympatric beetle populations in Arizona or an allopatric beetle population in California (Figure 2). However, the authors did support their findings that *Entomocorticium* sp. A, and *C. ranaculosus* had different optimal growth rates in a Mississippi population of *D. frontalis* (Figure 2). One explanation for this pattern is overall variability in daily temperature: daily temperature range is greater in Arizona and California than in Mississippi

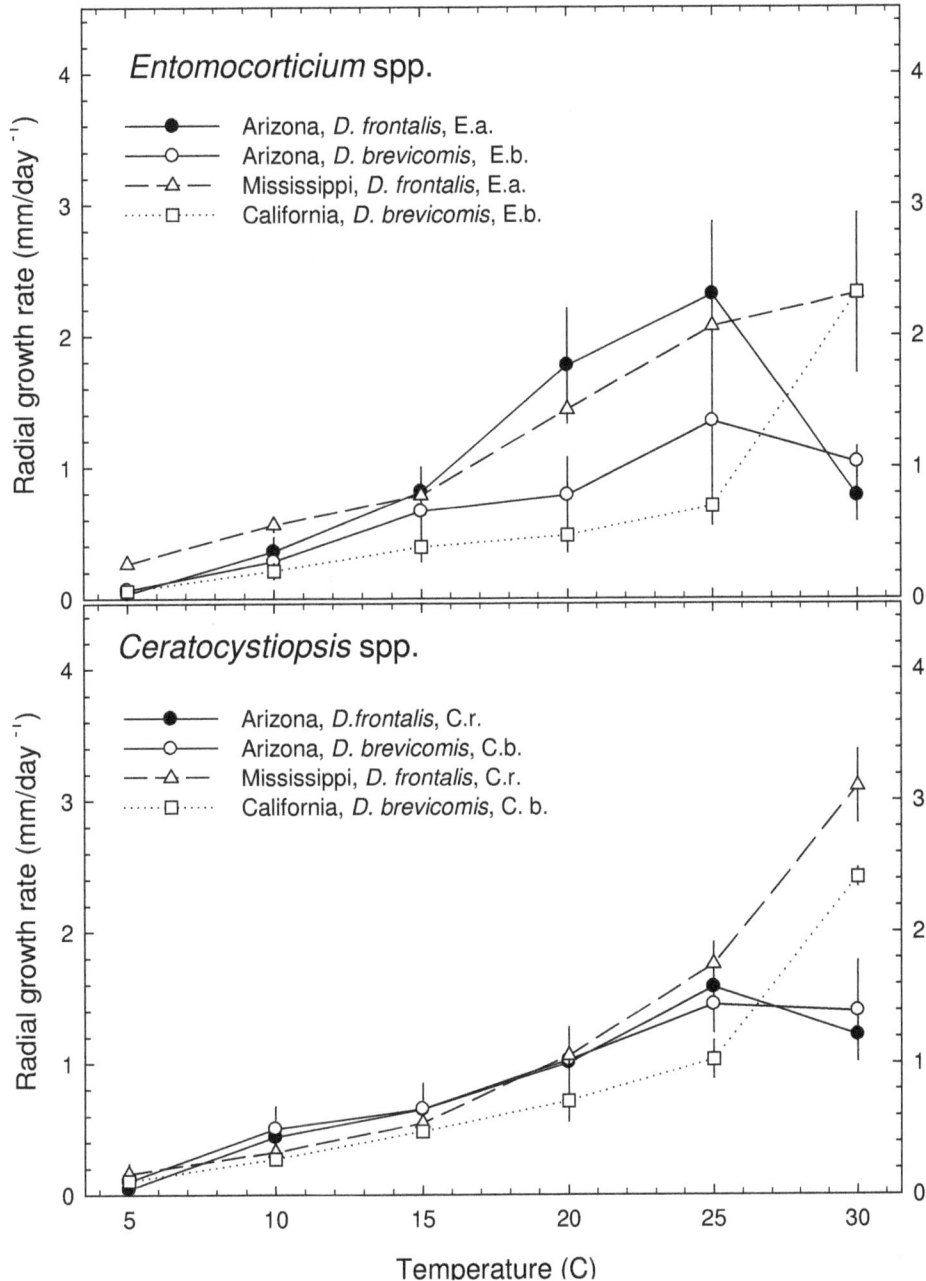

Figure 2. Growth rates of Ceratocystiopsis spp. and *Entomocorticium* spp. isolated from sympatric bark beetle populations in Arizona, an allopatric population of *D. brevicomis* in California, and an allopatric population of D. frontalis in Mississippi. Sympatric populations were pooled in this figure since there were no significant differences between radial growth rates of sympatric fungi. Error bars represent one standard error.

(Hereford, 2007; Southern Regional Climate Center, 2008; Western Regional Climate Center, 2008). Thus, optimal growth rate and species abundances of mycangial fungi may be influenced by daily temperature range, in addition to regional and seasonally mediated temperature differences. The rates of primary resource acquisition by sympatric fungi corresponded to the average maximum summer temperature in the region. In northern Arizona, climate records show that maximum

summer (June-September) temperatures average 25 °C (Hereford, 2007), which is the range where the greatest average growth rates were observed for sympatric fungi (Figure 2). In the experiments, an increase in ambient temperature of only 3 °C above the average summer maximum correlated with a dramatic decrease in growth rates of the sympatric fungi. In contrast, mycangial isolates from allopatric populations showed no apparent decrease in growth rates as ambient temperatures increased, and isolates from both populations grew optimally at 28°C. Unfortunately, no inferences can be made about primary resource acquisition by fungi beyond ambient temperatures of 28°C. However, previous work shows that the growth rates of mycangial fungi (both Entomocorticium and Ceratocystiopsis) isolated from *D. frontalis* in Mississippi declines once ambient temperatures exceeds 28°C (Hofstetter et al., 2007). If it is true that fungal growth rates correspond to regionally defined average maximum temperatures during summer months, then they would predict that isolates from allopatric *D. brevicomis* will grow optimally at 29.6°C, and isolates from allopatric *D. frontalis* will grow optimally at 30°C. Future studies with these organisms could benefit from assaying growth rates of mycangial fungi from both within and between beetle populations across a broader temperature range and at smaller intervals, since isolates appear to be highly sensitive to relatively small incremental variations in temperature.

Competitive resource acquisition

Fungi isolated from the sympatric beetle populations exhibited significantly greater frequencies of competitive acquisition and defense of media in a resource-limited environment. In sympatric beetle populations, *Ceratocystiopsis* species defended media resources from colonization by opposing isolates with significantly higher frequency than *Entomocorticium* species (Table 1). Data regarding competitive resource acquisition and resource defense by allopatric fungi were strongly asymmetric: in almost, no case (< 5%) was a fungal isolate from an allopatric beetle population able to colonize media occupied by a fungal isolate from a sympatric beetle population. Thus, competitive resource acquisition by allopatric isolates occurred almost exclusively in pairings with other allopatric isolates. The opposite was not true for sympatric fungi, which were able to competitively acquire media resources colonized by isolates from all beetle populations. However, all competition assays were performed at 25°C, where sympatric fungi exhibited their optimal growth. Thus, it is possible that the observed differences in competitive performance between sympatric and allopatric isolates were due to localized adaptations to temperature or the seasonality of collection (Six and Bentz, 2007). However, optimal growth rate of mycangial fungi from *D. frontalis* collected in Mississippi and Alabama is reported between 25 - 28°C (Klepzig et al., 2001; Hofstetter et al., 2007). The authors suggest that future studies related to competition between multiple mycangial species focus on testing interactions across a broader range of temperatures. In nature, many factors may contribute to the outcomes of competitive interactions between multiple fungi. For example, Licyayo et al. (2007) found that ammonia concentrations and pH had strong effects on interspecific

interactions between fungal species. Similarly, melanin and other pigments have been shown to strongly mediate competition between fungal species (Yuen et al., 1999; Klepzig, 2006). In a *D. frontalis* system, water availability was determined to play an important role in fungal competition (Klepzig et al., 2004). The studies only account for differences in competitive ability among fungi between regions of sympatry and allopatry, and by beetle species and fungal species within sympatric populations. However, the present study represents a first assessment of combative interactions between the mutualists of two or more bark beetle species, and suggests that fungi are likely to adapt to a competitive environment when multiple mycophagous beetle species inhabit a single plant host. Future studies of competitive interactions could benefit by testing interspecific fungal interactions across a gradient of host plant *Dendroctonus* beetles are frequently exposed to a terpenoid-saturated environment during the colonization of host tissues, and exposure to terpenoid compounds strongly impacts the growth performance of beetle – associated fungi (Paine and Hanlon, 1994; Hofstetter et al., 2005).

In conclusion, the mycangial fungi associated with *D. brevicomis* and *D. frontalis* were variable with respect to primary resource acquisition and competitive interactions. Growth rates of fungi did not vary by beetle species or fungal species when beetle populations were sympatric, however fungal growth rates did show substantial variation across regions. In sympatric populations, *Ceratocystiopsis* species were more likely to resist colonization by a competitor than *Entomocorticium* species. Mycangial isolates from sympatric beetle populations were more likely to competitively acquire resources. Interactions between sympatric isolates and allopatric isolates were asymmetric: sympatric isolates were better competitors and frequently colonized media resources inhabited by an allopatric isolate. Interestingly, competition also appeared to be symmetric among allopatric isolates.

These data reported here support the hypothesis that interactions between mycangial fungi of multiple *Dendroctonus* species are antagonistic (−, −), since fungi that occurred in sympatry were stronger competitors. Furthermore, these findings may be extendable to other systems where multiple insects with fungal associates colonize the same plant host and insect-fungal associations have been increasingly recognized as a central theme in arthropod ecology (Blackwell and Vega, 2005). The importance of these interactions for beetle larval performance and fungal-beetle relationships are still unknown and further experiments are needed to determine how competition between multiple fungal associates of *Dendroctonus* species affect beetle fitness.

ACKNOWLEDGEMENTS

The author thanked Amanda Garcia, Sherri Smith, and

Danny Cluck for capturing and sending bark beetles to their lab. The author appreciated the efforts of the USDA Forest Service, Southern Research Station laboratory technicians and the USDA Forest Service, Rocky Mountain Research Station for providing us with laboratory space. The author thanked Brandy Francis for assistance with sequencing efforts, Cindy Liu for access to unpublished fungal sequencing primers, and Laine Smith for laboratory assistance. The author also thanked one anonymous reviewer for comments that improved the quality of the manuscript. The author acknowledged the USDA Forest Service funding sources: Southern Research Station Cooperative Agreement 06-CA-11330129-046 and Rocky Mountain Research Station Joint-Venture Agreement 05-PA-11221615-104.

REFERENCES

Ayres MP, Wilkens RT, Ruel JJ, Lombardero MJ, Vallery E (2000). Nitrogen budgets of phloem-feeding bark beetles with and without symbiotic fungi. Ecology 81: 2198–2210

Barras SJ, Perry T (1971). Gland cells and fungi associated with prothoracic mycangium of *Dendroctonus adjunctus* (Coleoptera: Scolytidae). Ann. Ent. Soc. Am. 64: 123 – 126.

Blackwell M, Vega FE (2005). Seven wonders of the insect world. In: Vega FE, Blackwell M (Eds). Insect – fungal associations, Oxford, New York, 333 pages.

Boddy L (2007). Interspecific combative interactions between wood-decaying basidiomycetes. FEMS Microb. Ecol. 1-15.

Bridges JB (1983). Mycangial fungi of *Dendroctonus frontalis* (Coleoptera: Scolytidae) and their relationship to beetle population trends. Environ. Entomol.12: 858- 861.

Bronstein JL (1992). Seed predators as mutualists: ecology and evolution of the fig/pollinator interaction. In: Bernays E (Ed). Plant-insect interactions, Vol. 4. CRC, Boca Raton, FL, pp 1-44.

Bronstein JL (1994). Our current understanding of mutualism. Q. Rev. Biol. 69: 31-51.

Coppedge BR, Stephen FM, Felton GW (1995). Variation in female southern pine beetle size and lipid content in relation to fungal associates. Can. Entomol. 127: 145 – 153.

Davis TS, Hofstetter RW (2009). The effects of gallery density and species ratio on the fitness and fecundity of two sympatric bark beetles (Coleoptera: Curculionidae). Environ. Entomol. 33 639-650.

Harrington TC (2005). Ecology and evolution of mycophagous bark beetles and their fungal partners. In: Vega FE, Blackwell M (Eds). Insect – fungal associations, Oxford, New York, pp 257 – 291.

Hereford R. Climate Variation at Flagstaff, Arizona—1950 to 2007. Open-file report 2007 - 1410. USGS. M Meyers, director.

Hofstetter RW, Mahfouz JB, Klepzig KD, Ayres MP (2005). Effects of tree phytochemistry on the interactions among endophloedic fungi associated with the southern pine beetle. J. Chem. Ecol. 31: 551-572.

Hofstetter RW, Cronin J, Klepzig KD, Moser JC, Ayres MP (2006). Antagonisms, mutualisms, and commensalism affect outbreak dynamics of the southern pine beetle. Oecologia 147: 679-691.

Hofstetter RW, Dempsey TD, Klepzig KD, Ayres MP (2007). Temperature-dependent effects on mutualistic, antagonistic, and commensalistic interactions among insects, fungi and mites. Community Ecology 8: 47- 56.

Hsiau, PTW, Harrington TC (1997). *Ceratocystiopsis brevicomi* sp. nov., a mycangial fungus from *D. brevicomis* (Coleoptera: Scolytidae). Mycologia 89: 661-669.

Huth C, Pellmyr O (1997). Non-random fruit retention in *Yucca filamentosa*: consequences for an obligate mutualism. Oikos 78: 576-584.

Kenis M, Wermelinger B, Gregoire J (2004). Research on parasitoids and predators of Scolytidae – a review. In: Lieutier F, Day KR, Battisti A, Gregoire JC, Evans HF (Eds). Bark and Wood Boring Insects in Living Trees in Europe, a Synthesis. Kluwer Academic Publishers. Dordrecht, Netherlands, pp 237-290.

Kirisits T (2004). Fungal associates of European bark beetles with special emphasis on the Ophiostomatoid fungi. In: Lieutier F, Day KR, Battisti A, Gregoire JC, Evans HF (Eds). Bark and Wood Boring Insects in Living Trees in Europe, a Synthesis. Kluwer Academic Publishers. Dordrecht, Netherlands, pp 181-235.

Klepzig KD, Wilkens RT (1997). Competitive interactions among symbiotic fungi of the southern pine beetle. Appl. Environ. Microbiol. 63: 621-627.

Klepzig KD, Moser JC, Lombardero MJ, Hofstetter RW, Ayres MP (2001). Symbiosis and competition: complex interactions among beetles, fungi and mites. Symbiosis 30: 83-96.

Klepzig KD, Flores-Otero J, Hofstetter RW, Ayres MP (2004). Effects of available water on growth and competition of southern pine beetle associated fungi. Mycol. Res.108: 183-188.

Klepzig KD (2006). Melanin and the southern pine beetle-fungus symbiosis. Symbiosis 40: 137-140

Kopper BJ, Klepzig KD, Raffa KF (2004). Effectiveness of modified White's solution at removing ascomycetes associated with the bark beetle *Ips pini*. For. Pathol. 33: 237-240.

Licyayo DC, Suzuki A, Matsumoto M (2007). Interactions among ammonia fungi on MY agar medium with varying pH. Mycoscience 48: 20-28.

Lieutier F, Day KR, Battisti A, Gregoire J, Evans JF (2004). Bark and Wood Boring Insects in Living Trees in Europe, a Synthesis. Kluwer Academic Publishers, Boston Massachusetts. 569 Pages.

Murphy EA, Mitchell DT (2001). Interactions between *Tricholomopsis rutilans* and ectomycorrhizal fungi in paired culture and in association with seedlings of lodgepole pine and Sitka-spruce. For. Pathol. 31: 331-344.

Paine TD, Hanlon CC (1994). Influence of oleoresin constituents from *Pinus ponderosa* and *Pinus jeffreyi* on growth of mycangial fungi from *Dendroctonus ponderosae* and *Dendroctonus jeffreyi*. J. Chem. Ecol. 20: 2551 – 2463.

Paine TD, Raffa KF, Harrington TC (1997). Interactions among scolytid bark beetles, their associated fungi, and live host conifers. Ann. Rev. Entomol. 42: 179-206.

Palmer TM, Stanton ML, Young TP (2003). Competition and coexistence: exploring mechanisms that restrict and maintain diversity within mutualist guilds. Am. Nat. 162: 63-78.

Rayner ADM, Webber JF (1984). Interspecific mycelial interactions - an overview. Br. Mycol. Soc. Symp. 8: 383-417.

Schlyter R, Anderbrandt O (1993). Competition and niche separation between two bark beetles: existence and mechanisms. Oikos 68: 437-447.

Scott JJ, Oh DC, Yuceer MC, Klepzig KD, Clardy J, Currie CR (2008). Bacterial protection of a beetle-fungus mutualism. Science 322: 63.

Six DL (2003). Bark beetle – fungus symbiosis. In: Bourtzis K, Miller TA (eds) Insect symbioses, Vol. 3. CRC Press, Boca Raton, pp 97 – 110.

Six DL, Klepzig KD (2004). *Dendroctonus* bark beetles as model systems for studies on symbiosis. Symbiosis 37: 1 – 26.

Six DL, Bentz BJ (2007) Temperature determines symbiont abundance in a multipartite bark beetle-fungus ectosymbioses. Microb. Ecol. 54: 112-118.

Southern Regional Climate Center (2008). Climatological normals for Mississippi, 1971-2000. Accessed December 17 2008. U.S. Department of Commerce: National Oceanic and Atmospheric Administration, National Weather Service.

Tilman D (1982) Resource competition and community structure. Monogr. Pop. Biol.17: 1-282.

Tuininga AR (2005). Interspecific interaction terminology: from mycology to general ecology. In: Dighton J, White JF, Oudemans P (Eds). The Fungal Community: Its organization and role in the ecosystem. CRC Press, Boca Raton, FL, pp 265-283.

Western Regional Climate Center. (2008). Period of record monthly climate summary: Susanville, CA. Accessed December 17, 2008. Desert Research Institute: http://www.wrcc.dri.edu/cgi-bin/cliMAIN.pl?ca8702.

White TJ, Bruns T, Lee S, Taylor J (1990). Amplification and direct sequencing of fungal ribosomal RNA genes for phylogenetics. In: Innis MA, Gelfand DH, Sninsky JJ, White TJ, (Eds). PCR protocols: a guide to methods and applications. Academic Press, San Diego, CA, pp. 315–322.

Whitney HS (1982). Relationships between bark beetles and symbiotic organisms. In: Mitton JB, Sturgeon KB (Eds). Bark beetles in North American conifers: a system for the study of evolutionary biology. University of Texas Press, Austin, pp 183 – 211.

Wood SL (1982). The Bark and Ambrosia Beetles of North and Central America

(Coleoptera:Scolytidae), a Taxonomic Monograph. Great Basin Naturalist Memoirs, No.6. Brigham Young University Press. Provo, Utah.

Yuceer C, Hsu CY, Erbilgin N, Klepzig KD (2010). Ultrastructure of the mycangium of the southern pine beetle, Dendroctonus frontalis (Coleoptera: Curculionidae, Scolytinae): Complex morphology for complex interactions. Acta Zoologica 91:000-000.

Yuen TK., Hyde KD, Hodgkiss IJ (1999). Interspecific interactions among tropical and subtropical freshwater fungi. Microb. Ecol. 37: 257-262.

3

Identification of a restriction endonuclease (SacC1) from *Saccharomyces cerevisiae*

Mukaram Shikara

Biotechnology Division, Applied Sciences Department, University of Technology, Baghdad, Iraq.

SacC1 is a novel restriction endonuclease from *Saccharomyces cerevisiae* that recognizes the palindromic sequence 5'CTCGAC3' cleaving both DNA strands upstream and downstream of its recognition sequence and makes a staggered cut at the distance of five bases from the recognition sequence on the upper strand and at the seventh base on the complementary strand. It shares similar characteristics with Sac I from *Streptomyces achromogenes* as well as Sst1 from *Streptomyces Stanford* and Psp124B1 from *Pseudomonas* species. It has been purified by ammonium sulphate precipitation, dialysis, and gel filtration using phosphocellulose, DEAE-cellulose and Sephadex G-100 with an optimal pH range (7.5-8.5), active at 37°C and dependent on Mg^{+2} or Mn^{2+} which increases its activity by 4- and 2-folds, respectively, while other cations decrease its activity to some extents. Cleavage on both sides of the recognition sequence is characteristic of Type IIB systems but all IIB enzymes studied so far have been found to recognize discontinuous sites and a distinctive subunit/domain organization that is not present in the SacC1 enzyme. There are similarities between SacC1 and other homing endonucleases belonging to the LAGLIDADG family such as a requirement for Mg^{2+} (or Mn^{2+}) for cleavage to take place, optimal activity at alkaline pH and stimulation of the reaction by moderate concentrations of the monovalent cation.

Key words: Purification, recognition site, restriction enzyme, *Saccharomyces, Streptomyces*, Type IIB.

INTRODUCTION

Restriction enzymes have proved to be invaluable for the physical mapping of DNA. They offer unparalleled opportunities for diagnosing DNA sequence content and are used in fields as disparate as criminal forensics and basic research. In fact, without restriction enzymes, the biotechnology industry would certainly not have flourished as it has (Roberts, 2005). Restriction enzyme systems are aim to destroy foreign DNA without destroying their own bacterial or fungal DNA (Roberts and Macelis, 1994). The key feature of restriction endonucleases is not only their ability to cleave DNA at their recognition sequences, but rather their ability to avoid cleaving DNA at any other sequence. The ability of these enzymes promoted extensive screening of bacteria and fungi by biochemical assays and by genome analyzes (Murray, 2000). They have been classified into three main groups according to their cofactor requirements and the type of DNA cleavage (Wilson, 1991). The Type II restriction enzymes recognize specific DNA sequences and cleave at constant positions at or close to that sequence to produce 5'-phosphates and 3'-hydroxyls. Usually they require Mg^{2+} ions as a cofactor, although some have more exotic requirements (Richard et al., 2003). They are also the simplest ones with respect to other properties such as subunit structure and cleavage characteristics and are composed of two separate enzymatic activities. One is a restriction endonuclease that cleaves DNA at a specific recognition sequence. The second is a DNA methyltransferase, which is able to methylate the same sequence and render it refractive to cleavage by the corresponding endonuclease (Halford, 2001).

In Iraq, several separate attempts to purify restriction endonucleses from local sources were tried with the earliest attempts made in mid 1990s (Putrus, 1995; Al-Khafagi, 1999).

*Corresponding author. mukaramshikara2010@yahoo.com.

Table 1. The oligonucleotide duplexes used in SacC1 activity stimulation experiments.

Number of duplex	Duplex	Oligonucleotides	Stimulation of SacC1 activity
1	15-mer	5'-agCACGAGCTCGCta	Yes
		3'-tcGTGCTCGAGCGat	
2	13-mer	5'-tagCACCTGCtat	No
		3'-atcGTGGACGata	
3	18-mer	5'-atagCAGAGCTCCGtata	Yes
		3'-tatcGTCTCGAGGCatat	
4	18-mer	5'-atagCACGAGCTCTtata	Yes
		3'-tatcGTGCTCGAGAatat	
5	20-mer	5'-atagGAGCTCCGGCTTtata	Yes
		3'-tatcCTCGAGGCCGAAatat	
6	20-mer	5'-atagCACCTGCtataaagt	No
7	20-mer	3-'tatcGTGGACGatatttcag	

The present attempt (with the help of Al-Azytoonah University, Amman, Jordan) aims to purify a restriction endonuclease from *Saccharomyces cerevisiae*, This yeast is a member of the largest genus that produces antibacterials, antifungal, immunosuppressants, and industrial enzymes including restriction enzymes where several restriction enzymes were purified from it. In addition, establishing a line for production of this important enzyme is very important as new restriction enzymes are still required in order to increase the range of DNA manipulation.

MATERIALS AND METHODS

Sources of media and analytical chemicals

All analytical chemicals, Phage Lambda, plasmid DNAs, all enzymes, DNA markers and dialysis tubing cellulose membrane (diameter 6 mm and width 10 mm) were purchased from Sigma Aldrich. [α-^{33}P]dATP is from Amersham. Oligonucleotides (Table 1) and the DNA sequencing kit are from InterScience. Kd ladder is from Genomics, Agilent Technologie. pMXBIO control plasmid containing the endonuclease target sequence served as a double-strand template using the forward and reverse primer 1212 and 1233 are obtained from New England Biolabs.

Isolation of yeast

S. cerevisiae strain R-Z128 was used throughout the study. Yeast is grown in YPD medium (1% yeast extract, 2% Bacto-peptone and 2% dextrose) at 30ºC with constant shaking for 3 days and then the cells were lysed by sonication in ultrasonic bath (Sonicator Branson 5210) for 20 x 10s and broken cells were removed by centrifugation (2 min, 15,600 ×g). The supernatant was centrifuged at 3000 x g for 30 min at 4℃ and the supernatant has been used as the "crude extract" and the source of the enzyme.

Purification of the restriction endonuclease

Solid ammonium sulphate was added to the "crude extract" to form 0 - 30%, 30 - 50% and 50 - 80% saturation fractions, respectively. After centrifugation at 4000 g for 15 min, the pellet (of each

fraction) was suspended in 40 mM potassium phosphate (pH 7.5) containing 5 mM 2-mercaptoethanol, and 10% glycerol (buffer A) was then dialyzed with two changes against 4 L of the above buffer for 24 h and measured for endonuclease activity.

The 50 - 80% saturation fraction was found to have a high endonuclease activity, so it has been purified further by layered onto a 1.5 x 40 cm phosohocellulose column that was previously equilibrated with 4 L of buffer A. The endonuclease was eluted from the column with a linear gradient of 0 - 0.6N NaCl in buffer A. The peaked fractions (Fractions 45 - 50) were pooled together and dialyzed for 5 h against 4 L of 40 mM Tris-HCl, pH 7.5 containing 5 mM 2-mercaptoethanl and 10% glycerol (buffer B) and then loaded onto a DEAE-cellulose column (1.5 x 1m) that was previously equilibrated with buffer B. Two peaks were obtained, the first peak with a very small activity was eluted in the washing region (fractions 36 - 50), while the second peak with most endonuclease activity was eluted at 0.76 - 0.84 N NaCl (Fractions 113 - 121) (Figure 1). The peaked fractions were pooled, dialyzed for 5 h and loaded into Sephadex G-100 column (Figure 2) and eluted with buffer B (see Methods). The top fractions (82 - 85) were used as the purified restriction enzyme. The homogeneity of the enzyme was determined by 10% SDS–PAGE. The total yield of SacC1 protein was 2.5 mg.L^{-1} of induced culture. All steps were carried out at 4℃. Endonuclease activity, protein and carbohydrate concentrations were determined for all fractions.

Determination of endonuclease activity

Endonuclease activity was assayed according to Brown and Smith method (1980) by incubation of 0.3 pmol lambda DNA with 3.0 pmol of the purified endonuclease enzyme in a final volume of 20 µL containing 40 mM Tris–HCl (pH 7.5), 10 mM MgCl$_2$ and 0.1 mg.mL^{-1} bovine serum albumin for 1 h at 37℃. The reaction mixture was stopped by cooling at 0ºC and with the addition of 20 mM EDTA. The cleavage products electrophoresed on a 0.8% agarose gel and DNA was visualized by staining with ethidium bromide. One unit of the enzyme was defined as the amount of the enzyme that can digest 1 µg of Lamba DNA for 1 h at 37℃ under standard conditions.

Determination of the recognition sequence and the cleavage site

The recognition sequence of SacC1 was determined by mapping of the recognition sites on phage λDNA. The fragments predicted by

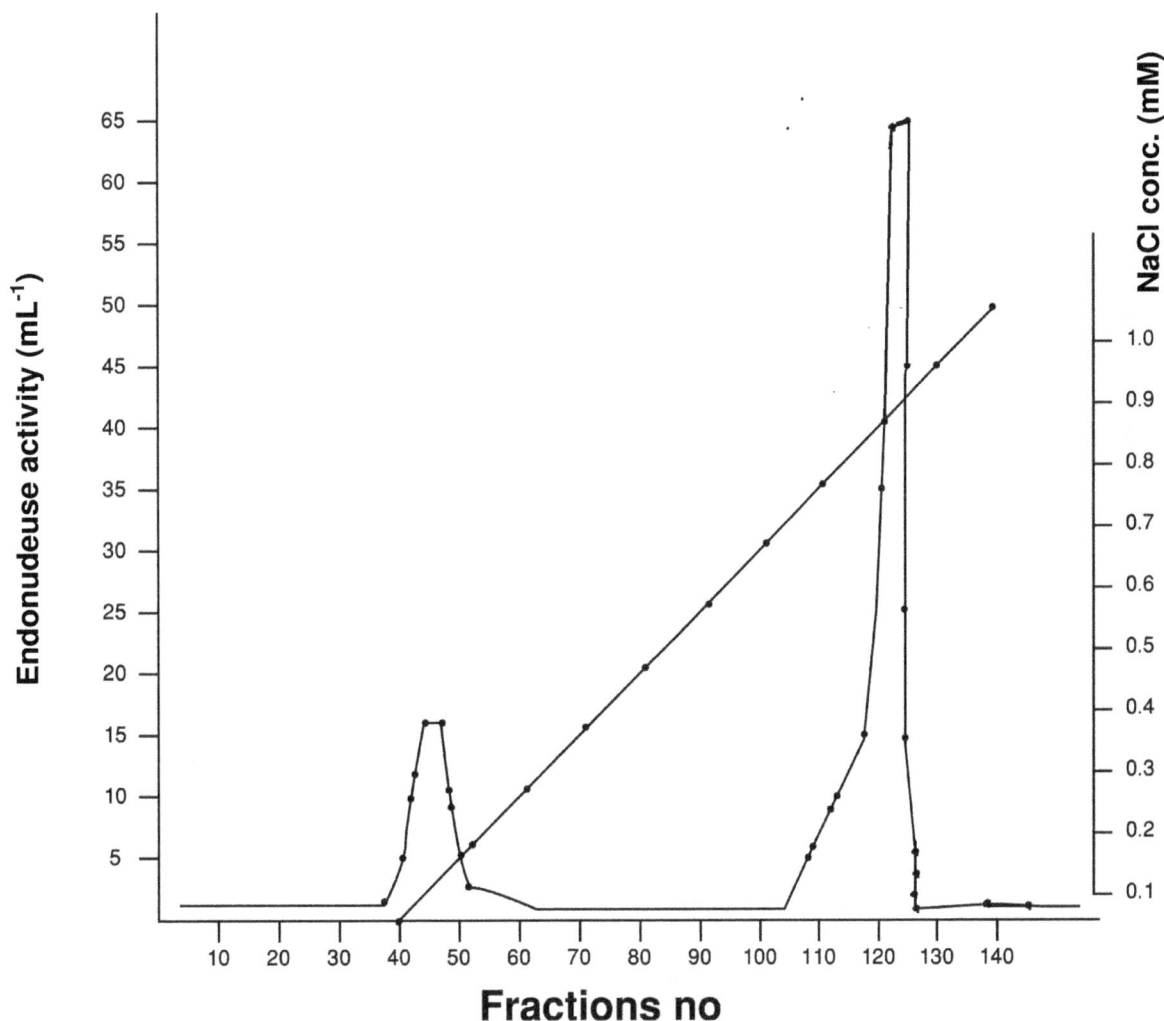

Figure 1. DEAE-cellulose chromatography of SacC1 endonuclease activity. The pooled active fractions from phosphocellulose column were eluted through the DEAE-cellulose column with 300 NaCl gradients (0-1M) as described in "Methods".

cleavage of the inferred recognition sites were compared with the observed fragments from SacC1 cleavage of different DNAs, while λDNA was used as a template to characterize the cleavage site of SacC1. A 20 mer oligodeoxyribonucleotide complementary to λDNA was used in direct sequencing through the SacC1 recognition site. Four dideoxy sequencing reactions using $[\alpha^{33}P]$ dATP and a DNA sequencing kit were carried out. The same primer and template were used in an extension reaction, which also included T7 DNA polymerase, dNTP and $[\alpha^{33}P]$ dATP. The extension reaction was heat inactivated, treated with SacC1 in the presence of 0.5 µM oligonucleotide duplex and the reaction mixture was divided into two. The aliquot was treated with T4 DNA polymerase in the presence of dNTP. All samples were diluted with sequencing dye solution and loaded on a standard sequencing gel along with the dideoxy sequencing reaction.

Determination of optimal pH, temperature, stability and metal ions

The influence of pH and the temperature of endonuclease activity were examined. The optimal pH was determined at 37°C in buffer

B, while the enzyme activity at various temperatures (20 - 80°C) were determined in buffer B, pH 8.0. The pH stability was studied by measuring the endonuclease activity at pH 8.0 after the enzyme was incubated at 37°C for 24 h. The effects of different metal ions were also determined.

Estimation of purity and the molecular weight

Sephadex G-100 was used according to the method of Andrews (1964). SacC1 samples (250 µg) were loaded onto a Sephadex G-100 column (1 x 50 cm) which was equilibrated with a buffer B. Separation was carried out at a flow rate of 0.2 mL.min^{-1}. The column was calibrated with dextran blue (>100 kDa), bovine serum albumin (66.2 kDa), Egg albumin (45 KDa), chymotrypsinogen A (25 kDa), lysozyme (14.4 kDa) and cytochrome C (12.4 kDa), Elution profiles were monitored by measuring absorbance at 280 nm and the specific Lamba DNA digestion profile by Sac I and SacC1. For the interpolation of unknown molecular mass, a linear dependence of the logarithm of the molecular mass on the elution time was assumed. Sodium dodecyl sulfate (SDS)-polyacrylamide gel electrophoresis (PAGE) was performed with 12%

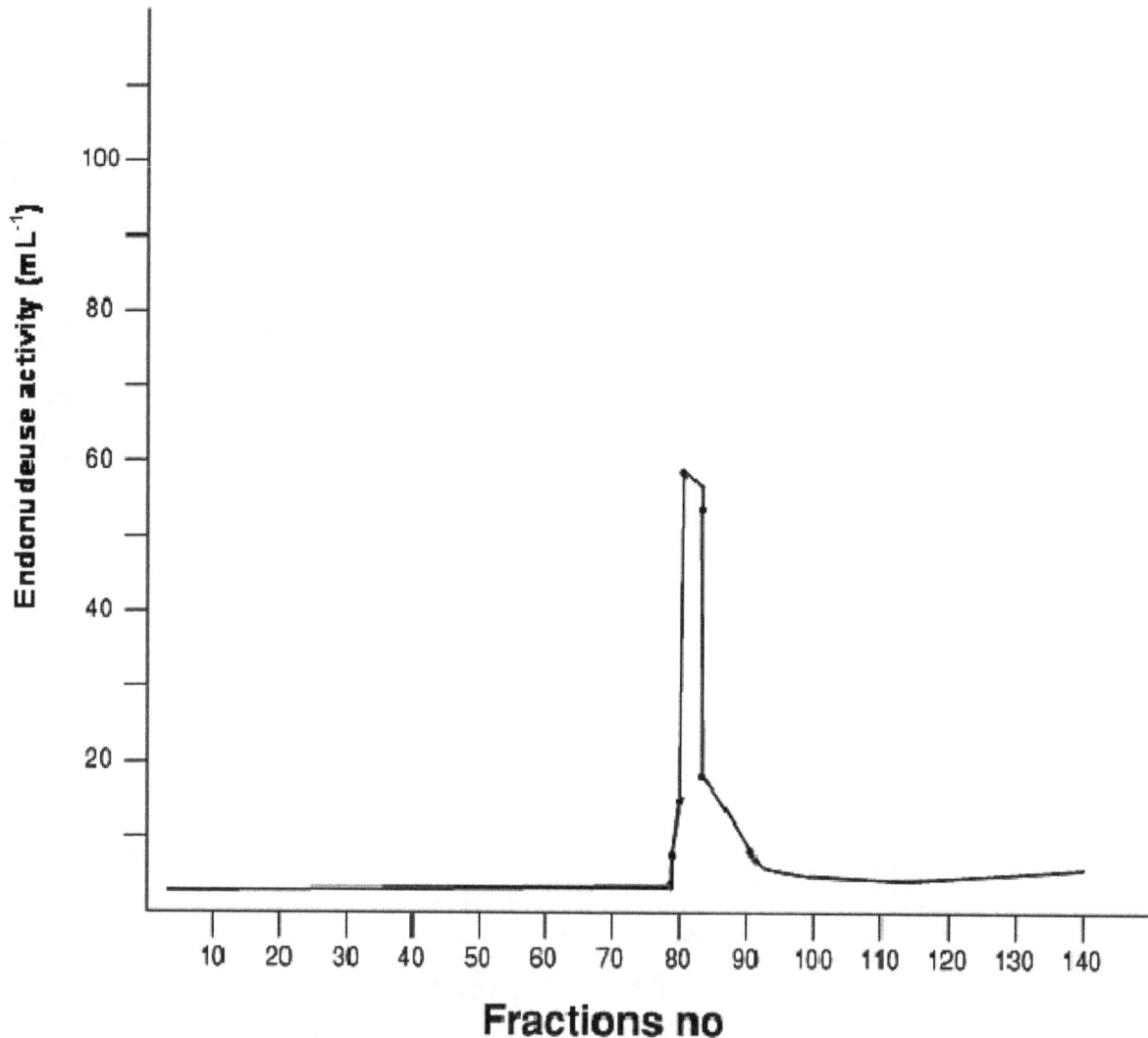

Figure 2. Sehadex G-100 Chromatography of SacC1 endonuclease activity. The pooled active fractions from DEAE-cellulose column were eluted through the Sephadex G-100 column as described in "Methods".

polyacrylamide gels as described by Laemmli (1970) and as modified by Maizel (1971). Molecular weight markers were obtained from Boehringer, Mannheim, Germany.

Estimation of protein and carbohydrates

Protein contents were determined by the method of Lowry et al. (1951) using Bovine serum albumin as a standard. Carbohydrates contents were determined by the method of Dubois et al. (1956) using glucose as a standard. In the case of electrophoresis, an endonuclease activity unit is defined as the ability of one volume of restriction enzyme to completely cut 1 ug of lambda DNA at 37°C for 1 h under standard conditions.

Determination of the optimal conditions for the *in vitro* endonucleolytic activity of the purified restriction enzyme

To determine the optimal conditions for the *in vitro* endonucleolytic activity of SacC1, different buffers were tested, taking into account ionic strength, pH, Mg^{2+} and ion concentrations.

RESULTS

Purification of the SacC1

Table 2 summarizes the purification steps.

Homogeneity

The homogeneity of the enzyme was determined by 10% SDS–PAGE. The total yield of purified enzyme's protein was $2.5\,mg.L^{-1}$ of induced culture.

Determination of the recognition site

To determine the substrate specificity of the enzyme, cleavage sites of SacC1 on λDNA were mapped by double digestion with SacI and SacC1 (Figure 3). A

Table 2. Purification steps of SacC1.

Step	Total volume (mL)	Total activity (units)	Total protein (mg)	Specific activity (unit.mg^{-1})	Purification	(%)
Crude extract	500	2345	7.61	308.2	1	100
50-80% (NH$_4$)$_2$SO$_4$ fraction	28	1495.2	3.58	417.7	1.4	63.7
Phosphocellulose	50	1296	0.61	2124.6	6.9	42.7
DEAE-Cellulose	30	827.6	0.18	4597.7	14.9	35.3
Sephadex G-100	16	464.6	0.007	66371.4	215.4	19.8

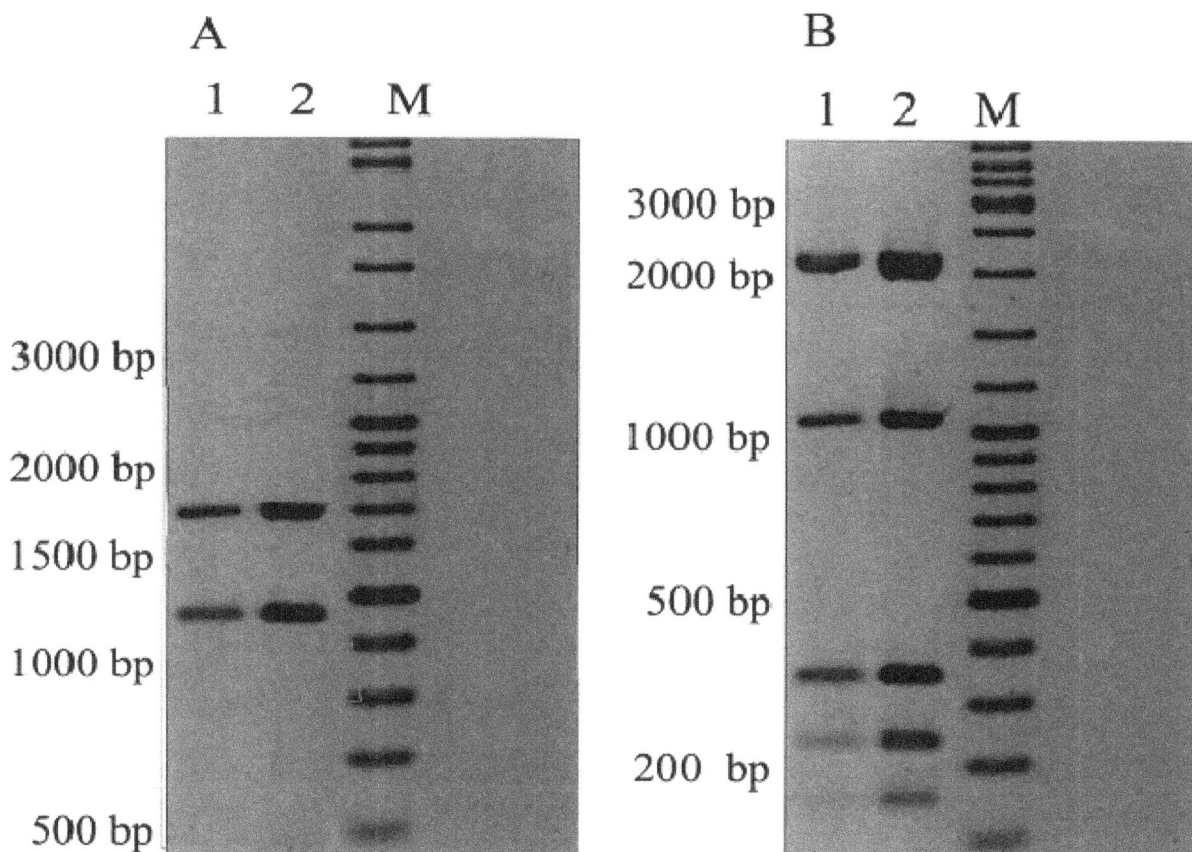

Figure 3. Determination of the recognition sequence of SacC1 a) Comparison of the plasmid DNA cleavage by SacC1 (lane 1) and SacI (lane 2).B) double digestion of the plasmid DNA with SacC1 and SacI. M) DNA molecular weight marker (100-10000).

computer-aided search of homologous nucleotide sequences revealed only one common sequence, 5'GAGCTC3', for the mapped SacC1 sites. The numbers of DNA fragments generated by SacC1 and SacI cleavage of λDNA and ϕX174 DNAs were 2 and 1 sites, respectively. pBR322, pUC19, pTZ19R, and M13mp18 were not cleaved since they do not contain 5'GAGCTC3' sequences.

The cleavage position was determined by the method of primer extension (Sanger et al., 1977) through comparison of the dideoxy sequence ladders with fragments generated by SacC1 cleavage and T4 DNA polymerase action on the digestion product (Figure 4). The fragment generated by SacC1 digestion co-migrates with the T band of the sequence ladder indicating that the cutting point is 5 nt away from the recognition sequence 5'-GAGCT/C3'. The single band obtained after treatment with T4 polymerase co-migrates with the C band (lane 1) of the sequence ladder, confirming that the cleavage point on the complementary DNA strand is 7 nt away from the recognition sequence. In summary, SacC1 cleaves to double-stranded DNA, generating fragments five and seven bases upperstream and downstream of its recognition sequence, respectively.

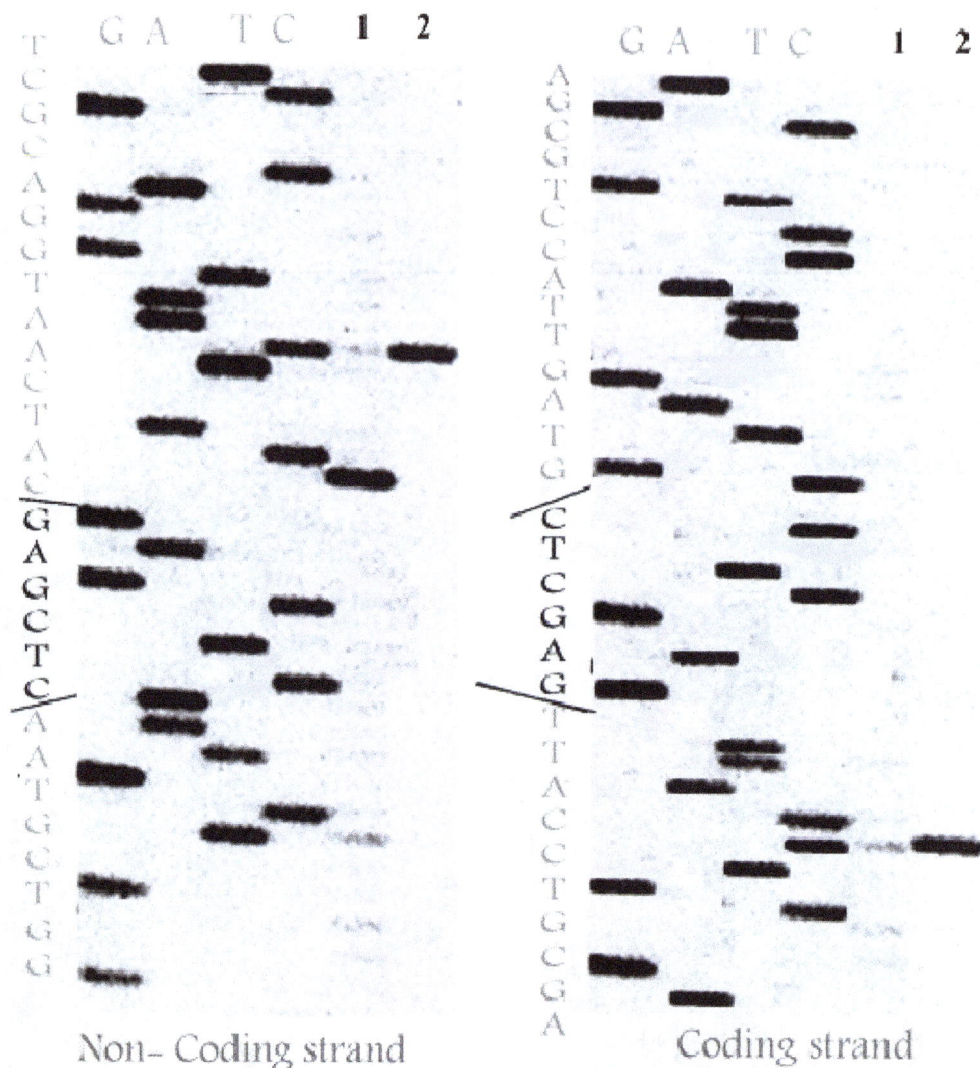

Figure 4. Determination of the cleavage site. Autoradiogram is showing the extension polymerization products digested or not with SacC1 and the corresponding sequence ladder. The cleavage sites are indicated in bold letters. Lane 1. T4 polymerase action on the SacC1digest, Lane 2. The product of the primed synthesis reaction cleaved with SacC1.

Estimation of molecular weight

The molecular mass of SacC1 was estimated by gel filtration on a Sephadex G-100 column to be 58,000 and by SDS-gel electrophoresis to be 64,000 (Figure 5).

Effects of pH and temperature

The optimum pH for the enzyme was determined by estimating the percent activity at different pH values. The pH optimum of the enzyme has a range of (7.5 - 8.5), while the optimum temperature was determined by examining the enzyme activity at different temperatures. Preliminary experiments showed its stability at room

temperature for 30 min, but for 6 h at 4 °C and more than three months at -20 °C (Table 3).

Metal ion requirements

The enzyme appears to be dependent on Mg^{+2} and Mn^{2+} 4- which increases its activity by 4- and 2-folds, respectively, while other cations decrease its activity to some extents. Salts such as sodium or potassium chlorides have no effect at all, while the presence of B-mercaptoethanol increases the activity slightly. ATP has no effect on it (Table 4). The maximum activity of the enzyme was observed at 37 °C in the presence of 40 mm Tris–HCl (pH 7.5), 10 mM $MgCl_2$ and 0.1 mg.mL^{-1} BSA in

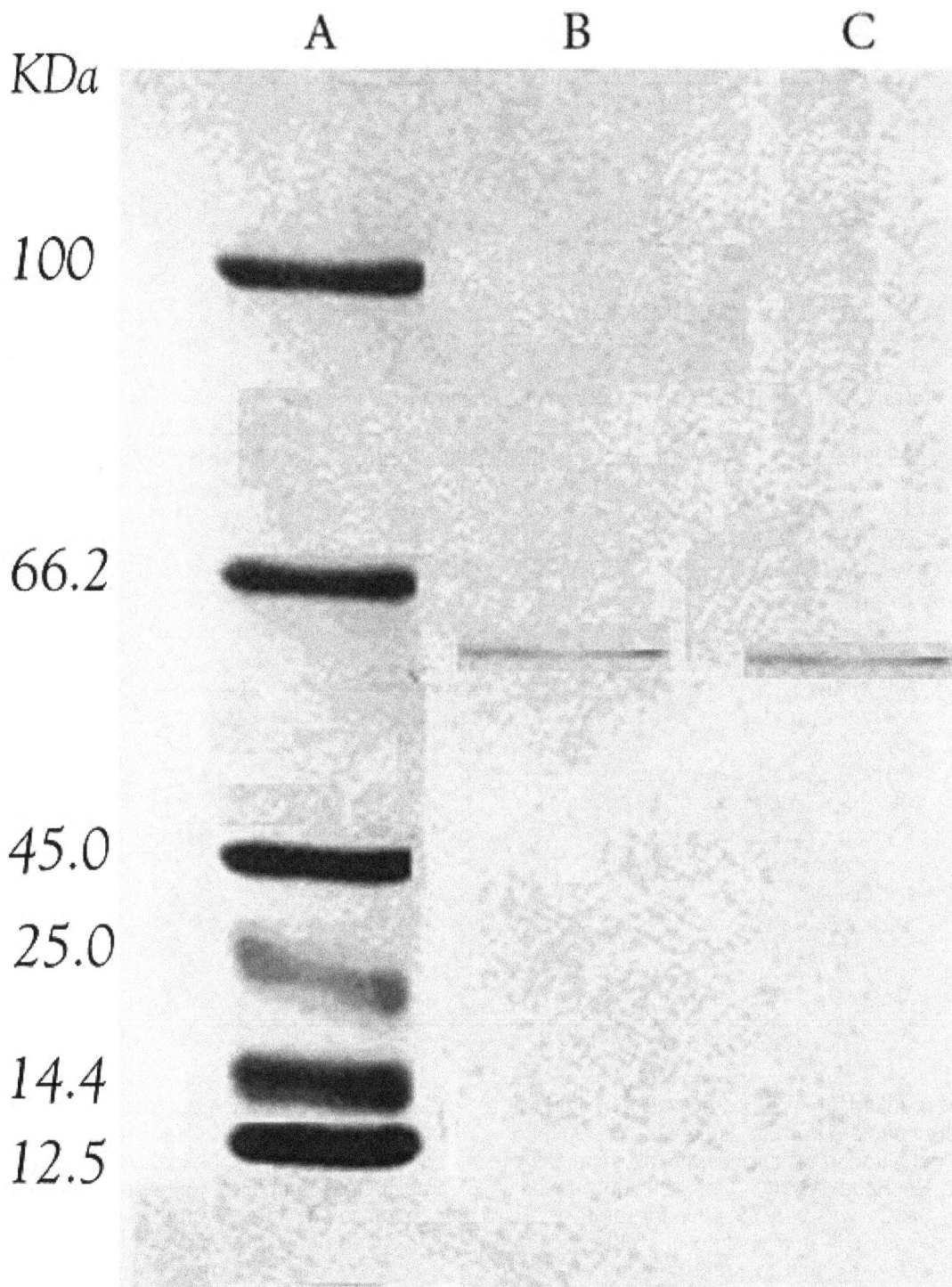

Figure 5. Determination of the molecular weight of SacC1 (B and C) while (A) is the protein ladder (10-100 kDa) was used as the protein molecular weight marker.

the absence of any salt or cofactors (Table 4).

DISCUSSION

SacC1 is a novel restriction endonuclease from Saccha-romyces cerevisiae that recognizes the palindromic sequence 5'CTCGAC3' cleaving both DNA strands upstream and downstream of its recognition sequence and makes a staggered cut at a distance of five bases from the recognition sequence on the upper strand and at the seventh base on the complementary strand.

Table 3. Effect of the temperatures on SacC1.

pH	Relative activity (%)	Temperature (°C)	Relative activity (%)
6.0	14	15	15
6.5	32	20	17
7.0	80	30	60
7.5	100	37	100
8.0	100	45	55
8.5	100	50	35
9.0	16	55	30
9.5	4	60	3

Table 4. Influence of different metal ions and reagents on SacC1.

Addition	Concentration (mM)	Relative activity (%)
No addition	None	100
Mg^{2+} ($MgCl_2$)	10.0	420
Ca^{2+} ($CaCl_2$)	10.0	98
Mn^{2+} ($MnCl_2$)	10.0	240
Zn^{2+} ($ZnCl_2$)	5.0	30
Fe^{3+} ($FeCl_3$)	5.0	120
Hg^{2+} ($HgCl_2$)	10.0	40
Ni^{2+} ($NiCl_2$)	10.0	51
Cu^{2+} ($CuCl_2$)	10.0	50
Cd^{2+} ($CdCl_2$)	1.0	78
EDTA	5.0	24
Na^+ (NaCl)	1.0	110
K^+ (KCl)	1.0	100
2-mercaptoethanol	0.5	112
Dithiothreitol	0.5	60
Sodium dodecyl sulfate	1.0	20
ATP	1.0-10.0	0
Urea	1.0	0
H_2O_2	1.0	0

SacC1 was purified from *S. cerevisiae* 215 fold with 20% recovery by ammonium sulphate precipitation, dialysis, and gel filtration using phosphocellulose, DEAE-cellulose and Sephadex G-100. The technique used for purification of SacC1 was very effective for removal of all interacting proteins. The purified fractions obtained from Sephadex G-100 were able to break lambda DNA into smaller sizes, making this enzyme a restriction enzyme.

The enzyme has an optimal pH range (7.5-8.5) and is active at 37°C. Preliminary experiments showed its stability at room temperature for 30 min, but for 6 h at 4°C and more than three months at -20°C. The enzyme is dependent on cations and Mg^{2+} (or Mn^{2+}) which increases its activity by 4- and 2- folds, respectively.

SacC1 shares similar characteristics with Sac I from *Streptomyces achromogenes* (Zhuravleva et al. 1987) as well as Sst1 from *Streptomyces Stanford* and Psp124B1 from *Pseudomonas* species. They all recognize 5'CTCGAC3' and have optimal temperatures of 37°C and optimal pH around 8.0. Sac I has a molecular weight of 50,000 while SacC1 has a molecular weight of 64,000. EcoRI has completely different characteristics from the above two enzymes (Tamerler et al., 2001; Dai, 2007).

Cleavage on both sides of the recognition sequence is a characteristic of Type IIB systems but all IIB enzymes studied so far have been found to recognize discontinuous (usually asymmetric) sites and a distinctive subunit/domain organization that is not present in the SacC1 enzyme. On the other hand, excision of small fragments by SacC1 bears some similarities to the action of *Bcgl* (Kong et al., 1993) and *HaeIV* (Piekarowicz et al., 1999) belonging to Type IIB system. However, conversely to these enzymes, SacC1 recognizes non-interrupted sequences and ATP do not have any influence on its activity.

Type IIB enzymes excise DNA fragments by cutting

either side of their sites and some of them, especially *R. BpII* and *R. AloI* recognizes palindromic sequences (Roberts, 2005). Some of the Type IIB systems need no cofactor other than Mg^{2+} to cut DNA, like SacC1.

The characteristics of the cleavage reaction catalyzed by SacC1 are reminiscent of those already reported for other homing endonucleases belonging to the LAGLIDADG family (Belfort and Roberts, 1997; Wende et al., 1996). The similarities are a requirement of Mg^{2+} (or Mn^{2+}) for cleavage to take place, optimal activity at alkaline pH and stimulation of the reaction by moderate concentrations of the monovalent cation. SacC1 catalyzes, also, a double-strand cleavage in the vicinity of the insertion site, creating 3'-overhangs of 5 nt, a feature common to all the LAGLIDADG homing endonucleases studied so far (Jurica and Stoddard, 1999). The cleavage occurs 7 nt downstream from the insertion site for the coding and non-coding strands, respectively, a pattern also observed for *I-PorI* by Andersen et al. (1994).

ACKNOWLEDGEMENTS

The author wishes to thank Mr. Jalal Ahmad Odeh and Miss Ghadir Al-Badawi, Faculty of Pharmacy, University of Al-Zaytoonah, Amman, Jordan for their great contributions in this research.

REFERENCES

Al-Khafagi KAM (1999). Extraction of a restriction enzyme from a local isolate of the thermophilic bacteria *Bacillus stearothermophilus*. MSc thesis, College of Science, University of Baghdad, Baghdad, Iraq.

Andrews P (1964). Estimation of the molecular weights of proteins by Sephadex gel-filtration. Biochemi. J. 91: 222-233.

Andersen JL, Thi-Ngoc HP, Garrett RA (1994). DNA substrate specificity and cleavage kinetics of an archaeal homing-type endonuclease from *Pyrobaculum organotrophum*. Nucleic Acids Research 22(22): 4583-4590.

Belfort M, Roberts RJ (1997). Homing endonucleases: keeping the house in order. Nucleic acid Res. 25(17): 3379-3388.

Brown NL, Smith M (1980). A general method for defining restriction enzyme cleavage and recognition site. Methods in Enzymology 65: 391-404.

Dai X (2007). Purification of wild type EcoRI endonuclease and EcoRI endonuclease RS187 crystal growth of WT EcoR I endonuclease-DNA complex and endonuclease RS187-DNA complex. MSc Thesis, University of Pittsburgh.

Dubois M, Gilles KA, Hamilton JK, Rebers RA, Smith F (1956). Colorimetric method for determination of sugars and related substances. Anal. Chem. 28: 350-360.

Halford SE (2001). Hopping, jumping and looping by restriction enzymes. Biochem. Soc. Trans. 29(4): 363-373.

Jurica MS, Stoddard BL (1999). Homing endonucleases: Structure, function and evolution. Cellular and molecular life sciences 55(10): 1304-1326.

Kong H, Morgan RD, Maunus RE, Schildkraut I (1993). A unique restriction endonuclease, BcgI, from *Bacillus coagulans*. Nucleic Acids Res. 21: 987-991.

Laemmli UK (1970). Cleavage of structural proteins during the assembly of the head of Bacteriophage T4. Nature 227: 680-685.

Lowry OH, Rosebrough NJ, Furr AL, Randall RJ (1951). Protein measurement with the Folin phenol reagent. J. Biol. Chem. 193: 1209-1220.

Maizel JV (1971). Polyacrylamide gel electrophoresis of viral protein. In: Maramorosch K, Koprowshi H (eds), Methods in Virology 5, Academic Press, New York, pp. 179-246.

Murray NE (2000). Type I Restriction Systems Sophisticated Molecular Machines. Microbiology and Molecular Biology Reviews 64: 412-434.

Piekarowicz A, Golaszewska M, Sunday AO, Siwinska M, Stein DC (1999). The HaeIV restriction modification system of *Haemophilus aegyptius* is encoded by a single polypeptide. J. Mole. Biol. 293: 1055-1065.

Putrus SG (1995). Extraction and characterization of an isoschizomer for the restriction enzyme Hind III from locally isolated *Haemophilus influenza*. M.Sc thesis, College of Science, University of Baghdad, Baghdad.

Richard JR, Belfort M, Bestor T, Bhagwat AS, (2003). Survey and Summary: A nomenclature for restriction enzymes, DNA methyltransferases, homing endonucleases and their genes. Nucleic Acids Res. 31(7): 1805-1812.

Roberts RJ (2005). How restriction enzymes became the workhorses of molecular biology. Proceedings of the National Academy of Science 102(17): 5905-5908.

Roberts RJ, Macelis D (2007). REBASE – restriction enzymes and methylases. Nucleic Acids Res. 29: 268-269.

Roberts RJ, Vincze T, Posfai J, Macelis (2005). D. REBASE – restriction enzymes and DNA methyltransferases. Nucleic Acids Research 33: D230-D232.

Sanger F, Nicklen S, Coulson AR (1977). DNA sequencing with chain terminating inhibitors. Proceedings of the National Academy of Science 74: 5463-5467.

Tamerler C, Onsan ZI, Kirdar B (2001). A comparative study on the recovery of EcoRI endonuclease from two different genetically modified strains of *Escherichia coli*. Turk. J. Chem. 25: 63-71.

Wende W, Grindl W, Christ F, Pingoud A, Pingoud V (1996). Binding, Bending and Cleavage of DNA Substrates by the Homing Endonuclease PI-SecI. Nucleic Acids Res. 24(21): 4123-4132.

Wilson GG (1991). Organization of restriction modification systems. Nucleic Acids Res. 19: 2539-2566.

Zhuravleva LI, Oreshkin EN, Bezborodov AM (1987). Isolation and purification of restriction endonuclease Sac I from *Streptomyces achromogenes* ATCC 12767. Prikladnaia Biokhimiia I Mikrobiologiia 23(2): 208-215.

4

Overview on *Echinophora platyloba,* a synergistic anti-fungal agent candidate

Avijgan Majid[1]*, Mahboubi Mohaddesse[2], Darabi Mahdi[2], Saadat Mahdi[3], Sarikhani Sanaz[1] and Nazilla Kassaiyan[1]

[1]Iranian Traditional Medicine research Center, Isfahan University of Medical Sciences, Isfahan, Iran.
[2]Barij Essence pharmaceutical Company, Isfahan, Kashan, Iran.
[3]Shahr-e-kord University of Medical Sciences, Shahr-e-kord, Iran.

Echinophora Platyloba DC. is one of the four native species of this plant in Iran. The aim of this five-step study was to investigate the antimicrobial properties of this plant. Dried aerial parts of the plant were extracted by ethanol 70% in percolator. The antimicrobial activity of ethanolic extract was evaluated against dermatophytes, *Candida (C) albicans* and gram positive bacteria by agar dilution method and microbroth dilution assay. Finally, the synergistic effect of Amphotericin B plus 5% ethanolic extract against *C. albicans* was determined by measuring MIC (minimum inhibitory concentration) and MLC (minimum lethal concentration) values. Gram positive bacteria were resistant to the extract according to measurement of zones of inhibition; *Trichophyton schenlaini* and *Trichophyton verucosum* were sensitive to concentrations $\geq$ 35 mg/ml, while other dermatophytes showed various susceptibilities to extract. MIC value of 5% ethanolic extract was 2 mg/ml against *C. albicans* using broth micro dilution method. In synergism assay, there was a 50% reduction in MIC and a 75% reduction in MLC values of the mixture of Amphotericin B and 5% ethanolic extract against *C. albicans* in comparison to Amphotericin B alone. Regarding this study, some degrees of synergy was recorded in the combination of Amphotericin B plus *E. platyloba* extract covering *C. albicans* which represented promising finding in antifungal therapy.

Key words: Amphotericin B, *Candida albicans*, *Echinophora platyloba*, ethanolic extract, traditional medicine.

INTRODUCTION

Historically, plants have provided a source of inspiration for novel drug compounds, as plant derived medicines have made large contributions to human health and well being (El Astal et al., 2005). Natural products have served as a major source of drugs for centuries and about half of the pharmaceuticals in use today are derived from natural products (Clark, 1996). Also, another study reports that 25 - 50% of current pharmaceuticals are derived from plants (Cowan, 1999). Microbiologists are combing the earth for phytochemicals which could be developed for treatment of infectious diseases. Plants are rich in a wide variety of secondary metabolites, such as tannins, terpenoids, alkaloids and flavonoids, which have been reported to have *in vitro* antimicrobial properties (Cowan, 1999). A study demonstrates the antifungal activity of extracts of some Thai medicinal plants which can be excellent candidates for the development of remedy for opportunistic fungal infections in AIDS sufferers (Phongpaichit et al., 2005). Another study reports the activity of traditional medicinal herbs from Balochistan, Pakistan against *C. albicans*, *Bacillus subtilis* and *Bacillus cereus* (Zaidi et al., 2005).

In Iran, traditional medicine has a major therapeutic role and for thousands of years, traditional healers have been using different plants to treat patients. *E. platyloba DC.* is one species of *Echinophora* genus (Rechinger, 1987). The 10 different species of this plant has been defined as: *Echinophora tenuifolia*, *E. platyloba* DC., *Echinophora sibthorpiana Guss*, *Echinophora anatolica Boiss*, *Echinophora cinera*, *Echinophora vadiaus Boiss*,

*Corresponding author. E-mail: irtradmed@mui.ac.ir, avijgan@yahoo.com.

Figure 1. Picture of *Echinophora platyloba* in the field.

Echinophora orientalis Hedge and *Lamond, Echinophora tournefotii joub, Echinophora trichophylla Sm, E. spinosa.* Four of these species are native to Iran: *Echinophora orientalis Echinophora* sibthorpiana, *Echinophora* cinerea and *E. platyloba* (Vanden et al., 2002). *E. platyloba* is mainly used for food seasoning in Iran (Chaharmahal va Bakhteyari province) (Sadrai et al., 2002), rather than preventing tomato paste and pickles from mold. The hypothesis of these serial studies was based on the plant's specific characteristic as food preserver which might have been due to its antimicrobial properties. During past 7 years, the authors did several studies on antimicrobial activity of *E. platyloba* ethanolic extract. This article is a review of all previous studies indicating the effectiveness of ethanolic extract of *E. platyloba* against *C. albicans,* dermathophytes and some gram positive bacteria, in addition to its significant synergy with Amphotericin B against *C. albicans* (Figure 1).

MATERIALS AND METHODS

Plant material

The plant was collected from the southwestern parts of Iran (Shahr-e-kord). A voucher specimen of plant was deposited in the Herbarium at the Faculty of Sciences, Isfahan University, Isfahan, Iran. The aerial parts of the plant were separated, shade dried and grinded into powder using mortar and pestle. The prepared powder was kept in tight containers protected completely from light.

Method for preparation of plant extract

Extraction of ethanolic extract was carried out by macerating 100 g of powdered dry plant in 500 ml of 70% ethanol (Istelak, Iran) for 48 h at room temperature. Then, the macerated plant material was extracted with 70% ethanol solvent using percolator apparatus (2 liter volume) at room temperature. The plant extract was removed from percolator, filtered through Whatman filter paper (NO.4) and dried under reduced pressure at 37°C with rotator evaporator before being added to ethanol as the solvent. Three different concentrations of ethanolic extracts (4, 5.2, and 11%) were prepared according to amount of evaporation.

Microorganisms

Cultures of Staphylococcus aureus ATTC 25923, *Staphylococcus epidermidis* and *Steptococcus pyogones* ATTC 19615 were obtained from Biotechnology Research Center of Iran, Tehran, Iran. C. albicans ATCC 10231 and dermathophytes *(T. schenlaini, T. verucosum, Trichophyton rubrum, Trichophyton mentagrophyte, Trichophyton violaceum, Epidermophyton flucosum, Microsporum gypsum, Microsporum canis* were obtained from Fungi and Parasitological department, Faculty of Medicine, Isfahan University of Medical Sciences, Isfahan, Iran. Bacterial suspensions were made in Brain Heart Infusion (BHI) broth to a concentration of approximately 10^8 CFU/ml using standard routine spectrophometrical methods. Suspensions of fungi were made in Sabouraud dextrose broth. Subsequent dilutions were prepared from the above suspensions, to be used in the tests.

4- Antimicrobial screening

Study No.1

This initial study was performed to evaluate the probability of any antimicrobial potency of the plant against ethanolic extracts (5 and 11%) in addition to the plant which derived essential oil. Essential oil was obtained by hydrodistillation method using a Clevenger apparatus. The yield was 0.3%. In brief, microbial suspension containing 10^8 CFU/ml of bacteria was swabbed and spread on Muller- Hinton agar. Three sterile blank discs (Padtan Tab Co, Tehran, Iran), impregnated with 20 μl of 5 and 11% ethanolic extracts, and essential oil and were placed on the inoculated agar. Kefline 30 mcg/disc, Cloxaciline 30 mcg/disc and Penicillin G 100 u/disc (Himedia Company in Iran) were used as positive control standards to determine the susceptibility of gram positive bacteria. The inoculated plates were incubated at 37˚C for 24 h. The antimicrobial activity was evaluated by measuring the diameter of zone of inhibition against the test microorganisms (Mahboobi et al., 2006).

The minimal inhibitory concentrations (MICs) of extracts against different bacteria were determined by micro broth dilution assay. The extract was twofold serially diluted with 10% Dimethylsulfoxide (DMSO) which contains 25 - 0.39 mg of extract per testing well. Muller Hinton broth was used as broth media exception of *S. pyogones* that was used as MHB supplemented with 3% horse lysed blood. After shaking, 100 μl of the extract was added to each well. The suspension of each organism was adjusted to 10^4 - 10^5 CFU ml $^{-1}$ and then 100 μl was added to each well and cultivated at 37˚C. MIC was defined as the first well with no visible growth after 24 h. Minimal bactericidal concentration (MLCs) were determined as the lowest concentration resulting in no growth on subculture (M7-A7, NCCLS, Wayne, PA, 2006).

Study No.2

The activity of 0, 35, 50, 150 and 250 mg/ml of ethanolic extract of E. platyloba was tested against dermatophytes by agar dilution method using sabouraud dextrose agar (SDA). Inocula were prepared by growing isolates on SDA slopes. All dermatophytes were incubated for 7 days at 30˚C. Slopes were flooded by normal saline. The final concentrations of dermatophytes were 2.5×10^3 - 2.5×10^4 Cfu ml^{-1}. 20 μl of inoculums were spread onto each plate.

The plates were incubated at 30˚C for 5 - 7 days. The MIC was the lowest concentration at which there was no visible fungal growth after incubation (Mustafa NK et al., 1999).

Study No. 3

The antifungal activity of ethanolic extract of E. platyloba ATCC 10231 was evaluated against C. albicans. The activity of 0, 35, 50, 150 and 250 mg extract ml^{-1} was tested against C. albicans ATCC 10231 by agar dilution assay using sabouraud dextrose agar (SDA) and Inocula 0.4×10^4 - 5×10^4 CFU ml^{-1}. The MIC was defined after 48 h.

Study No.4

A single colony of the strain of C. albicans ATCC10231 to be tested was grown overnight at 35˚C in sabouraud dextrose broth. The inoculum was prepared by diluting the overnight growth with 0.9% NaCl to obtain a turbidity of 0.5 McFarland (1×10^5 to 1×10^6 CFU/ml). The suspension was swabbed on sabouraud dextrose agar. To test the antifungal activity of extract, 20 μl of ethanolic extract (5, 4 and 11%) and 5, 10, 15 and 20 μl of essential oil were placed onto the inoculated plates. Disk containing Amphotericin B (Himedia Mumbai, India) and disk impregnated with DMSO were used as controls. The plates were incubated at 35˚C for 48 and the diameter of inhibitory zones (mm) was measured (Griggs et al., 2001).

Minimal inhibitory concentration: The minimal inhibitory concentrations (MICs) of extract and Amphotericin B against C. albicans ATCC 10231 were determined by micro broth dilution method. The E. platyloba extract and Amphotericin B were two fold serially diluted separately with 10% DMSO which contain 25 - 0.39 mg/ml extract and 256 - 1 μg/ml Amphotericin B per testing well. Sabouraud dextrose broth was used as broth media. After shaking, 100 μl of agent solutions was added to each well. The suspension of C. albicans was adjusted to 10^4 - 10^5 CFU ml $^{-1}$ and then 100 μl was added to each well and cultivated at 35˚C. MIC was defined after 24 h for yeast (Griggs et al., 2001).

Study No.5

The synergy between ethanolic extract of E. platyloba and Amphotericin B: The Amphotericin B was two fold serially diluted with 10% DMSO which contain 16 - 0.125 μg/ml Amphotericin B per testing well also each well contain 0.78 mg/ml of ethanolic extract (< MIC of E. platyloba extract against C. albicans found in study no 4). The inoculum was adjusted to 10^4 - 10^5 CFU ml $^{-1}$ and was added to each well and cultivated at 35˚C. MIC was defined after 24 h.

RESULTS

Result of study NO 1 (Avijgan et al., 2005)

MIC and MLC values for 5% extract against S. aureus, S. epidermidis, and S. Pyogones are shown in Table 1. The diameter of zone of inhibition for antibiotics varies from 30 - 56 mm and the average measurements for 5% ethanolic extract, 11% ethanolic extract and the essential oil were 8, 8 and 12 mm, respectively.

Table 1. MIC and MLC values of 5% ethanolic extract of *E. platyloba* against gram positive bacteria.

Names measurements (mg/ml)	S. aureus	S. epidermidis	S. pyogones
MIC	3.1	1.5	3.1
MLC	12.5	6.2	6.1

Table 2. The susceptibility for some dermatophyts to 5% ethanolic extract of *Echinophora platyloba*.

Concentrations of extract (mg/ml) fungi	35	50	150	250
T. schenlaini	S	S	S	S
T. verucosum	S	S	S	S
T. rubrum	R	R	R	R
M. canis	R	R	R	S
M. gypsum	R	R	R	R
T .violaceum	R	R	S	S
T .mentagrophyte	R	R	R	S
E. flucosum	R	R	R	S

S = Sensitive; R = Resistant.

Table 3. The susceptibility of *C. albicans* ACTT 10231 to different types and concentrations of *E. platyloba* and Amphotericin B by disc diffusion method.

DISC content	Ethanolic extract 4%	Ethanolic extract 5%	Ethanolic extract 11%	Amphotericin B	Ethanolic extract 5% + Amphotericin B	Aqueous extract	ETHANOL 70%
Zone of Inhibition (mm)	8	13	12	18	22	0	7
MIC (mg/ml)	NC	1569	NC	2	1	NC	NC
MLC(mg/ml)	NC	3125	NC	8	2	NC	NC

NC= not checked.

Results of study NO.2: (Avijgan et al., 2006)

The susceptibility of dermathophytes to 5% ethanolic extract are shown in Table 2.

Result of study NO 3 (Avijgan et al., 2006)

In this study the authors tested the susceptibility of *C. albicans* to 5 different concentrations (0, 35, 50, 150 and 250 mg/ml) of 5% ethanolic extract. After 24 h of incubation period, the yeast grew only in the tube used as the control; while no growth was recorded in the other 4 tube containing 35, 50, 150 and 250 mg/ml of extract.

Result of study NO 4 (Avijgan et al., 2006)

C. albicans grew in two tubes in this study, the one used as control tube and the one containing 1 mg/ml of the extract. No growth was recorded in tubes containing 2, 4, 8, 16, 32, 64, 128 and 256 mg/ml concentrations of extract.

Result of study NO.5 (Mahbobi et al., 2009)

The results of this part of the study are shown in Table 3.

DISCUSSION

The medicinal properties of plant species have made an outstanding contribution in the origin and evolution of many traditional herbal therapies. These traditional knowledge systems have started to disappear with the passage of time due to scarcity of written documents and relatively low income in these traditions. Over the past few years, medicinal plants have regained a wide recognition due to an escalating faith in herbal medicine in view of its lesser side effects compared to allopathic medicine in addition, the necessity of meeting the requirements of medicine for an increasing human population (Chandra et al., 2006).

E. platyloba is one of the four species of the plant native to Iran (Rechinger, 1987). As a traditional herb, It is used for food seasoning (Sadrai et al., 2002), rather than preventing the pickles and tomato past from mold. Various studies have expressed the anti fungi properties of plants (Na et al., 2003; Quiroga et al., 2004; Shin et al., 2003; Akagawa et al., 1996; Nwosu et al., 1998; Phongpaichit et al., 2005; Zaidi et al., 2005). According to previously done surveys, E. platyloba is an enriched source of saponin, alkaloid and flavonoid (Nourozi, 1989), while there are studies demonstrating that, these three substances have significant antifungal activity (Renault et al., 2003; Mel'nichenko et al., 2003; Kariba et al., 2002; Quiroga et al., 2004). The hypothesis of these serial studies was based on the plant's specific characteristic as food preserver which might have been due to its antimicrobial properties.

In the first study, the 11 and 5% ethanolic extracts and the essential oil, 3% did not exhibit any inhibitory activity against S. aureus, S. epidermidis and S. pyogones. According to Table 1, MIC and MLC values of 5% extract against bacteria were much higher than MIC and MLC values of antibiotics used in this test. Also, diameters of zones of inhibition of 5% extract, 11% extract, and essential oil of plant in comparison to diameters of zones of inhibition of used antibiotics, revealed the fact that the plant dose not have effective antibacterial properties. In the second trial, as it is apparent in Table 2, T. schenlaini and T. verucosum were sensitive to concentrations ≥ 35 mg/ml and they grew only in the plate used as control. T. rubrum and M. gypsum showed resistance to all concentrations. T. mentagrophyte, M. canis and E. flucosum were resistant to 35, 50 and 150 mg/ml but sensitive to 250 mg/ml. T. violaceum was resistant to 35 and 50 mg/ml but sensitive to 150 and 250 mg/ml. according to this result, the 5% ethanolic extract showed antifungal activity against T. schenlaini and T. verucosum. Further studies are definitely needed to evaluate the susceptibility of T. schenlaini and T. verucosum to lower concentrations of the extract. In the third study, the 5% ethanolic extract showed antifungal activity against C. albicans, in concentrations ≥ 35 mg/ml (35, 50, 150 and 250 mg/ml). Yeast grew only in the tube used as control.

In the forth study C. albicans growth was inhibited by concentrations ≥ 2 mg/ml of extract (2, 4, 8, 16, 32, 64, 128 and 256 mg/ml). The last study was done on the base of anticandidal activity of the plant that was revealed in the third study. According to the results in Table 3, there was a 50% reduction in MIC and a 75% reduction in MLC values of the mixture of Amphotericin B and 5% ethanolic extract against C. albicans in comparison to Amphotericin B alone. The zone of inhibition of the mixture showed 22% increase in diameter in compa-

rison to that of Amphotericin B alone. In this test, the most potent antifungal agent was the mixture of ethanolic extract 5% plus Amphotericin B, followed by Amphotericin B, ethanolic extract 5%, ethanolic extract 11%, ethanolic extract 4% and ethanol 70% in order. Aqueous extract did not show any antifungal activity. Also, the results showed that 5% ethanolic extract was slightly stronger than 11% ethanolic extract.

A series of dose dependent side effects are related to usage of Amphotericin B. A report highlights hypokalemia due to usage of Amphotericin B as a rare cause of rhabdomyolysis. So that patients under treatment with Amphotericin B should be checked for this life-threatening complication regularly (da Silva et al., 2007). Amphotericin B is widely used for severe life threatening fungal infections. Its use is limited by a dose-dependent nephrotoxicity manifested by a reduction in glomerular filtration rate and tubular dysfunction. An elevated creatinine associated with Amphotericin B is not only a marker for renal dysfunction but is also links to a substantial risk for the use of hemodialysis and a higher mortality rate. Several manipulations have been proposed to try and minimize Amphotericin B induced nephrotoxicity (Deray et al., 2002). Mechanisms to prevent nephrotoxicity include the use of lipid formulations such as Amphotericin B lipid complex, Amphotericin B colloidal dispersion and liposomal Amphotericin B and the concurrent use of volume repletion (Goldman et al., 2004).

Antifungal combination may increase the magnitude and rate of microbial killing in vitro, shorten the total duration of therapy prevent the emergence of drug resistance, expand the spectrum of activity and decrease the drug related toxicities by allowing the use of lower doses of antifungal.

Multiple compound therapies along the disease pathway may need to be manipulated simultaneously from an effective treatment. When one drug is used ,the required high dosage for efficacy often produce bioavailability problems and unwanted side effects in addition to drug resistance (Zhang et al., 2007). There are several arguments that justify the strategy of combining anti fungal drugs to optimize therapy such as the in vitro data showing the potential for a synergistic effect, broader spectrum of activity and decreased risk of emergence of resistant strains and absence of a negative or harmful effects of monotherapy (Kontoyiannis et al., 2004; Marr, 2004; Ramesh et al., 2008; Chendrasekur et al., 2002; Steinbach, 2005; Baddely et al., 2005). E. platyloba has been used for ages traditionally and effectively in Iran, so, it is presumed that side effects should be less. The synergistic combined mixture of Amphotericin B and the extract detected in this in vitro study need further in vivo studies to evaluate its actual effect. This will represent promising finding in antifungal therapy and enable the use of the local, rich plant heritage as an effective medicine with probably fewer side effects.

Conclusions

Regarding this study, it is clear that *E. platyloba* indeed exhibits a potent antifungal and a weak antibacterial activity. Its inhibitory action against *C. albicans* was the highest followed by *T. schenlaini* and *T. verucosum*. Some degrees of synergy was recorded in combination of Amphotericin B plus *E. platyloba* 5% ethanolic extract covering *C. albicans* .The synergistic combined mixture in this *in vitro* study need further *in vivo* studies to evaluate its actual effect.

ACKNOWLEDGMENTS

Great thanks to Shahr-e-kord University of Medical Sciences, Infectious Diseases Research Center of Isfahan University of Medical Sciences, Skin Diseases and Leishmanisis Research Center of Isfahan University of Medical Sciences, Barij essence pharmaceutical Company, Isfahan, Kashan, Iran, for their supports and also financial co-operation.

REFERENCES

Akagawa G, Abe S, Tansho S, Uchida K, Yamaguchi H (1996). Protection of C3H/HE J mice from development of Candida albicans infection by oral administration of Juzen-taiho-to and its component, Ginseng radix: possible roles of macrophages in the host defense mechanisms. Immunopharmacol Immunotoxicol., 18(1): 73-89.

Avijgan M, Mahbobi M, darabi M and Nilforoshzadeh MA (2005). The In vitro assessment of effect of Echinophora platyloba Extract On *S. epidermidis, S. aureus, St. pyogenes* in comparison to brand anti-biotics. Abstract Book, The 2end national congress of update indermatology Diseases and Leishmaniasis. Isfahan, Iran. pp. 75-76. (In Farsi)

Avijgan M, Saadat M, Nilforoshzadeh MA, Hafizi M (2006a). Anti fungal effect of echinophora Platyloba extract on some Common Dermathophytes. J. Med. plants. 5(18): 56-62. (In Farsi)

Avijgan M, Hafizi M and Saadat M (2006b). Anti fungal effect of Hydroalcohlic Extract of echinophora Platyloba DC. On Candida albicans. J. Med. Aromatical Plants Iran, 21(4): 545-552. (In Farsi)

Avijgan M, Saadat M, Nilforoshzadeh MA and Hafizi M (2006c). Antifungal effect of Echinophora Platyloba's Extract against Candida albicans. Iranian J. Pharmacol. Res., 4: 285-289.

Baddely JW, Pappas PG (2005). Antifungal combination therapy: Clinical potential. Drugs, 65: 1461-1480.

Chandra Prakash Kala, Pitamber Prasad Dhyani, Bikram Singh Sajwan (2006). Developing the medicinal plant sector in northern India: challenges and opportunities. J. Ethnobiol. Ethnomed., 2: 32.

Chendrasekur PH, Cutright JL, Manavathu EK (2002). Efficacy of vericonazole in the treatment of acute invasive aspergillosis. Clin. Infect. Dis., 34: 563-571.

Clark AM (1996). Natural Products as a resource from new drugs. Pharm. Res., 13(8): 1133-44.

Cowan MM (1999). Plant products as antimicrobial agent. Clin. Microbial. Rev., 12(4): 564-82.

Da Silva PS, de Oliveira Iglesias SB, Waisberg J (2007). Hypokalemic rhabdomyolysis in a child due to amphotericin B therapy. Eur J. Pediatr., 166: 169-171.

Deray G, Mercadal L, Bagnis C (2002). Nephrotoxicity of amphotericin B. Nephrologie. 23(3): 119-22.

El Astal ZY, Ashour AERA and Kerrit AAM (2005). Antimicrobial Activity of Some Medicinal Plant Extracts in Palestine. Pak J. Med. Sci., 21(2) 187-193.

Goldman RD, Koren G (2004). Amphotericin B nephrotoxicity in children. J. Pediatr. Hematol. Oncol., 26(7): 421-6.

Griggs JK, Manandhar NP, TowersGH, Taylor RS (2001). The effects of storage on the biological activity of medicinal plants from Nepal. J Ethnopharmacol. 77: 247-252.

Kariba RM, Houghton PJ, Yenesew A (2002). Antimicrobial activities of a new schizozygane indoline alkaloid from Schizozygia coffaeoides and the revised structure of isoschizogaline. J. Nat. Prod., 65(4): 566-569.

Kontoyiannis DP, Lewis RE (2004).Toward more effective antifungal therapy: the prospects of combination therapy. Br. J. heamatol., 126(2): 165-175.

Mahbobi M, Avijgan M, Darabi M and Kassaiean N (2009). The synergistic effect of Echinophora Platyloba extract and Amphotricin B against *Candida albicans*. J. Med. Plants, 8. 30: 36-43.

Mahbobi M, Shahcheraghi F, and Feizabadi MM (2006): Bactericidal effects of essential oils from clove lavender and geranium on multi-drug resistant of *P. aeruginosa*. Iranian J. Biotech., 4(2): 137-140.

Marr k (2004).combination antifungal therapy: where are we now and where are we going? Oncology, 18(S7): 24-29.

Mel'nichenko EG, Kirsanova MA, Grishkovets VI, Tyshkevich LV, Krivorutchenko IuL (2003). Antimicrobial activity of saponins from Hedera taurica C a r r. Mikrobiol. Z, 65(5): 8-12.

Mustafa NK, Tanira MOM, Dar FK, Nsanze H (1999). Antimicrobial activity of Acacia nilotica subspp. Nilotica fruit extracts. Pharm. Pharmacol. Communicat., 5:583-6.

Na M, Li Ly, Yang YD (2003). Anti-fungal test of composite agastache lotion on seven pathogenic fungi and its clinical application] Zhongguo Zhong Xi Yi Jie He Za Zhi. 23(6): 414-6.

National Committee for Clinical laboratory Standard (2006). Method for dilution antimicrobial susceptibility test for bacteria that grow aerobically, Approved standard, seventh edition. M7-A7, NCCLS, Wayne, PA,.

Nourozi M (1989). Evaluation of photochemical and antimicrobial effect of Echinophora platyloba: PhD Thesis. Faculty of Pharmacy of Tehran University of Medical Sciences. Pharmacogenosy Department. 35-4. (In Farsi)

Nwosu MO, Okafor JI (1998). Preliminary studies of the antifungal activities of some medicinal plants against Basidiobolus and some other pathogenic fungi. Mycoses. 1995 May-Jun; 38(5-6):191-5.

Phongpaichit S, Subhadhirasakul S, Wattanapiromsakul C (2005). Antifungal activities of extracts from Thai medicinal plants against opportunistic fungal pathogens associated with AIDS patients. Mycoses. 48(5): 333-8.

Quiroga EN, Sampietro AR, Vattuone MA (2004). In vitro fungitoxic activity of Larrea divaricata cav. extracts. Lett. Appl. Microbiol., 39(1): 7-12.

Ramesh Putheti, Okigbo, RN (2008).Effects of plants and medicinal plant combinations as anti-infective. Afr. J. Pharm. Pharmacol., 2(7): 130-135.

Rechinger K.H (1987). Flora Iranica. Akademische Druke: U. Uerlagsantalti; Graz, Austria. 72.

Renault S, De Lucca AJ, Boue S, Bland JM, Vigo CB, Selitrennikoff CP (2003). CAY-1, a novel antifungal compound from cayenne pepper. Med. Mycol., 41(1): 75-81.

Sadrai H, Asghari G, Yaghobi K (2002). Evaluation of ethanolic extract and essential oil of Echinophora platyloba on isolated ileum of rat. Pajouhesh in Pezesheki, 7: 150-155. (In Farsi)

Shin S, Kang CA (2003). Antifungal activity of the essential oil of Agastache rugosa Kuntze and its synergism with ketoconazole. Lett. Appl. Microbiol., 36(2): 111-5.

Steinbach W (2005). Combination antifungal therapy for invasive aspergillosis utilizing new targeting strategies. Curr. Drug Targets Infect. Disord., 5: 203-210.

Vanden D, vancechoottee M (2002). Susceptibility testing of fluconazole by NCCLS Broth Macrodilotion Method, E-Test, and disk Diffusion for application in therovtine laboratory. J. Clin. Microbiol., 40(3): 918-921.

Zaidi MA, Crow SA Jr (2005). Biologically active traditional medicinal herbs from Balochistan, Pakistan. J. Ethnopharmacol., 4; 96(1-2): 331-4

Zhang L, Yan K, Zhang Y, Huang R, Bian J, Zheng C, Sun H, Chen Z, Sun N, An R, Min F, Zhao W, Zhao Y, You J, Song Y, Yu Z, Liu Z, Yang K, Gao H, Dai H, Zhang X, Wang J, Fu C, Pei G, Lei J, Zhang S, Goodfellow M, Jiang Y, Kuai J, Zhou G, Chen X (2007). High throughout synergy screening identifies microbial metabolites as combination agents for the treatment of fungal infections, Proc. Natl. Acad. Sci. USA, 104(11): 4606-4611.

5

Studies on the mycoflora associated with sugarcane factory waste and pollution of River Nile in upper Egypt

A.H.M. El-Said[1]*, T. H. Sohair[1] and A. G. El-Hadi[2]

[1]Department of Botany, Faculty of Science, South Valley University, Qena, Egypt.
[2]Department of Botany, Faculty of Science, Az-Zawiyah University, Libya.

Sixty-nine species and four varieties which belong to twenty eight genera of terrestrial fungi were recovered from polluted and nonpolluted water and mud samples on glucose and cellulose-Czapek's agar at 28°C. The most common species from the two substrates on the two types of media were *Aspergillus flavus, Aspergillus fumigatus, Aspergillus niger, Cladosporium cladosporioides, Fusarium oxysporum, Mycosphaerella tassiana* and *Penicillium chrysogenum.* Twenty-six species belonging to 14 genera were isolated from polluted (26 species and 14 genera) and nonpolluted (17 and 10) mud samples on sabouraud's dextrose agar at 28°C. The most prevalent species were *Acremonium retiulum, Alternaria alternata, A. flavus, Aphanoascus fulvescens, Aspergillus terreus, Aphanoascus sp., Penicillium funiculosum* and *Stachybotrys chartarum.*

Key words: Pollution, River Nile, terrestrial fungi and keratinophilic fungi.

INTRODUCTION

Freshwater fungi including those of strictly aquatic and those of terrestrial habitats have been continuously studied for about one century and commonly found in pools, ponds, lakes, rivers, streams and bogs, as well as in marginal soils. They live as saprophytes or parasites on plants and animals. Many investigations were carried out on terrestrial fungi in the world (Park, 1974; El-Hissy, 1979; Moustafa and Khosrawi, 1982; Bettucci and Rodriguez, 1989; El-Hissy et al.,1989; 1990b; Bettucci et al., 1990; El-Nagdy and Abdel-Hafez, 1990; Moharrum et al., 1990; Rodriguez et al., 1990; Khallil and Abdel-Sater, 1992; Bettucci et al., 1993; Hyde and Goh, 1999; El-Hissy et al., 2001; Ho et al., 2003; Cai et al., 2002; 2003; 2006; Petra et al., 2005; Jiao et al., 2006; Mudur et al., 2006; Cai and Hyde, 2007; Jiang et al., 2008; Mongkol and Kevin, 2008). The present investigation is aimed to study the effect of sugarcane factory pollutants on the occurrence and distribution of terrestrial fungal population in the water and submerged mud of the River Nile.

The occurrence and distribution of keratinophilic and related fungi of different mud habitats have been investigated (Hassan, 1982; 1991; Mangiarotti and Caretta, 1984; Miyoshi et al., 1985; Hassan and Batko, 1986; Chabasse, 1988; Hassan and Shoulkamy, 1991; Soon, 1991; Abdullah and Dina, 1995; Abdullah and Hassan, 1995; Ulfig et al., 1996; 1997; Ali-Shtayeh et al., 1999; Ali-Shtayeh and Rana, 2000).

MATERIALS AND METHODS

Thirty samples of polluted and nonpolluted water and submerged mud were collected from River Nile during the working season of Nag Hamady sugarcane factory from December - April 2007 - 2008. The water samples were analyzed chemically for the estimation of total soluble salts and organic matter contents (Jackson, 1958). A pH-meter was used for pH determination of water and submerged mud.

Determination of terrestrial fungi

The dilution plate method (Johnson and Curl, 1972) was used for the estimation of fungi in submerged mud samples. For the recovery of

---
*Corresponding author. E-mail: husseinelsaid@yahoo.com.

terrestrial fungi from the aggregated water, 1 ml of each water sample was transferred into each of five petri-dishes. A Modified Czapek's agar medium in which glucose (10 g/L) or powder cellulose (20 g/L) were used for isolation of glucophilic and cellulose-decomposing fungi, respectively. Streptomycin (20 u/ml) and Rose Bengal (30 ppm) were added as bacteriostatics agents. The plates were incubated at 28°C for 7 days and the developing fungi were counted and identified (Morphologically, based on macro- and microscopic characteristics). The fungal numbers were calculated per 1 ml water in every sample.

Determination of keratinophilic and related fungi

The dilution plate method (Johnson and Curl, 1972) was used for the estimation of fungi in mud samples on Sabouraud's dextrose agar medium (Moss and McQuown, 1969), which was supplemented with chloramphenicol (0.5 mg / ml medium) and cycloheximide (0.5 mg / ml medium). The plates were incubated at 28°C for 4 to 6 weeks and the developing fungi were counted and identified.

RESULTS AND DISCUSSION

The total soluble salts of polluted water and submerged mud samples ranged between 0.1 to 1.9 and 1 to 9% and their contents of organic matter fluctuated between 0.003 to 0.58 and 2.3 to 51.9%, respectively. El-Hissy et al. (2001) found that the total soluble salts and organic matter contents in water and submerged mud of the River Nile polluted with Kom Ombo sugar cane factory fluctuated between 125 to 540 mg/L, 0.262 - 0.678% and 0.34 to 6.54 mg/L, 0.88 to 2.85%, respectively. The pH values of the water and submerged mud samples were alkaline and ranged between 7.3 to 8.8 and 8.5 to 9.8, respectively. These results are in agreement with the results of El-Hissy et al. (2001) who found that pH ranged between 5.4 to 8.19 and 8.05 to 8.6, respectively, in a similar context using similar methods. Also, the total soluble salts of nonpolluted water and submerged mud samples tested fluctuated between 0.05 to1.5 and 1 to 7% and their contents of organic matter ranged between 0.003 to 028 and 2.3 to 43.3%, respectively. The pH values of the water and submerged mud samples ranged between 7.6 to 9.1 and 8.8 to 10.9, respectively.

Polluted terrestrial fungi

Sixty-five species and four varieties belonging to 28 genera were isolated from water (21 genera and 49 species plus 3 varieties) and submerged mud (25 genera and 57 species plus 2 varieties) samples on glucose- and cellulose-Czapek's agar at 28°C (Table 1). In this respect, El-Hissy et al. (2001) isolated 41 species which belong to 32 genera of terrestrial fungi from water and mud samples polluted by industrial effluents of Kom Ombo sugar cane factory on glucose-and cellulose-Czapek's agar at 28°C. The most common genera from the two substrates on the two types of media were *Acremonium, Aspergillus, Cladosporium, Fusarium,*

Gibberella, Mucor, Mycosphaerella, Penicillium and *Trichoderma*. They occurred in 40 to 100, 26.7 to 93.3, 33.3 to 100 and 26.7 to 100% of the samples comprising 1.4 to 32.9, 2.9 to 20.5, 1.7 to 38.7 and 0.5 to 46.9% of total fungi, respectively. From the above genera the most prevalent species were *Acremonium strictum, Aspergillus flavus, Aspergillus fumigatus, Aspergillus niger, Cladosporium cladosporioides, Cladosporium sphaerospermum, Fusarium oxysporum, Gibberella fujikurio, Mucor circinelloides, Mycosphaerella tassiana, Penicillium chrysogenum, Penicillium oxalicum, Penicillium puberulum* and *Trichoderma harzianum*. They were encountered in 13.3 to 93.3, 26.7 to 93.3, 26.7 to 100 and 26.7 to 100% of the samples comprising 0.2 to 17.5, 0.7 to 12.8, 0.7 to 20.2 and 0.3 to 24% of total fungi, respectively (Table 1). Ali-Shtayeh and Rana (2000) found that the most common fungi in polluted field soils and raw city sewage in Jordan were *Alternaria alternate, Aspergillus candidus, Geotrichum candidium* and *Paecilomyces lilacinus*.

Most of the above species were encountered previously but with various numbers and frequencies from water and mud of Ibrahim canal (Abdel-Hafez and Bagy, 1985), River Nile (El-Hissy et al., 1990a), Aswan High Dam Lake (El-Hissy et al., 1990b), some ponds of Kharga Oases (El-Nagdy and Abdel-Hafez, 1990), water, soil and air polluted by the Manquabad Superphosphate (Khallil and Abdel-Sater, 1992), and River Nile polluted with industrial effluents of Kom Ombo sugar cane factory (El-Hissy et al., 2001) in Egypt. Also, the previous fungal species were reported from the world (Barlocher and Kendrick, 1974; El-Hissy, 1979; Hiremath et al., 1985; Bettcci and Roquebert, 1989; Bettucci et al., 1990; 1993; Hyde and Goh, 1999; El-Hissy et al., 2001; Cai et al., 2002; 2003; 2006; Ho et al., 2003; Petra et al., 2005; Jiao et al., 2006; Mudur et al., 2006; Cai and Hyde, 2007; Jiang et al., 2008; Mongkol and Kevin, 2008). The remaining genera and species were moderate or less frequent (Table 1).

Nonpolluted terrestrial fungi

Forty-six species and 2 varieties in addition to 20 genera were collected from water (10 genera and 29 species) and submerged mud (18 and 39 plus 2 varieties) samples on glucose and cellulose-Czapek's agar at 28°C (Table 1). The most prevalent genera from the two substrates on the two types of media were *Aspergillus, Cladosporium, Fusarium, Mycosphaerella, Penicillium* and *Trichoderma*. They were found in 26.7 to 93.3, 26.7 to 86.7, 26.7 to 86.7 and 26.7 to 86.7% of the samples constituting 1.6 to 37.7, 1.2 to 58.7, 2.1 to 37.7 and 1.02 to 37.1% of total fungi, respectively. From the above genera the most common species were *A. flavus, A. fumigatus, A. niger, C. cladosporioides, F. oxysporum, Penicillium chrysogenum* and *Penicillium corylophilum*.

Table 1: Total counts (TC, calculated per g in all samples) number of cases of isolation (NCI, out of 15) and occurrence remarks (OR) of fungal genera and species recovered from polluted and nonpolluted water and mud on glucose and cellulose–Czapek's agar at 28°C.

Genera and species	Polluted — Water — Glucose TC	Glucose NCI and OR	Cellulose TC	Cellulose NCI and OR	Mud — Glucose TC	Glucose NCI and OR	Cellulose TC	Cellulose NCI and OR	Nonpolluted — Water — Glucose TC	Glucose NCI and OR	Cellulose TC	Cellulose NC and OR	Mud — Glucose TC	Glucose NCI and OR	Cellulose TC	Cellulose NCI and OR
Acremonium	2750	6M	1200	5M	4250	9H	2050	5M	250	2L			150	1R	700	5M
A.cerealis (Karst.)W.Gams	450	4M			500	3L	350	2L	100	1R					50	1R
A.furcatum F.& V. Moreau	100	1R					150	1R	50	1R						
A.murorum (Corda) W.Gams					250	2L										
A.retiulum W.Gams	1400	3L			1700	3L	600	2L							100	1R
A.strictum W.Gams	800	6M	1200	5M	1800	8H	950	4M	100	2L			150	1R	550	5M
Alternaria					600	4M	1850	7M			50	1R	2050	8H	3400	7M
A.alternata (Fries) Keissler					600	4M	1850	7M			50	1R	1750	8H	3000	7M
A.chlamydospora Mouchacca													150	1R	100	1R
A.tenuissima (Kunze:Pers.) Wiltshire													150	1R	300	1R
Aspergillus	20050	15H	8250	14H	33850	15H	34450	15H	9050	14H	15250	13H	18600	13H	18200	13H
A.candidus Link	50	1R			200	2L			950	1R			800	3L		
A.carneus (V.Tiegh.)Blochwitz	150	2L			150	2L							400	3L		
A.flavus Link	10950	14H	1100	14H	13800	15H	17650	15H	4700	14H	9400	13H	8450	13H	9750	12H
A. flavus var. colmnaris Rapper and Fennell																
A.fumigatus Fresenius	3450	13H	4150	12H	7300	15H	4800	13H	2050	10H	2150	10H	4000	12H	3750	12H
A.niger Van Tieghem	2850	13H	2200	9H	4550	12H	2300	13H	800	8H	2850	8H	1850	10H	3000	8H
A.ochraceus Wilhelm	150	2L	350	3L	650	5M			50	1R			100	1R	400	4M
A. sydowii (Bain and Sart.) Thom and Church					2250	8H	250	2L					1600	5M	450	1R
A. terreus Thom	350	3L			0600	7M	6650	8H	500	5M	850	2L	1200	6M	850	2L
A. terreus var aureus Thom and Raper	200	2L	450	2L	2300	7M	2800	6M					200	4M		

Table 1. Contd.

Species	Col 1	Col 2	Col 3	Col 4	Col 5	Col 6	Col 7	Col 8
A.ustus Fennell and Raper	100			550 / 5M				100 / 1R
A.versicolor (Vuill.)Tiraboschi	1R			500 / 4M				2L
Botryotrichum atrogriseum Van Beyma				100 / 1R				100 / 2L
Chaetomium								
C.globosum Kunze				1050 / 3L				
				800 / 3L				
C.spirales Zoph				250 / 1R				
Cladosporium	1850 / 4M	5000 / 8H	500 / 4M	1750 / 7M	2700 / 9H	7050 / 11H	1200 / 4M	1550 / 7M
C.cladosporioides (Fres.)de Vries	1850 / 4M	2400 / 8H	500 / 4M	1550 / 7M	1950 / 9H	5100 / 11H	900 / 4M	1450 / 6M
C.sphaerospermum Penzig	2600 / 4M		200 / 2L	750 / 5M	1950 / 5M	300 / 4M	100 / 2L	
Cochliobolus spicifer Nelson							100 / 1R	
Emericella nidulans (Edidam)Vuillemin			150 / 2L	100 / 1R	100 / 1R		100 / 1R	
Epicoccum nigrum Link					100 / 1R	100 / 1R	100 / 1R	
Fenniellia flavipes Wiley and Simmons		500 / 2L				200 / 1R	200 / 1R	
Fusarium	5450 / 8H	2300 / 6M	300 / 4M	400 / 4M	4500 / 13H	5100 / 7M	5250 / 9H	5250 / 13H
F.moniliforme Sheldon							1250 / 7M	
F.oxysporum Shelecht	3650 / 8H	650 / 4M	300 / 4M	300 / 4M	4400 / 13H	3100 / 7M	3800 / 9H	4400 / 13H
F.poae (Peck) Wollenweber						200 / 2L	200 / 2L	200 / 2L
F.semitectum Berk.& Rav.	1800 / 2L	1650 / 3L	50 / 1R	50 / 1R	100 / 1R	1100 / 6M	400 / 3L	400 / 3L
F.tricinctum (Corda) Sacc.					700 / 3L	250 / 2L	250 / 2L	
Gibberella	150 / 1R	300 / 2L	100 / 2L	250 / 2L	800 / 5M	1500 / 5M	4400 / 8H	850 / 7H
G.acuminata Wollenweber					100	200 / 3L	200 / 3L	200 / 3L
G.avenacea Cooke						100 / 2L	100 / 2L	100 / 2L
G.fujikuroi (Sawada) Ito		150 / 2L	150	150 / 1R	800 / 5M	1300 / 5M	4400 / 8H	500 / 5M
G.intricans Wollenweber		150 / 2L	100 / 1R			100 / 1R		
G.zeae (Schwabe)Petch							50 / 1R	50 / 1R
Humicola grisea (Tassi) Goid					1250 / 4M	200 / 3L		
Mucor		850 / 5M		600 / 4M	600 / 4M	1900 / 7M	1950 / 4M	2200 / 6M
M.circinelloides Van Tieghton		600 / 5M		600 / 5M	400 / 4M	1050 / 6M	600 / 5M	1050 / 5M

Table 1. Contd.

Taxon	C1	C2	C3	C4	C5	C6	C7	C8
M.racemosus Fresenius	1150 4M	1350 3L	850 7M	200 2L			250 3L	6300
Mycosphaerella tassiana (Albertini and Schweinti)	2500 8H	1400 4M	7600 13H	400 4M	2650 11H	450 8M	8600 11H	
Mycothecium verrucaria (Alb. and Sch.) Dit.				100 1R		100 1R		
Nectria	2100 8H	2300 9H	1500 4M	7600 11H			400 3L	
N.haematococca Berkeley and Brown	2100 8H	2300 9H	1400 4M	7500 11H			400 3L	
N.viridescens C.Booth			100 1R	100 1R			100 1R	
Paecilomyces variotii Bainier								
Penicillium	20150 13H	9050 12H	19400 15H	12350 12H	9550 10H	7450 9H	8750 13H	9900 12H
P.aurantiogriseum Dierckx	100 2L		100 1R				100 1R	
P.brevicompacium Dierckx			150 2L				100	
P.chrysogenum Thom	10800 13H	5150 12H	4750 13H	7200 12H	2600 9H	5150 9H	3650 11H	6650 12H
P.citrinum Thom	1150 7M	750 5M	1400 8H		650 5M		700 5M	200 3L
P.coryloohilum Dierckx	700 7M	350 1R	3900 7M		1950 5M	1950 6M	850 5M	400 4M
P.duclauxii Delacroix	500				50 1R			
P.funiculosum Thom			300 2L		3500 7M			
P.oxalicum Currie andThom	2150 8H	1100 4M	6600 6M	2900 4M	800 2L	100 2L	900 4M	750 2L
P.purberulum Bainier	4700 12H	1700 7M	2200 8H	2250 5M	250 5M	250 5M	2450 5M	1900 7M
P.viridicatum Westling	50 1R		50 1R					
Phoma glomerata (Corda) Woolenweber and Hochapfel	900 3L	1350 4M	50 1R	450 2L	700 2L		100 1R	900 3L
Rhizopus stolonifer (Ehrenb.)Link	50 1R		50 1R	400 3L				
Scopulariopsis brevicaulis (Sacc.)Bainier		100 1R	200 2L					
Scytalidium lignicola Pesante.		150 1R	950 3L	400 2L				
Stachybotrys							500 4M	1400 2L
S.atra var. microspora Mathur and Sankhla				100 1R				1350 2L

Table 1. Contd.

Species									
S. chartarum (Ehrenb: Lindt) Hughes	850		300	2L			500	4M	50 / 1R
S.cylindrospora C.W.Jensen									
Torula herbarum (Pers.) Link	100 / 50	1R / 1R	250	1R	50	1R	50		
Trichoderma	2200 9H	3550 8H	2200 6M	1650 6M	2200 8H	750 5M	1500 8H	1050 4M	500
T..hamatum (Bonord.)Bain	350 3L		1600 6M		1600 6M	50		50 1R	
T.harzianum Rafai	1700 9H	550 9H	600 4M	350 4M	600 4M	150 3L	250 3L	550 4M	500
T.pseudokoingii Rafai	150 1R	500 1R						500	
T.viride Pers	2500 8H		1300 6M	700 3L	550 5M		1250 8H	500 3L	500 3L
Trichothecium roseum (Pers.) Link: Gary	50 1R								
Trimmatostroma salicis Corda	150 1R		150 1R	350 2L	150 1R				
Total counts	61250	40250	87400	73500	25300	26000	49300	49000	49000
Number of genera : 28	21	25			10	18			
Number of species : 69 + 4 Varieties	49 +3 varieties	57 +2 varieties	29						

They emerged in 26.7 to 93.3, 26.7 to 86.7, 26.7 to 86.7 and 26.7 to 80% of the samples comprising 1.2 to 18.6, 1.2 to 36.2, 1.3 to 17.1 and 0.8 to 19.9% of total fungi, respectively (Table 1).

Ali-Shtayeh and Rana (2000) found that the most common fungi in non-polluted field soils and raw city sewage in Jordan were A. alternate, A. candidus, G. candidium and P. lilacinus. All the above fungi were isolated from all over the world by several researchers (Barlocher and Kendrick, 1974; Abdel-Hafez and Bagy, 1985; Hiremath et al., 1985; Bettucci et al., 1990; 1993; El-Hissy et al., 1990a; b; El-Nagdy and Abdel-Hafez, 1990; Khallil and Abdel-Sater, 1992; Bettcci and Roquebert, 1995; Hyde and Goh, 1999; El-Hissy et al., 2001; Cai et al., 2002, 2003; 2006; Ho et al., 2003; Petra et al., 2005; Jiao et al., 2006; Mudur et al., 2006; Cai and Hyde, 2007; Jiang et al., 2008; Mongkol and Kevin, 2008). The remaining fungal genera and species were less common (Table 1). Numerous species were isolated only from polluted water and submerged mud on glucose or cellulose agar medium: Acremonium mucorum, Aspergillus ustus, Aspergillus versicolor, Botryotrichum atrogriseum, Chaetomium spirales, Cochliobolus spicifer, Epicoccum nigrum, Fusarium poea, Gibberella acuminata, Gibberella avenacea, Humicola grisea, Nectria viridescens, Penicillium funiculosum, Penicillium viridicatum, Rhizopus stoloifer, Scopulariopsis brevicaulis, Scytalidium lignicola, Stachybotrys cylindrospora, Trichoderma pseudokoningii, Trichothecium roseum and Trimmatostroma salicis (Table 1).

Keratinophilic and related fungi

Twenty-six species belonging to 14 genera were isolated from polluted (26 genera and 15 species) and nonpolluted (17 genera and 11 species) mud samples on sabouraud's dextrose agar at 28°C (Table 2). The most common genera were Acremonium, Alternaria, Aspergillus, Aphanoascus, Penicillium and Stachybotrys (Table 2). They occurred in 26.7 to 100 and 26.7 to 80% of the samples comprising 1.9 - 46.3 and 2.4 to 35.8% of total fungi, respectively (Table 2). From the above genera the most prevalent species were Acremonium retiulum, A. alternate, A. flavus, Aphanoascus fulvescens, A. terreus, Aphanoascus sp., C. cladosporioides, P. funiculosum and Stachybotrys chartarum. They

Table 2. Total counts (TC, calculated per g mud in all samples), number of cases of isolation (NCI, out of 15) and occurrence remarks (OR) of fungal genera and species recovered from polluted and nonpolluted mud samples on Sabouraud's dextrose agar at 28°C.

Genera and Species	Polluted mud		Nonpolluted mud	
	TC	NCI&OR	TC	NCI&OR
Acremonium	4800	10H	2550	7M
A. cerealis	200	2L	--	--
A. retiulum	2600	6M	2550	7M
A. strictum	2000	6M	--	--
Alternaria alternata	550	4M	1000	4M
phanoascus A	13600	15H	6050	12H
A. fulvescens	10700	13H	3600	9H
A. terreus	1000	6M	600	5M
Aphanoascus sp.	1900	8H	1850	6M
Aphinisia queenslandica	150	1R	300	3L
Aspergillus	2000	8H	2200	8H
A. flavus	500	4M	900	7M
A. fumigatus	300	2L	400	3L
A.sydowii	800	6M	--	--
A. terreus	300	1R	--	--
A.ustus	100	1R	900	5M
Chrysosporium	500	3L	300	1R
C. luteum	150	1R	150	1R
C. pannorum	350	3L	150	1R
Cladosporium	1400	4M	850	6M
C. cladosporioides	1200	4M	850	6M
C. sphaerospermum	200	2L	--	--
Epicoccum nigrum	100	1R	--	--
Fusarium oxysporum	100	1R	--	--
Humicola grisea	500	1R	--	--
Mycosphaerella tassiana	1500	6M	550	3L
Nectria haematococca	100	1R	--	--
Penicillium	3100	7H	2700	9H
P. chrysogenum	300	3L	550	4M
P. corylophilum	1400	5M	300	3L
P. funiculosum	1400	5M	1850	10H
Stachybotrys chartarum	1000	5M	400	4M
Total count	8450		16900	
Number of genera 14	14		10	
Number of species 28	28		18	

recovered from 26.7 to 86.7 and 26.7 to 66.7% of the samples contributing 1.7 - 36.4 and 2.4 - 21.3% of total fungi, respectively (Table 2). Ali-Shtayeh and Rana (2000) found that the most common fungi in polluted and non-polluted field soils and raw city sewage in Jordan were *Microsporum gypseum*, *Trichophyton ajelloi*, *Arthroderma cuniculi*, *Arthroderma curreyi*, *Chrysosporium keratinophilum*, *Chrysosporium tropicum* and *pannorum*. Most of the above species were isolated previously, but with various numbers and frequencies

from the world (Garg, 1966; Ajello and Padhye, 1974; McAleer, 1980; Hassan, 1982; 1991; Cano et al., 1985; Hassan and Batko, 1986; Ogbonna and Pugh, 1987; Chabasse, 1988; Ulfig and Ulfig, 1990; Hassan and Shoulkamy, 1991; Soon, 1991; Abdullah and Dina, 1995; Abdullah and Hassan, 1995; Ulfig et al., 1996; 1997; Ali-Shtayeh et al., 1999).

Several authors have suggested that *A. fuluvescens* may be an opportunistic dermatophyte (Rippon et al., 1970; Marin and Campos, 1984), while Cano et al. (1990)

demonstrated that *A. fuluvescens* behaves as an internal opportunistic pathogen. Some species were isolated only from polluted mud such as: *Acremonium cerealis, Aspergillus strictum, Aspergillus sydowii, Aspergillus terreus, C. sphaerospermum, Epicoccum nigrum, F. oxysporum, Humicola grisea* and *Nectria haematococca*. The remaining genera and species were isolated in less frequents (Table 2). Some species were isolated only from polluted mud such as: *A. cerealis, A. strictum, A. sydowii, A. terreus, C. sphaerospermum, E. nigrum, F. oxysporum, H. grisea* and *N. haematococca*.

In conclusion, in this investigation the authors studied the effect of the pollutants of the Nag Hamady sugar cane factory on occurrence and distribution of aquatic and terrestrial fungal population in the water and submerged mud of the River Nile. This study shows that the numbers and frequencies of fungi in polluted water and submerged mud samples were higher than nonpolluted samples of water and submerged mud. This is due to the effect of wastes of the sugar cane factory in the River Nile, so pouring industrial factory wastes into River Nile must be avoided.

REFERENCES

Abdel-Hafez SII, Bagy MMK (1985).Survery on the terrestrial fungi of Ibrahimia canal water in Egypt. (Ismaillia Conf.). Proc.Egypt.Soc., p. 4.

Abdullah SK, Dina (1995). Isolation of dermatophytes and others keratinophilic fungi from surface sediments of the Shatt-Al-Arab River and its creeks at Basrah,Iraq. Mycoses, 34: 163-166.

Abdullah SK, Hassan DA (1995). Isolation of dermatophytes and other keratinophilic fungi from surface sediments of the Shatt Al-Arab River and its creeks at Basrah.Iraq. Mycoses, 38: 163-166.

Ali-Shtayeh MS, Jamous RMF, Abu-Ghdeib SI (1999). Ecology of cycloheximide- resistant fungi in field soils receiving raw city wastewater or normal irrigation water .Mycopathologia, 144: 39-54.

Ali-Shtayeh MS, Rana FJ (2000). Keratinophilic fungi and related dermatophytes in polluted soil and water habitats. Iberoamericana de Micologia, pp. 51-59.

Ajello L, Padhye A (1974). Keratinophilic fungi of the Galapagos Islands. Mykosen, 17: 239-243.

Barlocher F, Kenderick B (1974). Dynamics of the fungal population on leaves in a streame. J. Ecol., 62: 761-790.

Bettucci L, Rodriguez C (1989). Composition and organization of the *Penicillium* and its telemorphs taxocene of two grazing land soils in Uruguay. Cryptogamie, Mycol., 10 (2): 107-116.

Bettucci L, Rodriguez C, Indarte R (1993). Studies fungal communities of two grazing-land soils in Uruguay. Pedobiologia, 37: 72-82.

Bettucci L, Rodriguez C, Roqubert M (1990). Studies fungal communities of volcanic ash soils along an altitydinal gradient in Mexico. Pedobiologia, 34: 61-67.

Bettcci L, Roqubert M (1995). Studies micro-fungi from a tropical rain forst litter and soil : a preliminary study. Nova Hedwigia, 61(1-2): 111-118.

Cai L, Hyde KD (2007). New species of Clohiesia and *Paraniesslia* collected from freshwater habitats in China. Mycoscience, 48(3): 182-186.

Cai L, Tsui CKM, Zhang KQ, Hyde KD (2002). Aquatic fungi from Lake Fuxian,Yunnan, China. Fungal Divers., 9(1): 57-70.

Cai L, Tsui CKM, Zhang KQ, McKenzie EHC (2003). Freshwater fungi from bamboo and wood submerged in the Liput River in the Philippines. Fungal Divers., 13(1): 1-12.

Cai L, Ji KF, Hyde KD (2006). Variation between freshwater and terrestrial fungal communities on decaying bamboo culms. Antonie van Leeuwenhoek, 89(2): 293-301.

Cano J, Mayayo E, Guarro T (1990). Experimential pathogenicity of *Aphanoascus* spp. Mycoses, 33,41-45.

Cano J, Punsola I, Guarro J (1985). Geographic distribution of the genus *Chrysosporium* in catalonia according to climates and types of soils, Rev. Iber. Micol., 2 : 91-108.

Chabasse DC (1988). Taxonomic study of keratinophilic fungi isolated from soil and some mammals in France. Mycopathologia, 101: 133-140.

El-Hissy FT (1979). Aquatic and terrestrial fungi from the surface and casts of earthworms in Egypt.Bull. Fac. Sc. Assiut Univ., 8: 201-210.

El-Hissy FT, Khallil AM, El-Nagdy MA (1989). Aquatic fungi associated with seven species of Nile fishes (Egypt).Zentralbl Mikrobiol., 144: 305-314.

El-Hissy FT, Khallil AM, El-Nagdy MA (1990a). Fungi associated with some aquatic plants collected from freshwater areas at Assiut (Upper Egypt). J. Islamic Acad. Sci., 3: 298-304.

El-Hissy FT, Moharram AM, El-Zayat SA (1990b). Studies on the mycoflora of Aswan High Dam Lake, Egypt,monthly variation. J. Basic. Microbiol., 30 (2): 231-236.

El-Hissy FT, Mortada SMN, Khallil AM, Abdel-Motaal FF (2001). Aquatic fungi recovered from water and submerged mud polluted with industrial effluents. J. Bio. Sci., 1 (9): 854-858.

El-Nagdy MA, Abdel-Hafez SII (1990). Occurrence of zoosporic and Terrestrial fungi in some ponds of Kharga Oases, Egypt. J. Basic Microbiol., 30: 233-240.

Garg AK (1966). Isolation of dermatophytes and other keratinophilic fungi from soils in India. Sabouraudia, 4: 259-266.

Hassan SKM (1982). *Mitochytrium regale* sp. nov, a new keratinophilic water fungus from Poland. Acta Mycol., 18(2): 155-160.

Hassan SKM, Batko A (1986). *Nowakowskiella keratinophila* sp.now.,a keratinophilic fungus from the brackish water. Acta Mycol., 22(2): 193-196.

Hassan SKM (1991). Chytrids in Egypt: I-Saprophytic species of the cladchytriaceae from water streams. Cryptogamie Mycol., 12(3): 211-225.

Hassan SKM, Shoulkamy MA (1991). Chytridiaceous fungi from water streams in Upper Egypt.Zentralbl. Mikrobiol., 146: 509-523.

Hiremath AB, Prabhakar MN, Jayarj YM (1985). Fungi of waste waters and stabilizations pond. Plant Sci., 95 (4): 263-270.

Ho WH, To PC, Hyde KD (2003). Induction of antibiotic production of freshwater fungi using mix-culture fermentation. Fungal Divers., 12(1): 45-51.

Hyde KD, Goh TK (1999). Fungi on submerged wood from the River Coln, England. Mycol. Res., 103(12): 1561-1574.

Jackson ML (1958). Soil chemical analysis constable and Co. London.

Jiao P, Swenson DC, Gloer JB, Campbell J, Shearer CA (2006). Decaspirones A-E: New Bioactive Spirodioxynaphthalenes from the freshwater aquatic fungus *Decaisnella thyridioides*. J. Nat. Prod., 69: 1667-1671.

Jiang M, Wongsawas M, Wang HK, Lin FC, Liang YC (2008. Three new records of lignicolous freshwater hyphomycetes from Mainland, China. J. Agric. Technol., 4(1): 101-108.

Johnson LF, Curl EA (1972). Methods of research on ecology of soil – Borne pathogens.Burgess Publ. Co. Minneapolis, p. 247.

Khallil AM, Abdel-Sater AM (1992). Fungi from water , soil and air polluted by the industrial effluents of Manqubad Superphosphate factory (Assiut,Egypt). Int. Biodeterior. Biodegradation, 30: 363-386.

Mangiarotti AM, Caretta G (1984). Keratinoohilic fungi isolated from a small pool. Mycopathologia, 85: 9-11.

Marin G, Campos R (1984). Dermatofitosis por *Aphanoascus fulvescens*. Sabouraudia, 22: 311-314.

McAleer R (1980). Investigation of keratinophilic fungi for soils in Western Australia. Mycopathologia, 70: 155-165.

Miyoshi H, Matuura R, Hata Y (1985). An ecological survey of fungi in the mangrove estuary of Shiira River, Iriomote Island, Okinnawa. REP USA MAR BIOL INST KOCHI Univ., 7: 33-38.

Moharrum AM, El-Hissy FT, El-Zayat SA (1990). Studies on the mycoflora of Aswan High Dam,Egypt.Vertical fluctuations. J. Basic Microbiol., 30 (3):197-208.

Mongkol WHK, Kevin DHF (2008). New and rare lignicolous hyphomycetes from Zhejiang Province,China. J. Zhejiang Uni. Sci., B 9: 797-801.

Moss ES, McQuown AL (1969). Atlas of medical mycology.3rd edition. The Williams and Wikins Company. Baltimore, p. 366.

Moustafa AF, Khosrawi LK (1982). Ecological studies of fungi in the tidal mud-Flats of Kuwait. Mycopathologia, 79: 109-114.

Mudur SV, Swenson DC, Gloer JB, Campbell J, Shearer CA (2006). Heliconols A-C:Antimicrobial Hemiketals from the freshwater aquatic fungus *Helicodendron giganteum*. Org.Lett., 8:3191-3194.

Ogbonna CIC, Pugh GFJ (1987).Keratinophilic fungi from Nigerian soil. Mycopathologia, 99: 115-118.

Park D (1974). Accumulation of fungi by cellulose exposed in a River. Trans. Br. Mycol. Soc., 63:437-447.

Petra J, Gerd JK, Gudrum K (2005). Cadmium and zinc response of the fungi *Heliscus lugdunensis* and *Verticillium alboatrum* isolated from highly polluted water. Sci. Environ., 346: 274-279.

Rippon JW, Lee FC, McMillen S (1970). Dermatophyte infection caused by Aphanoascus fulvescens. Arch. Dermatol.,102: 552-553.

Rodriguez C, Bettucci L, Roqubert M (1990). Fungal communities of volcanic ash soils along an altitudinal gradient in Mexico.I-Composition and organisation. Pedobiologia, 34: 43-49.

Soon SH (1991). Isolation of keratinophilic fungi from soil in Malaysia. Mycopathologia, 113: 155-158.

Ulfig K, Guarro J, Cano J, Gene J, Vidal P, Figueras MJ, Lukasik W (1997). The occurrence of keratinolytic fungi in sediments of the river Tordera (Spain).FEMS Microbiol Ecol., 22: 111-117.

Ulfig K, Terakowski M, Paiza G, Kosarewicz O (1996). Keratinolytic fungi in sewage sludge. Mycopathologia, 136: 41-46.

Ulfig K, Ulfig A (1990).Short communication:Keratinophilic fungi in bottom sediments of surface water. J. Med. Vet. Mycol., 28: 419-422.

6

Influence of bacteria and protozoa from the rumen of buffalo on *in-vitro* activities of anaerobic fungus *Caecomyces* sp. isolated from the feces of elephant

Ravinder Nagpal[1*], Anil Kumar Puniya[1], Jatinder Paul Sehgal[2] and Kishan Singh[1]

[1]Dairy Microbiology Division, National Dairy Research Institute, Karnal 132001, (Haryana), India.
[2]Dairy Cattle Nutrition Division, National Dairy Research Institute, Karnal 132 001 (Haryana), India.

Anaerobic fungal isolates *Caecomyces* sp. from the feces of elephant and *Orpinomyces* sp. from buffalo rumen were co-cultured *in-vitro* with rumen bacterial and protozoal fractions collected from buffalo to observe the possible fate of these fungi in the rumen, if inoculated as microbial-feed supplements. When co-cultured together or separately with rumen bacteria and protozoa, *Caecomyces* sp. was adversely affected. However, bacterial and protozoal counts were higher, compared to the counts when grown alone. Similar patterns of results were observed when *Orpinomyces* sp. was grown in co-culture with bacteria and protozoa separately as well as together, indicating that it is possibly the inhibitory action of bacteria and protozoa, and not inter-species competition, that affects the growth of fungi preventing them from attaining their full fibre-degrading potential. Conversely, although fungal counts were lowered during their co-culturing with bacterial and protozoal fractions, their co-culturing increased the FPase activity of the co-cultured fraction which could be the apparent reason for enhanced fibre degradation.

Key words: Rumen microflora, anaerobic fungi, microbial interactions, fibre degradation.

INTRODUCTION

The rumen is a highly complex ecosystem that contains different microbial species. Ruminant's performance depends on the activities of their microorganisms to utilize the dietary feeds. The rumen microbial ecosystem comprised at least 30 bacterial (10^{10} to 10^{11}/ ml rumen fluid) (Stewart et al., 1997), 40 protozoa (10^5 to 10^7) (Williams and Coleman, 1997), and 6 fungal species ($<10^5$) (Ozkose et al., 2001; Nagpal et al., 2009b). Bacteria, fungi, and protozoa are responsible for 50 to 82% of cell-wall degradation (Lee et al., 2000). Although, substrate competition is high in the rumen, the synergism and symbiosis among different groups of microorganisms make the utilization of substrates more efficient. Many relationships are known to exist among microorganisms

in the rumen (Lee et al., 2000), and it is well established that anaerobic fungi actively participate in degradation of plant materials in ruminants, as these penetrate plant tissues better than bacteria or protozoa (Orpin and Joblin, 1988). Therefore, a considerable potential exists for the manipulation of fungal activity in the rumen to benefit the utilization of poor quality roughages by domesticated ruminants for the increased production responses; and one potential mean may involve inoculation of efficient fungal strains into the ruminants (Paul et al., 2004; Dey et al., 2004; Lee et al., 2000; Thareja et al., 2006; Tripathi et al., 2007; Nagpal et al., 2009a, b; 2010). The interactions of anaerobic fungi with other rumen microbes can be positive, negative or neutral, depending on the microbial groups involved and the type of substrate used. Since rumen fungi produce appreciable amounts of H_2, they can interact with H_2 utilizers that in turn alter their metabolite production. Methanogens are the principal H_2 utilizers in rumen; and stable co-cultures of fungi and methanogens

*Corresponding author. E-mail: nagpal511@gmail.com.

have been established *in-vitro* (Orpin and Joblin, 1997). In contrast, fibre degradation by *Neocallimastix frontalis* has been found to decrease in co-cultures with non-lactate utilizing *Selenomonas ruminantium*, a sugar fermenting H_2 consuming rumen bacterium, thus indicating the occurrence of interspecies hydrogen transfer (Richardson and Stewart, 1990).

The fungi release metabolites such as free sugars, which serve as energy sources for other bacteria. The fungi themselves may depend on the bacteria for vitamins, heme and amino acids (Williams et al., 1994). Co-culture of anaerobic fungi with rumen bacteria could also inhibit the activity of fungi (Dehority and Tirabasso, 2000), suggesting the role of bacteria in controlling fungal activities *in-vivo*. Moreover, since fungal zoospores are of small size, they are likely to be a prey for protozoa (Morgavi et al., 1994). Consequently, the rumen fungi do not appear to attain their full fibre-degrading potential in rumen due to the inhibition by other microbes. Therefore, the present investigation was aimed to study the effect of co-culturing with rumen bacteria and protozoa from buffalo on *in-vitro* activities of anaerobic rumen fungi *Caecomyces* sp. (from elephant feces), and was compared with that of *Orpinomyces* sp. (from buffalo rumen) to observe the possible fate of these fungi in the rumen, if exploited as direct-fed microbials/ animal feed additives.

MATERIALS AND METHODS

Anaerobic fungi *Caecomyces* sp. and *Orpinomyces* sp. were isolated from Indian elephant and Buffalo (Nagpal et al., 2009b, 2010), respectively, by following the method of Joblin (1981) with cellobiose as a carbon and energy source; and were characterized on the basis of number of flagella/ zoospore, thallus morphology (monocentric or polycentric), and rhizoid (filamentous or a vegetative cell) type (Trinci et al., 1994; Thareja et al., 2006; Nagpal et al., 2009b).

For co-culturing of fungal isolates with rumen bacteria and protozoa, a bacterial and protozoal fraction was prepared from rumen liquor of buffalo fed on a standard diet containing 10 kg green fodder maize, 1 kg concentrate mixture, and wheat straw *ad lib*, maintained at institute's cattle yard. Total rumen bacterial fraction was prepared by inoculating supernatant of strained rumen liquor in the basal anaerobic media containing cycloheximide (0.05 mg/ ml) and sodium lauryl sulpahte (0.01 mg/ ml). Total rumen protozoal fraction was prepared by inoculating the resuspended pellet of centrifuged rumen liquor into the media containing cycloheximide (0.05 mg/ ml), penicillin (0.10 mg/ ml) and streptomycin (0.10 mg/ ml). The rumen fungi and bacterial and protozoal fractions were grown anaerobically at 39 °C in basal anaerobic media (Obispo and Dehority, 1992), in the presence of antibiotics, cycloheximide or sodium lauryl sulphate to inhibit bacteria, fungi or protozoa, respectively. In co-cultures of fungi and bacteria, fungal broth and bacterial fraction was inoculated to the media and sodium lauryl sulphate was added; while for co-cultures of fungi and protozoa, fungal culture broth and protozoal fraction were added along with penicillin and streptomycin. Fungal and bacterial counts were taken as thallus forming units (tfu/ ml) and colony forming units (cfu/ ml), respectively, using roll-tube method

(Joblin, 1981). Protozoal counts were taken as direct microscopic counts (DMC/ ml) using methyl green as staining agent.

Filter paper cellulase (FPase) activities were estimated after incubation in Orpin's broth supplemented with 1% Whatman No. 1 filter paper (6 x 1 cm ≈ 50 mg) (Thareja et al., 2006; Tripathi et al., 2007; Nagpal et al., 2009b), keeping one un-inoculated set as control. Supernatants from incubated cultures were analyzed for estimation of reducing sugars (glucose) using dinitrosalicylic acid method (Miller, 1959). Reaction mixture, comprising 1.0 ml of 0.1 M phosphate buffer (pH 6.8), 0.5 gm of substrate and 0.5 ml of culture supernatant, was incubated at 39 °C for 1 h. A similar reaction mixture was prepared for control. The enzyme activities were calculated as IU, that is, µmol of glucose released per hour per ml of culture filtrate. All the data were statistically analyzed as per the method of Snedecor and Cochran (1980).

RESULTS AND DISCUSSION

When co-cultured with rumen bacteria and protozoa, the fungal population was found to be adversely affected (Table 1). During co-culturing of fungi and bacteria for 96 h, a reduction in fungal counts was observed. However, bacterial counts were found to be higher, compared to the counts when bacteria were grown alone (Table 1). Fungal counts were also found to be reduced, when isolate FE5 was co-cultured with protozoal fraction. On the other hand, when fungi, bacteria and protozoa were grown altogether, fungi could not survive after 48 h and even bacterial numbers were found to be negatively affected. Since the growth of isolate FE5 was found to be adversely affected during its co-culturing with bacteria and protozoa, it was assumed that, since the source of isolate FE5 was elephant, and rumen liquor for collecting bacterial and protozoal fractions, and for media preparation was taken from buffalo maintained at NDRI cattle yard, there could have been some inter-species interactions that were hampering the growth of fungi during its co-culturing with bacteria and protozoa. Hence, to verify this further, isolate RB_2, which was earlier isolated from buffalo, was also co-cultured with bacterial and protozoal fractions. In this case also, a similar effect was observed (Table 2).

Fungal counts were lower when co-cultured with bacteria and protozoa separately, and were further reduced when fungi, bacterial and protozoa were grown altogether. Therefore, it indicated that it was possibly because of the inhibitory action of bacteria and protozoa, and not inter-species competition, which was affecting the growth of fungi. Since fungi are slow-growers, this could have been the reason for their lowered counts during co-culturing with bacteria, because, by the time fungi started growing (72-96 h), bacterial population had already grown and produced enough metabolites that inhibited the growth of fungi.

Co-culturing of anaerobic fungi with rumen bacteria have been shown to inhibit the growth of the fungi (Bernalier et al., 1992; Roger et al., 1993; Dehority and Tirabasso, 2000). Stewart et al. (1992) and Bernalier et al.

Table 1. Counts of anaerobic fungus *Caecomyces* sp. (log tfu/ml), bacteria (log cfu/ml) and protozoa (log DMC) when grown alone and in co-cultures.

Treatment	Incubation period			
	0 h	48 h	72 h	96 h
Fungi, bacteria and protozoa grown alone				
Fungal counts	4.63 ± 0.3^a	4.67 ± 0.2^a	4.86 ± 0.5^b	4.94 ± 0.2^c
Bacterial counts	6.91 ± 0.3^a	7.07 ± 0.2^b	7.16 ± 0.1^c	7.07 ± 0.2^b
Protozoal counts	5.06 ± 0.1^a	5.28 ± 0.2^b	5.44 ± 0.3^c	5.46 ± 0.2^d
Fungi and bacteria grown in co-culture				
Fungal counts	4.63 ± 0.3^a	4.62 ± 0.1^a	4.70 ± 0.2^b	4.76 ± 0.4^b
Bacterial counts	6.91 ± 0.3^a	6.98 ± 0.1^a	7.21 ± 0.3^b	7.12 ± 0.5^c
Fungi and protozoa grown in co-culture				
Fungal counts	4.63 ± 0.3^a	4.61 ± 0.0^a	4.59 ± 0.2^a	4.44 ± 0.2^b
Protozoal counts	$5.06\pm o.1^a$	5.37 ± 0.3^b	5.42 ± 0.4^c	5.42 ± 0.2^c
Fungi, Bacteria and protozoa grown altogether				
Fungal counts	4.63 ± 0.3^a	4.59 ± 0.5^a	ND	ND
Bacterial counts	6.91 ± 0.3^a	6.96 ± 0.3^a	6.90 ± 0.5^a	ND
Protozoal counts	5.06 ± 0.1^a	5.12 ± 0.1^b	5.21 ± 0.2^c	5.36 ± 0.1^d

[a-d]: values with different superscripts in same treatment at different incubation periods differ significantly (P≤0.05).

Table 2. Counts of anaerobic fungus *Orpinomyces* sp. (log tfu/ml), bacteria (log cfu/ml) and protozoa (log DMC) when grown alone and in co-cultures.

Treatment	Incubation period			
	0 h	48 h	72 h	96 h
Fungi, bacteria and protozoa grown alone				
Fungal counts	4.65 ± 0.3^a	4.69 ± 0.3^a	4.83 ± 0.4^b	4.96 ± 0.3^c
Bacterial counts	6.91 ± 0.3^a	7.07 ± 0.2^b	7.16 ± 0.1^c	7.07 ± 0.3^b
Protozoal counts	5.06 ± 0.1^a	5.28 ± 0.4^b	5.44 ± 0.3^c	5.46 ± 0.2^d
Fungi and bacteria grown in co-culture				
Fungal counts	4.65 ± 0.3^a	4.66 ± 0.2^a	4.76 ± 0.5^b	4.89 ± 0.2^b
Bacterial counts	6.91 ± 0.3^a	7.01 ± 0.5^a	7.18 ± 0.6^b	7.12 ± 0.2^c
Fungi and protozoa grown in co-culture				
Fungal counts	4.65 ± 0.3^a	4.67 ± 0.0^a	4.76 ± 0.1^a	4.57 ± 0.3^b
Protozoal counts	5.06 ± 0.1^a	5.34 ± 0.3^b	5.45 ± 0.2^c	5.45 ± 0.4^c
Fungi, Bacteria and protozoa grown altogether				
Fungal counts	4.65 ± 0.3^a	4.61 ± 0.3^a	4.49 ± 0.4^b	ND
Bacterial counts	6.91 ± 0.3^a	6.91 ± 0.2^a	6.95 ± 0.6^a	6.71 ± 0.3^b
Protozoal counts	5.06 ± 0.1^a	5.18 ± 0.3^b	5.29 ± 0.3^c	5.11 ± 0.2^d

[a-d]: Values (means ± SD; n = 3) with different superscripts in same treatment at different incubation periods differ significantly (P≤0.05).

(1993) also found an extracellular, thermo-labile protein produced by ruminococci, which inhibited the activities of anaerobic fungi. Dehority and Tirabasso (1993) also found that mixed rumen bacteria produce a heat stable compound *in-vitro*, which inhibits growth of the rumen fungi. Activities of *N. frontalis* were also found to decrease in co-cultures with *Selenomonas ruminantium*, a sugar fermenting H_2 consuming rumen bacterium, indicating the occurrence of interspecies hydrogen transfer (Richardson and Stewart, 1990). Thus, the rumen

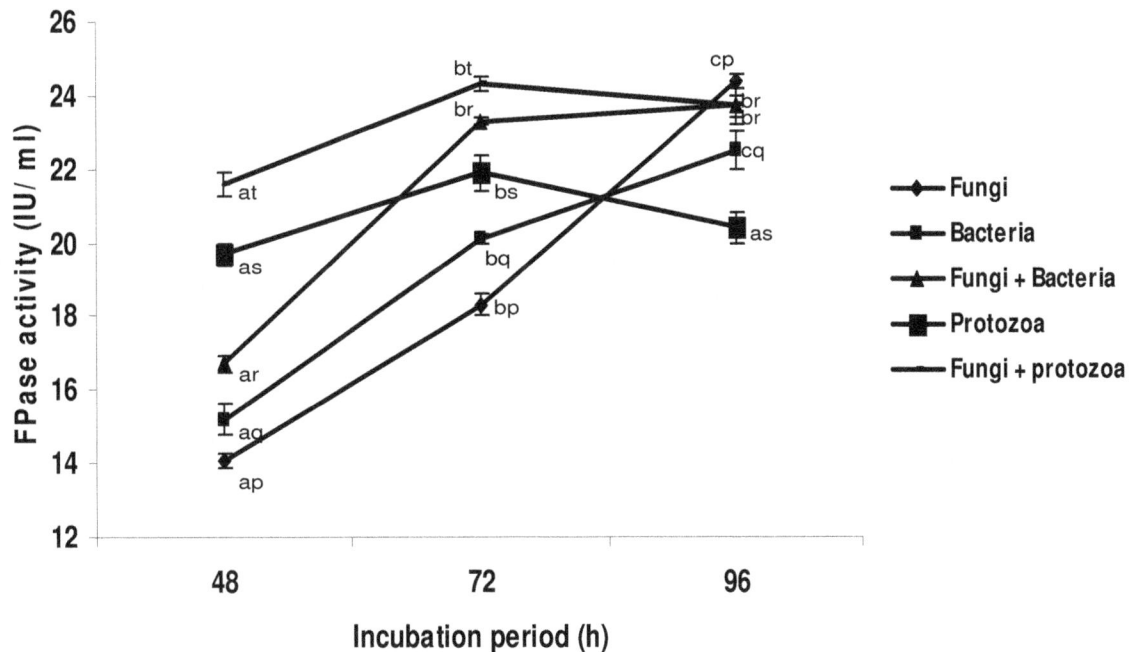

Figure 1. FPase activity of fungi, bacteria and protozoa grown alone and in co-cultures. [a-c]: Values (means ± SD; n = 3) with different superscripts in same treatment at different incubation periods differ significantly (P≤0.05). [p-t]: Values (means ± SD; n = 3) with different superscripts in different treatment at same incubation periods differ significantly (P≤0.05).

fungi do not appear to attain their full fibre-degrading potential in rumen due to the inhibition by bacteria. Since these fungi reproduce through small-sized zoospores which act as food for protozoa, hence, fungal counts were lowered during their co-culturing with protozoa due to the predatory action of protozoa over fungal zoospores. Co-incubation of protozoa with fungi have earlier also shown that the protozoa are able to ingest and digest fungi (Morgavi et al., 1994). The fungal growth was negatively affected by rumen protozoa, certainly because of protozoal predation on zoospores, or possibly due to the degradation of fungal sporangia by protozoal chitinolytic enzymes (Morgavi et al., 1994). Lee et al. (2000) also observed an inhibition of fungi as well as bacteria when co-cultured with protozoa. Moreover, since rumen is a continuous culture system, the co-culturing experiment carried under in-vitro conditions could not give clear picture of microbial interactions that take place inside the rumen.

On the other hand, when supernatant from co-cultured samples were analyzed for FPase activity using filter paper as substrate, FPase activity of co-cultured samples was found to be higher than that of samples from single fraction (Figure 1). FPase activity of isolate FE5 was 14.1, 17.8 and 19.3 IU/ ml after 48, 72 and 96 h, respectively; and in case of bacterial fraction, the activity was 12.3, 72.7 and 25.3 IU/ ml. When fungi and bacterial fraction were co-cultured, the FPase activity of the

supernatant was 16.7, 23.3 and 23.7 IU/ ml after 40, 72 and 96 h, respectively. An increase in the rate and extent of cellulose degradation during co-cultures of fungi and methanogens has also been observed by Fonty and Joblin (1991); Orpin and Joblin (1997); Wood et al. (1986); Joblin (1989) and Bernalier et al. (1991). Similarly, FPase activity of protozoal fraction was 23.1, 21.9 and 19.3 IU/ ml after 48, 72 and 96 h, respectively; and it increased to 21.2, 24.3 and 23.7 IU/ ml, when protozoal fraction was co-cultured with isolate FE5. Onodera et al. (1988) and Lee et al. (2000) also reported higher cellulolytic enzyme in co-cultures of rumen protozoa and bacteria than in fungal monoculture. Hence, although fungal counts were lowered during their co-culturing with bacterial and protozoal fractions, their co-culturing was found to increase the FPase activity of the co-cultured fraction. And this could have been one of the reasons for enhanced fibre degradation/ utilization in the rumen when fungi are fed to the animal.

Conclusion

It is possibly the inhibitory action of bacteria and protozoa, and not the inter-species competition or cross-species adjustments which affects the growth of fungi and prevents them from attaining their full fibre-degrading potential in rumen. Furthermore, the enhanced populations of protozoa and bacteria during co-culturing with fungi, and higher FPase activity could be the probable reason for enhanced fibre degradation in the rumen when fungi are fed to the animal.

ACKNOWLEDGEMENTS

The authors are grateful to NDRI, Karnal, and ICAR, New Delhi, India, for providing fellowships and necessary facilities required for the research work.

REFERENCES

Bernalier A, Fonty G, Gouet P (1991). Cellulose degradation by two rumen anaerobic fungi in mono-culture or in co-culture with rumen bacteria. Anim. Feed Sci. Tech., 32: 131-136.

Bernalier A, Fonty G, Bonnemoy F, Gouet P (1992). Degradation and fermentation of cellulose by the rumen anaerobic fungi in axenic cultures or in association with cellulolytic bacteria. Curr. Microbiol., 25: 143-148.

Bernalier A, Fonty G, Bonnemoy F, Gouet P (1993). Inhibition of the cellulolytic activity of Neocallimastix frontalis by Ruminococcus flavefaciens. J. Gen. Microbiol., 139: 873-880.

Dehority BA, Tirabasso PA (1993). Antibiosis between rumen bacteria and fungi. 22nd Biennial Conference on Rumen Function, Chicago, IL. November, 9-11, p. 6.

Dehority BA, Tirabasso PA (2000). Antibiosis between ruminal bacteria and ruminal fungi. Appl. Environ. Microbiol., 66: 2921-2927.

Dey A, Sehgal JP, Puniya AK, Singh K (2004). Influence of an anaerobic fungal cultures (Orpinomyces sp.) administration on growth rate, ruminal fermentation and nutrient digestion in calves. Asian-Aust. J. Anim. Sci., 17: 820-824.

Fonty G, Joblin KN (1991). Rumen anaerobic fungi: Their role and interactions with other rumen microorganisms in relation to fiber digestion. In: Tsuda T, Sasaki Y and Kawashima R (eds) Physiological Aspects of Digestion and Metabolism in Ruminants. Proceedings of 7th International Symposium on Ruminant Physiology, Academic Press, Inc. SDC, USA, pp. 665-680.

Joblin KN (1981). Isolation, enumeration and maintenance of rumen anaerobic fungi in roll-tubes. Appl. Environ. Microbiol., 42: 1119-1122.

Joblin KN (1989). Physical disruption of plant fibre by rumen fungi of the Sphaeromonas group. In: Nolan JV, Leng RA, Demeyer DI (eds). The roles of protozoa and fungi in ruminant digestion. Australia: Penambul Books, Armidale, pp. 1-2.

Lee SS, Ha JK, Cheng KJ (2000). Influence of an anaerobic fungal culture administration on in vivo ruminal fermentation and nutrient digestion. Anim. Feed Sci. Tech., 88: 201-217.

Miller GL (1959). Use of dinitrosalicylic acid reagent for determination of reducing sugars. Anal. Chem., 31: 426-427.

Morgavi DP, Sakurada M, Mizokami M, Tomita Y, Onodera R (1994). Effects of ruminal protozoa on cellulose degradation and the growth of an anaerobic ruminal fungus, Piromyces sp strain OTS1, in-vitro. Appl. Environ. Microbiol., 60: 3718-3723.

Nagpal R, Puniya AK, Griffith G, Goel G, Puniya M, Sehgal JP, Singh K (2009a). Anaerobic rumen fungi: potential and applications. In: Khachatourians G, Arora DK, Rajendran TP, Srivastava AK (eds.) Agriculturally important microorganisms: An international multi-volume annual review series Academic World International, Bhopal, India, 1(17): 375-393.

Nagpal R, Puniya AK, Sehgal JP, Singh K (2010). In-vitro fibrolytic potential of anaerobic rumen fungi from ruminants and non-ruminant herbivores. Mycosci. (in-press; DOI:10.1007/s10267-010-0071-6).

Nagpal R, Puniya AK, Singh K (2009b). In-vitro fibrolytic activities of the anaerobic fungus, Caecomyces sp., immobilized in alginate beads. J. Anim. Feed Sci., 18:758–768.

Obispo NE, Dehority BA (1992). A most probable number method for enumeration of rumen fungi with studies on factors affecting their concentration in the rumen. J. Microbiol. Methods, 16: 259-270.

Onodera R, Yamasaki N, Murakami K (1988). Effect of inhibition by ciliate protozoa on the digestion of fibrous materials in vivo in the rumen of goats and in an in vitro rumen microbial ecosystem. Agric. Biol. Chem., 52: 2635-2637.

Orpin CG, Joblin KN (1988). The rumen anaerobic fungi. In: Hobson PN (eds). The rumen microbial ecosystem. London: Elsevier, App Sci., pp. 129-150.

Orpin CG, Joblin KN (1997). The rumen anaerobic fungi. In: Hobson PN, Stewart CS (eds). The rumen microbial ecosystem. London: Chapman and Hall, pp. 140-195.

Ozkose E (2001). Morphology and molecular ecology of anaerobic fungi. PhD dissertation, University of Wales, Aberystwyth, UK.

Paul SS, Kamra DN, Sastry VRB, Sahu NP, Agarwal N (2004). Effect of anaerobic fungi on in-vitro feed digestion by mixed rumen microflora of buffalo. Reprod. Nutr. Dev., 44: 313–319.

Richardson AJ, Stewart CS (1990). Hydrogen transfer between Neocallimastix frontalis and Selenomonas ruminantium grown in mixed culture. In: Belaich JP, Bruschi M and Garcia JL (eds). Microbiology and biochemistry of strict anaerobes involved in interspecies hydrogen transfer. Plenum Press, New York: pp. 463-466.

Roger V, Bernalier A, Grenet E, Fonty G, Jamot J, Gouet P (1993). Degradation of wheat straw and maize stem by a monocentric and a polycentric rumen fungi, alone or in association with rumen cellulolytic bacteria. Anim. Feed Sci. Tech., 42: 69-82.

Snedecor GW, Cochran WG (1980). Statistical Methods. 7th Ed. The Iowa State University Press, Iowa, USA.

Stewart CS, Duncan SH, Richardson AJ (1992). The inhibition of fungal cellulolysis by cell-free preparations from ruminococci. FEMS Microbiol. Lett. 97: 83-88.

Stewart SC, Flint HJ, Bryant MP (1997). The rumen anaerobic fungi. In: Hobson, P.N. and Stewart, C.S. (eds) The Rumen Microbial Ecosystem. Kluwer Academic Publishers, Inc. Book News, Portland, pp. 10-72.

Thareja A, Puniya AK, Goel G, Nagpal R, Sehgal JP, Singh P, Singh K (2006). In-vitro degradation of wheat straw by anaerobic fungi from small ruminants. Arch. Anim. Nutr. 60: 412-417.

Trinci APJ, Davies DR, Gull K, Lawrence MI, Bonde-Nielsen B, Rickers A, Theodorou MK (1994). Anaerobic fungi in herbivorous animals. Mycol. Res., 98: 129-152.

Tripathi VK, Sehgal JP, Puniya AK, Singh K (2007). Hydrolytic activities of anaerobic fungi from wild blue bull (Boselaphus tragocamelus). Anaerobe, 13: 36-39.

Williams AG, Coleman GS (1997). The rumen anaerobic fungi. In: Hobson, P.N. and Stewart, C.S. (eds) The Rumen Microbial Ecosystem. Kluwer Academic and Publishers, Book News, Inc., Portland, pp. 73-139.

Williams AG, Withers SE, Naylor GE, Joblin KN (1994). Effect of heterotrophic ruminal bacteria on xylan metabolism by the anaerobic fungus Piromyces communis. Lett. Appl. Microbiol., 19: 105-109.

Wood TM, Wilson CA, McCrae SI, Joblin KN (1986). A highly active extracellular cellulase from the anaerobic rumen fungus Neocallimastix frontalis. FEMS Microbiol. Lett., 34: 37-40.

7

Isolation and random amplified polymorphic DNA (RAPD) analysis of wild yeast species from 17 different fruits

Prashant Kumar Lathar*, Arti Sharma and Isha Thakur

Department of Biotechnology, Jay Pee University of Information Technology, Waknaghat 173234, Solan (H. P), India.

The purpose of this study was to isolate the wild yeast strains present on different fruits and performing random amplified polymorphic DNA (RAPD) analysis to know the genetic inter relationship between different isolated species. Seventeen different fruit namely mango, apple, banana, black grapes, sapotae, orange, plum, jamun, pear, cherry, dates, pomegranate, figs, papaya, pineapple, green grapes and raisins were used as natural sources for yeast isolation. Amplicon fingerprints for the isolated species were obtained by RAPD assay using five different primers. Among them, only three primers allowed discrimination among the 17 isolated species. Jaccard's genetic similarity coefficient varied from 0.00 to 0.98 (as isolate S9 did not show any relatedness with the other clads).

Key words: Random amplified polymorphic DNA, polymerase chain reaction, amplicon, dendrogram, wild yeast strains.

INTRODUCTION

Yeast is a unicellular eukaryotic fungus, very common in the environment and is mostly saprophytic. It has been classified as ascomycetes or basidomycetes under fungi taxonomy Kreger-van et al. (1984) and there are about 1500 species of yeast Kurtzman et al. (2006), Barnett et al. (1990). The commercial importance of strains of yeast species *Saccharomyces cerevisiae* has made it a model organism of study on both research and industrial importance (Legras et al., 2007). Fermenting wild yeast species are being isolated from the natural sources for over decades and is being used in various fermentation processes. Yeast has been isolated from variety of natural sources like leaves, flowers, fruits etc (Spencer and Spencer, 1997; Davenport et al., 1980; Tourna, 2005; Li et al., 2008). Being a sugar-loving micro-organism, it is usually isolated from sugar rich materials. Fruits contain high sugar concentration so yeast species are naturally present on them and can be easily isolated

from them. Distinct wild yeast species are supposed to be present and associated with different fruits in natural environments (Spencer et al., 1997). As because of yeast fermentative characteristic, there is always a need for yeast strains with better features of fermentation especially high ethanol tolerance for production of ethanol as biofuel on commercial scale (Colin et al., 2006). Moreover, besides *S. cerevisiae*, there is always a search on for new wild/non toxic-fermentative yeast species for their further industrial exploitation in fermentation industry, baking industry, therapeutic production etc (Legras et al., 2007). Traditional methods like morphological, physiological and biochemical studies used for taxonomic identification of yeast isolates (Kurtzman et al., 2006; Barnett et al., 1990; Rosa et al., 2006). But being laborious and time consuming techniques, they are not appropriate for routine identification (Couto et al., 1994). Molecular biology based methods have been developed and can be applied to the field of yeast taxonomy in routine identification works (Quesada and Cenis.1995; Loureiro Malfeito-Ferreira., 2003). One such method is a variant of the polymerase chain reaction (PCR) technique based on the amplification of

*Corresponding author. E-mail: latharpk@gmail.com.

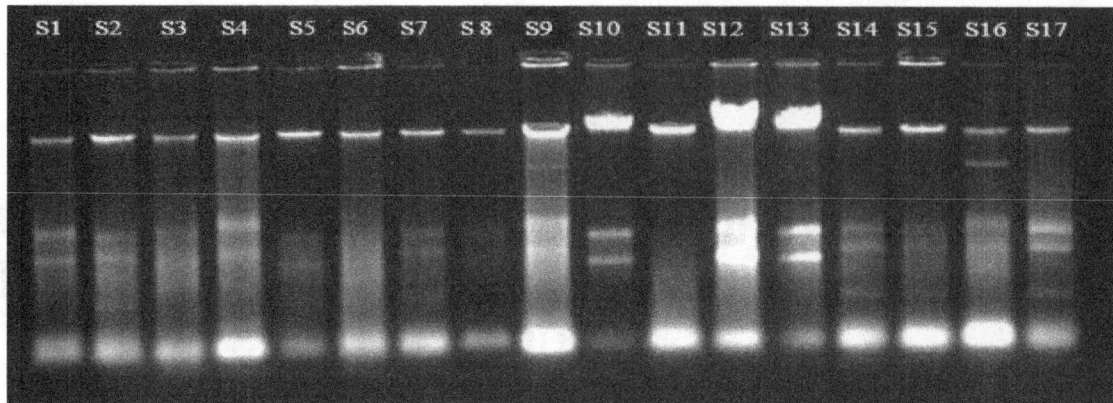

Figure 1. DNAquality testing of different yeast isolates.

random fragments of DNA (RAPD) (Xufre et al., 2000).

This technique utilizes short (5–15 mer) oligonucleotide primers of arbitrary sequence at low annealing temperature that hybridize at the loci distributed at random throughout the genome, allowing the amplification of polymorphic DNA fragments (Quesada et al., 1995).

MATERIALS AND METHODS

All chemicals used in the experiment were of highest purity and were obtained from Sigma Chemical Company (St. Louis, MO), Merck limited (Mumbai, India) and Hi Media (Mumbai, India).

Isolation and medium

Fruit samples were obtained from local sources of Himachal Pradesh and mostly naturally decaying/fermented samples were preferred. 100 g of each fruit sample was taken in a sterile mortar and crushed to a fine paste by mixing with sterile water. Then mixture was kept for overnight at normal room temperature so that natural wild yeast present on fruit samples might grow and develop to enrichment culture. A loopful of liquid portion from each sample was streaked (Quaternary streaking) in plate (with replica) containing MYPG medium (yeast extract 0.3%, malt extract 0.3%, peptone 0.5%, glucose 1% and agar 3%), pH 6.4 (phosphate buffer system) and incubated at 26°C for 2 days.

DNA isolation

For isolation of chromosomal DNA from the isolated yeast species, single isolated colony was selected from each plate and grown in test tube with 20 ml of MYPG broth (pH=6.4) at 26°C for 24 h. DNA was isolated from 24 h old broth culture with following materials: Yeast lysis buffer (pH-8.0), TE buffer (pH-8.0), zymolyase enzyme, phenol-chloroform-isopropyl alcohol solution and ethanol (100%). DNA pellet suspended in TE buffer are stored at 4°C for further PCR amplification (Rosa et al., 2006).

PCR amplification and gel run

The PCR reaction mixture (15 μl) contains 2 μl (~40 to 60 ng)

sample DNA, 1X PCR Buffer (10 mM) pH: 8.0, MgCl2 (3mM), dNTP mix (0.2 mM), Primer 1 (0.5 μM) and Taq DNA Polymerase (0.5 units, Intron technologie, USA).

Amplification of isolated DNA was done using random primers namely RAPD primer-OPA12, RAPD primer-OPB09, RAPD primer-OPC06, with PCR conditions: Initial denaturation at 94°C for 4 min, 39 cycles of: 45 s at 94°C, 45 s at 35°C, 1 min 30 s at 72°C and followed by final extension at 72°C for 5 min (Bio-Rad-My Cycler Thermal Cycler, USA). Gel electrophoresis (2% agarose) of amplified DNA done under standard electrophoresis procedure.

RAPD analysis

Arbitrary primers were obtained from OPC (Operon Technologies, California, USA). The DNA fragments were visualized with ethidium bromide, photographed and analyzed. A 100-bp DNA ladder (Pharmacia Biotech, Uppsala) used as the size standard. The binary data generated here were used to estimate levels of polymorphism by dividing the polymorphic bands by the total number of scored bands. The polymorphism information content (PIC) was calculated by the formula: $PIC = 2 Pi (1-Pi)$ where, Pi is the frequency of occurrence of polymorphic bands in different primers. Pair-wise similarity matrices were generated by Jaccard's coefficient of similarity (Jaccard, 1908) by using the SIMQUAL format of NTSYS-pc (Rohlf, 2002). Jaccard coffiecient was calculated by converity data into binary form. 1 for presence of band and 0 for absence. Matrix was generated for calculating similarity or dissimilarity between yeast strains. A dendrogram constructed by using the unweighted pair group method with arithmetic average (UPGMA) with the SAHN module of NTSYS-pc to show a phenetic representation of genetic relationships as revealed by the similarity coefficient (Sneath and Sokal, 1973).

RESULTS AND DISCUSSION

17 yeast species were isolated, purified and morphological as well as genetic diversity analysis using RAPD method were successfully carried out. DNA quality testing is illustrated in Figures 1 - 5. Yeast specific defined media used was able to inhibit the growth of bacterial population in all the cultures. In some fungal contamination, fungus growth showed after 3 days of incubation of primary cultures and pure cultures were easily made before fungal

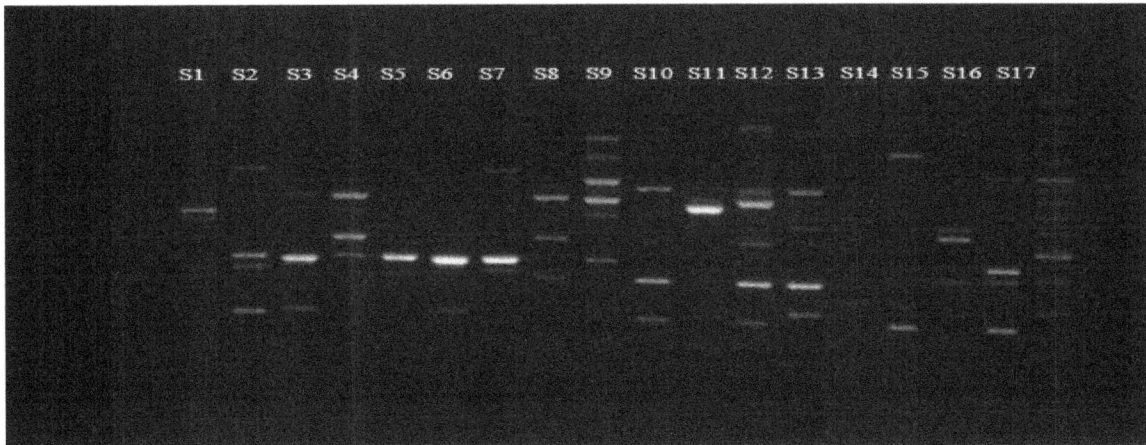

Figure 2. Amplification of genomic DNA of various yeast isolates by PCR using RAPD primer.. OPA12.

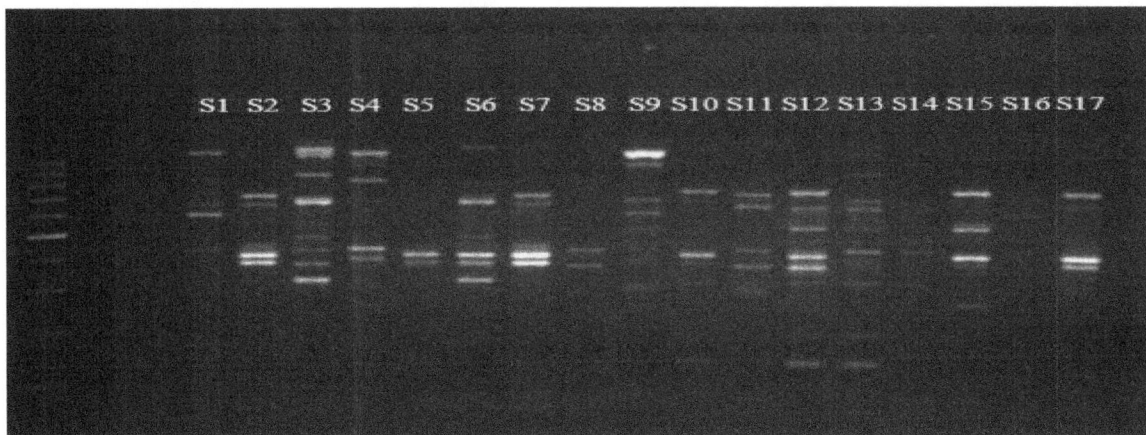

Figure 3. Amplification of genomic DNA of various yeast isolates by PCR using RAPD primer.. OPB02.

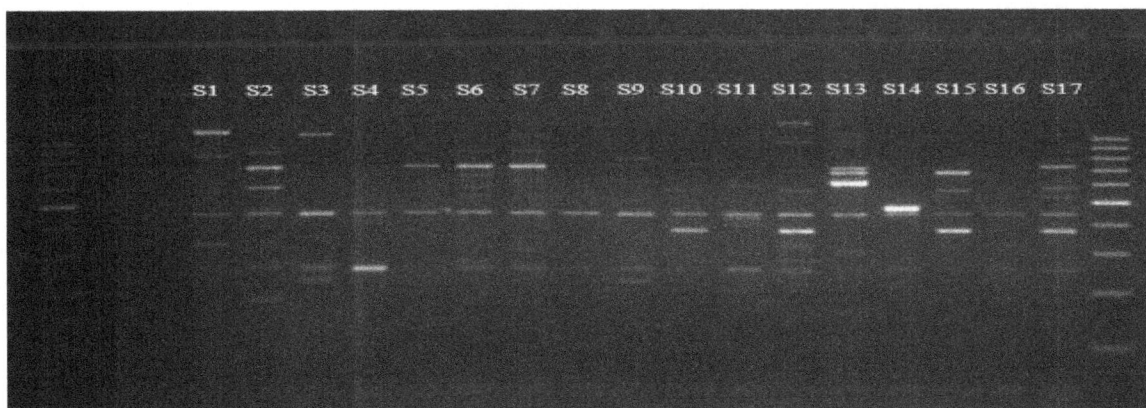

Figure 4. Amplification of genomic DNA of various yeast isolates by PCR using RAPD primer.. OPC06.

Growth became dominant. Studies related to determination of type of fungal contamination were not made as pure cultures were easily made by following the enrichment culturing. Although, in the primary cultures with inoculums from banana, sapota, orange, mango, pineapple and pear fruits more than single type of yeast

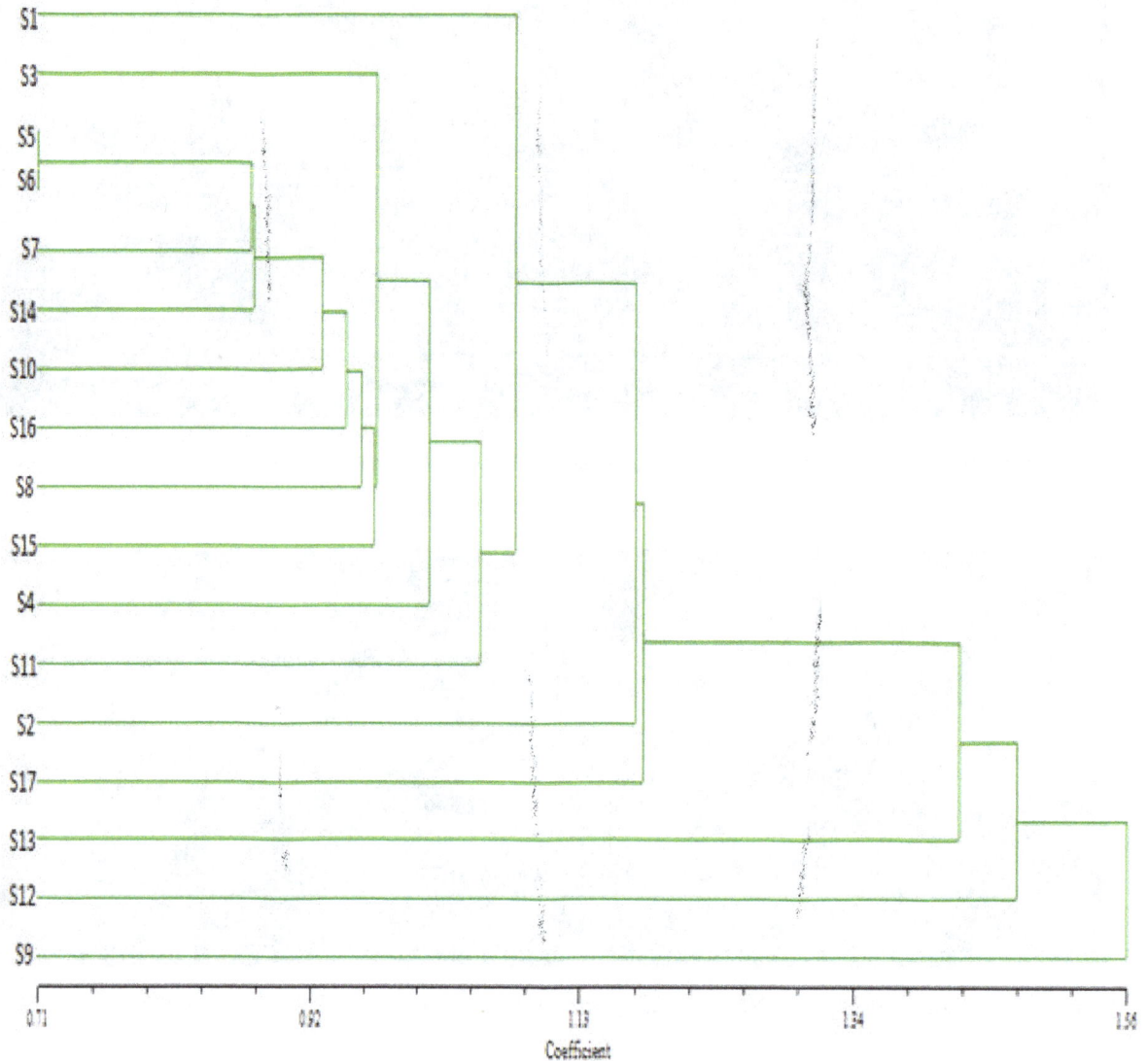

Figure 5. Dendrogram depicting genetic diversity in isolated yeast. The data is on RAPD profiles generated by three primers(OPA-12, OPB-09 and OPC-06).

culture in form of mixed culture existed. From such mixed cultures distinct single isolated colony were picked based on maximum growth among the whole culture and pure cultures of those selected colonies were made and maintained. Morphological data based on different parameters of colony characteristics are tabulated in Table 1. Yeast isolated from fruits namely dates and figs shared some common morphological characteristics. Identical colony characteristics were observed among yeast colonies from green grapes, raisins and black grapes. Yeast isolated from banana, sapota, pear and mango shared some of the common morphological characteristics with maximum growth during incubation period as compared to other isolated species. Few viable isolated colonies were obtained from Papaya under the similar incubation and growth media conditions. Fungal contamination was also

observed and mostly in plates with inoculums from banana, mango, jamun and orange.

Jaccard's genetic similarity coefficient varied from 0.00 to the highest genetic similarity coefficient (0.98) observed between S5 and S7 that is yeast isolated from green grapes and black grapes as sources. UPGMA cluster analysis of the Jaccard's similarity coefficient generated a dendrogram that illustrated the overall genetic relationship among the genotypes surveyed. Clusters are formed on the basis of distance. The dendrogram clearly showing the sufficient distance to form apart clusters. Cluster analysis indicated four distinct clusters. Cluster1 includes S1, S3, S5, S6, S7, S14, S10, S16, S8, S15, S4 and S11. Cluster 2 includes S2 and S17, Cluster 3 with S13 and S12, only S9 in Cluster 4. The difference between the 1st and 2nd cluster is lesser than the 3rd and 4th,

Table 1. Yeast Isolate code, corresponding fruit source and colony morphological data of different yeast isolated from 17 fruit sources.

Isolate name	Corresponding fruit source		Colony characteristics					
	Common Indian name	Scientific/ botanical name	Color	Shape	Surface	Elevation	Edge	Consistency
S1	Orange	Citrus sinensis	Cream	Circular	Dry, Rough	Convex	Serrate	Dull, matte
S2	Date	Phoenix dactylifera	Cream	Oval	Glistening	Flat	Curled	Mucoid
S3	Apple	Malus pumila	White	Circular	Wrinkled	Bulged	Curled	Stringy
S4	Pear	Pyrus communis	Cream	Evenly circular	Shiny	Raised	Undulate	Butyrous
S5	Green Grapes	Vitis vinifera	Off-white	Circular	Smooth	Raised	Entire	Butyrous
S6	Raisins	Vitis vinifera	Off white	Circular	Smooth, glistening	Raised	Entire	Butyrous
S7	Black grapes	Vitis vinifera	Cream	Spheriodal	Smooth glistening	Raised	Entire	Butyrous
S8	Mango	Mangifera indica	Chalky white	Irregular	Rough	Convex	Undulate	Brittle
S9	Jamun (Black Plum)	Eugenia jambolana or Syzygium cumini L	White	Small circular	Smooth	Raised	Curled	Stringy
S10	Sapota / Zapote (Sapodilla)	Casimiroa edulis	Chalky white	Filamentous	Granular	Raised	Serrate	Dry
S11	Pineapple	Ananas comosus	Chalky white	Small circular	Mucoid	Raised	Entire	Butyrous
S12	Papaya	Carica papaya	Dull White	Spheriodal	Smooth	Umbonate	Entire	Butyrous
S13	Cherry	Prunus avium	Off white	Circular	Smooth	Raised	Undulate	Viscous
S14	Plum	Prunus cultivar	White	Circular	Smooth	Convex	Curled	Butyrous
S15	Pomegranate	Punica granatum	Cream	Evenly circular	Smooth glistening	Raised	Entire	Viscous
S16	Banana	Musa acuminate	Chalky white	Irregular	Wrinkled, dry	Umbonate	Undulate	Dry brittle
S17	Figs	Ficus carica	Cream	Spheriodal	Shiny	Pulvinate	Entire	Viscous

but its sufficient to form another clad As isolated yeasts were from different sources, high variability both at phenotypic (as seen in colony morphologies) and genotypic (from band pattern of Gel run of randomly amplified DNA generated by 3 primers) was observed. The polymorphism information content (PIC) calculated from the frequency of polymorphic bands in primer OPA12 was 0.63, in primer OPB09 was 0.49 and in primer OPC06 was 0.53. The greater genetic similarity coefficient between yeast isolates from green grapes, raisins and black grapes as well as identical morphological characteristics indicates association of same strains of particular yeast specie with these fruits. Likewise, it is also known that S. cerevisiae is naturally presented on grapes

Spencer et al. (1995).

The basis of our experiment is the association of different yeast strains with the corresponding fruit and so they have distinct fermentative properties. Further analysis regarding characterization, fermentative properties, ethanol tolerance and subsequent selection of high ethanol yielding species of yeast from different fruits (Colin et al., 2006) were not being evaluated in the study.

Presence of yeast species on fruits can also be dependent on the geographical factors as well as the place from where the fruits are obtained (Rosa et al., 2006) and their further post harvesting treatment. In other manner, analysis variability in yeast species isolated from fruits that are of different geographical origins can provide a

procedure to trace the origin in import-export products. Yeast also undergoes a number of changes in accordance with the environment. Isolation of yeast species from different fruit sources and further the RAPD assay is randomized so the present study cannot be compared with any previous work. But it is clearly proved in the dendrogram that yeast isolated from different sources are very much different from each other, being, may be of different species or of different strains.

Conclusion

Presence of genetic diversity using RAPD analysis as well as different observed colony morphological characteristics in the yeast strains from different fruit sources, it purposes a methodology for easy and quick isolation of yeast strains for both research and industrial analysis.

REFERENCES

Barnett JA, Payne RW, Yarrow D (1990). "Yeasts: Characteristics and identification". Cambridge University Press, Cambridge, 2nd. Ed., p.1002.

Couto MMB, Vossen JMBM, Hofstra H, Huis Veld JHJ (1994). "RAPD analysis: a rapid technique for differentiation of spoilage yeasts". Intern. J. Food Microbiol., 24(1-2): 249-260.

Jaccard P (1908): "Nouvelles recherches sur la distribution florale". Société Vaudoise des Sciences Naturelles 44: 223–270.

Kreger-van Rij NJW (1984). "The Yeasts: A Taxonomic Study". Elsevier Science Publishers BV, Amsterdam, 3rd ed., pp. 893–905.

Kurtzman CP, Fell JW, Rosa CA, Peter G (2006). "Yeast systematics and phylogeny - implications of molecular identification methods for studies in ecology". The Yeast Handbook. Germany: Springer-Verlag, Berlin-Herdelberg, pp. 11-30.

Kurtzman CP, Piskur J (2006). Taxonomy and phylogenetic diversity among the yeasts. In Comparative genomics using fungi as models. Edited by: Sunnerhagen P and Piškur J. Heidelberg, Springer Verlag; [Hohmann S (Series Editor): Topics Curr. Genet., 15:29-46.

Legras JL, Merdinoglu D, Cornuet JM, Karst F (2007). "Bread, beer and wine: Saccharomyces cerevisiae diversity reflects human history". Mol. Ecol., 16 (10): 2091–2102.

Li H, Veenendaal E, Shukor NA, Cobbinah JR, Leifert C (2008). "Yeast populations on the tropical timber tree species Milicia excels". Lett. Appl. Microbiol., 21(5): 322 – 326

Loureiro V, Malfeito-Ferreira M (2003). "Spoilage yeasts in the wine industry". Int. J. Food Microbiol.,86 (1-2): 23–50.

Colin McBryde, Jennifer M, Lopes de B, Miguel JV (2006). "Generation of Novel Wine Yeast Strains by Adaptive Evolution". Am. J. Enol. Vitic., 57: 423–30.

Davenport RK, Mossel DA, Skinner FD, Passinfre SM (1980). "Experience with some methods for the enumeration and identification of yeasts occurring in foods. In: Biology and activities of yeasts". London: Acad. Press, 9: 279.

Quesada MP, Cenis JL (1995). "Use of random amplified polymorphic DNA (RAPD)–PCR in the characterization of wine yeasts". Am. J. Enol. Vitic. 46, 204-208.

Rohlf FJ (2002). "NTSYS-pc numerical taxonomy and multivariate analysis system, version 2.2. Exeter Software", New York, USA: Exeter Software.

Sneath PH, Sokal RR (1973): "Numerical taxonomy- The principles and practice of numerical classification". Freeman Press, San Francisco, California, USA, p. 573.

Spencer JFT, Spencer DM (1997). "Yeasts in Natural and Artificial Habitats". Springer-Verlag, Berlin- Heidelberg, p. 381.

Tournas VH (2005). "Moulds and Yeasts in fresh and minimally processed vegetable and sprouts". Int. J. Food Microbiol., 99: 71-77.

Xufre Angela, Fernanda Simões , Francisco Gírio, Alda Clemente, M. Teresa Amaral-Collaço (2000). "Use of RAPD Analysis for Differentiation among Six Enological Saccharomyces spp. Strains". Food Technol. Biotechnol., 38 (1): 53–58.

8

Heat resistance of genus *Byssochlamys* isolated from bottled raphia palm wine

E. I. Eziashi*, I. B. Omamor, C. E. Airede, C. V. Udozen and N. Chidi

Nigerian Institute for Oil Palm Research (NIFOR), Plant Pathology Division, P. M. B. 1030, Benin City, Edo State, Nigeria.

Bottled *raphia* palm wine was cultured in a laboratory medium amended with 0.5% acetic acid (pH 4.8). Two cultures of identified heat resistant yeast (HRY) and one unidentified yeast species were isolated. Cultures of the isolates grown on potato dextrose agar for 10 days at 26°C, survived pasteurization temperature at 80°C for 20 min and 85°C for 15 min. Of these HRY identified were *Byssochlamys nivea*, *Byssochlamys zollerniae* and one unidentified yeast species. To determine the source of contamination, fresh un-pasteurized *Raphia* palm wine was cultured. Result revealed that, colonies of the three HRY were higher compared with the pasteurized *Raphia* palm wine. Frequencies of occurrence at 80°C, 85°C and in un-pasterurized raphia palm wine were *B. nivea* 15.2, 6.1 and 24.2%; *B. zollerniae* 6.1, 3.0 and 12% and yeast species 9.1, 6.1 and 18.2% respectively. The thermal destruction time were *B. nivea* 90°C for 15 min, *B. zollerniae* 90°C for 5 min and yeast species 90°C for 10 min. The result indicates they are acid tolerant and thermophilic yeasts with *B. nivea* having the highest frequency of occurrence.

Key words: Culture, pasteurization, spoilage, identification, thermophilic.

INTRODUCTION

Raphia hookeri is the most economically important plant among the eight raphia species indigenous to Nigeria (Okolo, 2008; Otedoh, 1978). The exploitation of *Raphia* for the sap (palm wine) and other products of socio-economic importance such as pissava, fibre, oil edible grubs, poles, thatch etc are mainly from the wild (Udom, 2000). Wine is tapped from the panel which consists of the base of short spear leaves and the apical emerging terminal inflorescence axis (Tuley, 1965). The wine is rich in vitamins, carbohydrates and yeast (Obahiagbon, 2007).

Filamentous fungi are morphologically complex micro-organisms exhibiting different structural forms through out their life cycles (Adrio and Demain, 2003). The life cycles of filamentous fungi starts and ends in the form of spores. In submerged cultures, these fungi have different morphological forms ranging from dispersal of mycelial filaments to densely mycelial masses as pellets (Xu and Yank 2007). Microorganisms are an important part of our environment and are a principal cause of food spoilage. When food is contaminated by harmful microorganism, the products can cause severe human food-borne diseases, either due to the organisms themselves or the toxins released by them (Laplace-Buihe et al., 1993). The presence of these microorganisms in the products, even at low concentration may severely affect their quality (Laplace-Buihe et al., 1993).

Fruit juices contain various concentration of sucrose, which constitutes a very important component of the medium for the growth of fungi (Palou et al., 1998). Microbial spoilage is a serious problem for the food industry as fungal contamination can occur during processing as well as handling of the end products. Since yeast can generally resist extreme conditions better than bacteria, they are often found in products with low pH and in those containing preservatives (Macrae et al., 1993). Especially yeast spoilage has increased in recent years as a result of lower doses of preservatives and milder preservation

---

*Corresponding author. E-mail: eziashius@yahoo.com.

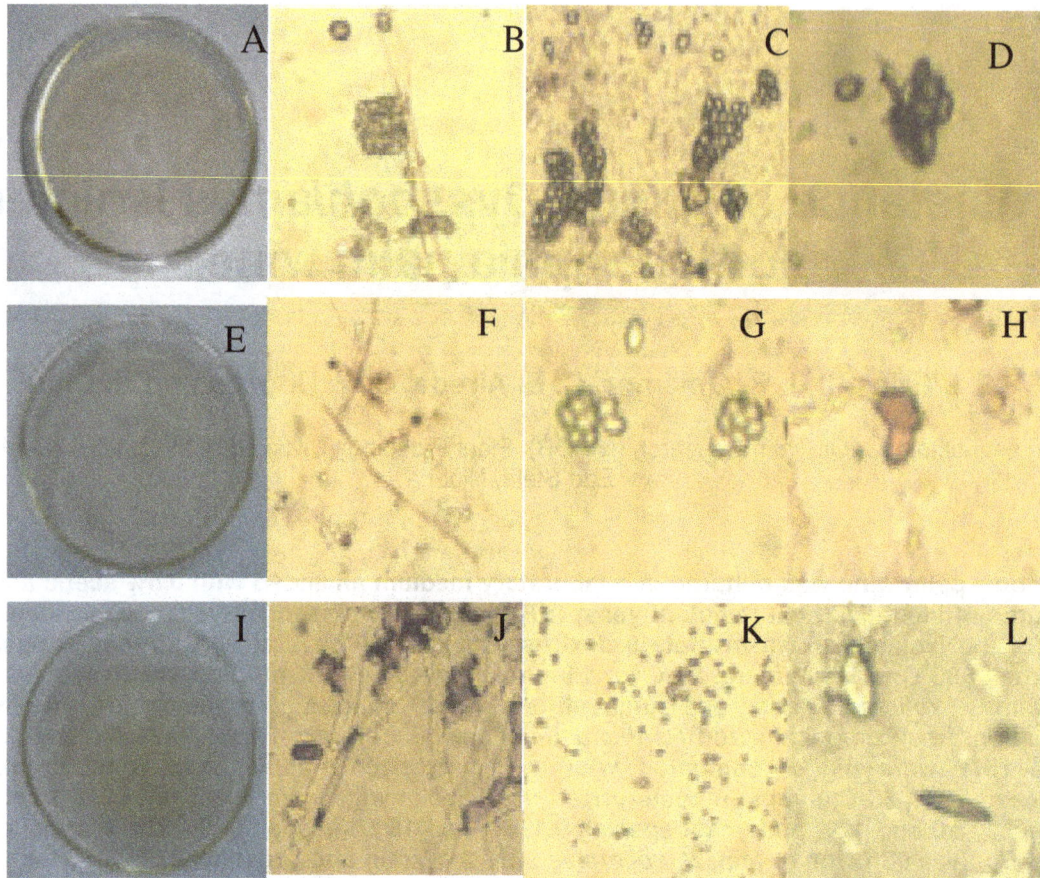

Plate 1 (a-l). Light photograph and microscope of *Byssochlamys* grown on PDA 10 days after Incubation.(a) Pure culture of *Byssochlamys nivea*; (b) Conidiophore (c). Conidia (d) Ascospores. (Conidia-scale bars = 10 μm). (e) Pure culture of *Byssochlamys zollerniae*; F. Conidiophore; G. Conidia; (h) Ascospores. (Conidia-scale bars = 10 μm). (i) Pure culture of Yeast species; (j) Conidiophore; (k) Conidia; L. Ascospores.

preservation processes required for higher standards of food quality (Beuchat, 1989). Spices, herbs and plants are essential oils added to foods primarily as flavoring agents have been shown to possess a broad range of antimicrobial activities (Palou et al., 2002).

The genus *Byssochlamys* contains two economically important species, *Byssochlamys nivea* and *Byssochlamys fulva*. Both species cause spoilage of processed fruit products and are among the most commonly encountered fungi associated with spoilage of heat processed fruits in countries worldwide (Tournas, 1994). *Byssochlamys* species produce ascospores which are heat resistant and survive considerable periods of heat above 85°C (Beuchat and Rice, 1979; Splittstoesser, 1987). In addition to their heat resistance, *Byssochlamys* species can grow under very low oxygen tension (Taniwaki, 1995) and can form pectinolytic enzymes. The combination of these three physiological characteristics makes *Byssochlamys* species very important spoilage fungi in pasteurized and canned fruits. *Byssochlamys* has a *Paecilomyces* anamorph (Samson et al., 2009). Patulin,

a toxic secondary metabolite can be produced by *B. nivea* and *B. fulva*, as well as several species of *Penicillium* and *Aspergillus* (Jackson and Dombrink-Kurtzman, 2006). The objective of this study was to isolate and identify yeasts causing spoilage of the pasteurized *Raphia* palm wine with the view to determine their thermal destruction temperatures (Figure 1).

MATERIALS AND METHODS

Experiment

Different bottled *Raphia* palm wines (BRPW) were collected as samples from four different locations in Edo state of Nigeria. Eight out of the twenty-two BRPW stored for four months from each of the four locations were spoiled. The medium was constituted according to Chu and Chang (1973). Different aqueous dilutions (10^{-2}, 10^{-3} and 10^{-4}) of the suspension were applied onto plates and 20 ml of melted medium at around 50°C was added. After gently rotating, the plates were incubated at 26°C for 10 days. Isolated colonies of yeasts were transferred from mixed cultures of the plates onto respective agar plates (Waksman, 1961) and incubated aerobically

Table 1. Yeasts isolated from pasteurized and un-pasteurized raphia palm wine and their frequencies of occurrence.

T °C	Yeast contaminants	Pasteurized mean number of yeast colonies	% Frequencies
80 °C – 20 min	B. nivea	5	15.2
	B. zollerniae	2	6.1
	Yeast species	3	9.1
85 °C – 15 min	B. nivea	2	6.1
	B. zollerniae	1	3.0
	Yeast species	2	6.1
90 °C-15 min	B. nivea	-	-
	B. zollerniae	-	-
	Yeast species	-	-
Unpasteurized	B. nivea	8	24.2
	B. zollerniae	4	12.1
	Yeast species	6	18.2

at 26 °C for 10 days. Each BRPW sample was treated with four plates for each aqueous dilution, totaling twelve plates for each sample. Their mean colonies and frequencies of occurrence were determined according to Omamor (2007). The first yeast isolate with the highest frequency of occurrence was sent to CABI identification services Surrey United Kingdom for morphological and molecular identifications, the second yeast isolate was identified according to Samson et al. (2009) while the third yeast isolate was left unidentified.

Morphological characterization

Purified isolates of yeasts were identified to the generic level by comparing their morphology of spore-bearing hyphae with the entire conidiophores and structure of ascospores with the Byssochlamys morphologies as described by Samson et al. (2009). This was done by using the cover-slip method in which an individual culture was transferred to the base of cover slips buried in potato dextrose agar (PDA) (Duarte and Archer, 2003) for photomicrographs. They were incubated for ten days at 26 °C ± 2. Structures of conidiophores and ascospores were visually estimated by using a motic microscope attached to a motic digital camera connected to a computer. The experiment was repeated three times.

RESULTS AND DISCUSSION

B. nivea, Byssochlamys zollerniae and one unidentified yeast species were isolated from both pasteurized and un-pasteurized Raphia palm wine. B. nivea with the highest frequencies of occurrence at 80, 85 °C and in un-pasteurized Raphia palm wine (Table 1), which was sent to CABI identification services Surrey United Kingdom for morphological and molecular identification, was identified as B. nivea Wasting IMI No. 396923. Both B. nivea, B. zollerniae and the yeast species survived pasteurization temperatures at 80 °C for 20 min and 85 °C for 15 min. However, B. nivea at 90 °C for 15 min, B. zollerniae at 90 °C for 5 min and Yeast species at 90 °C for 10 min

were eliminated. Their colonies spread averagely on PDA at 26 °C and covering the petri plates within 10 days (Plates 1a, e and i). Frequencies of occurrence at 80, 85 °C and in un-pasteurized Raphia palm wine were B. nivea 15.2, 6.1 and 24.2; B. zollerniae 6.1, 3.0 and 12, and yeast species 9.1, 6.1 and 18.2 respectively (Table 1).

The genus Byssochlamys is morphologically well-defined and characterized by ascomata in which crosiers and globose asci are formed with ellipsoidal ascospores. The ascomatal initials consist of swollen antheridia and coiled ascogonia. Using light micrograph, the conidiophores, conidia and ascospores were seen. The conidium measured 3.1 - 4.3 × 2.6 - 3.2 µm (Plate 1c) and was characterized with smooth wall. The conidium of B. zollerniae measured 3.1 - 3.8 × 2.5 - 3.2 µm (Plate 1g).

The introduction of B. nivea, B. zollerniae and the yeast species into the bottled and pasteurized Raphia palm wine might have been due to inadequate pasteurization, contaminants from the host palm, sub-standard condition of the palm wine tapping panel and the bottling unit. This was supported by Odutayo et al. (2004).

The thermal destruction temperature at 90 °C for 15 min in this study will most likely result in loss of desirable fresh flavor, vitamins, carbohydrates and other nutritional substances, although food scientists and the food industry are searching for novel methods that may destroy undesired microorganisms with less adverse effects on product quality (Rosenthal and Silva, 1997). Thermally pasteurized fruit juices are often characterized by a loss of desirable fresh flavor characteristics (Butz and Tauscher, 2002). The thermal pasteurization employed in this study has eliminated the acid tolerant and thermophilic yeasts causing spoilage of the Raphia palm wine, thus increase the shelve life and reduce the superior quality. This agrees with Butz and Tauscher (2002) reports that un-pasteurized products perceived by customers are to be of superior quality but its shelf life is very limited.

The present study on morphological identification and features of *B. nivea* confirms the presence of ascospores. It is responsible for the spoilage of bottled *Raphia* palm wine. This agrees with Beuchat and Rice (1979), reported that *Byssochlamys* spp. Produce ascospores which are heat resistant and survive considerable periods of heat above 85°C. The same was supported by Splittstoesser (1987). In addition to their heat resistance, *Byssochlamys* species can grow under very low oxygen tensions (Taniwaki, 1995) and can form pectinolytic enzymes. The combination of these three physiological characteristics makes *Byssochlamys* species very important spoilage of pasteurized and canned fruits (Samson et al., 2009).

Pasteurization temperature for fresh palm wine is between 70 - 80°C. At this temperature *B. nivea* is still viable. There must be alternative means to ensure that, these yeasts are eradicated in-order to ensure good quality *Raphia* palm wine with the reduction of thermal destruction time. This study did not investigate the product quality after the thermal destruction time.

REFERENCES

Adrio JL, Demain AL (2003). Fungal biotechnology. Int'l Microbiol 6:191-199.

Beuchat LR, Golden DA (1989). Antimicrobials occurring naturally in foods. Food Technol., 43:134-142.

Beuchat LR and Rice SL (1979). Byssochlamys spp. and processed fruits. Adv. Food Res., 25:237–288.

Butz P, Tauscher B (2002). Emerging Technologies: Chemical Aspects. Food Res. Int., 35:279

Chu FS, Chang CC (1973). Pectolytic enzymes of eight Byssochlamys fulva isolates. Mycologia 65:920-925

Duarte MIR, Archer SA (2003). In vitro toxin production by Fusarium solani f. sp. piperis Fitopatologia Brasilcira 28:229 – 235.

Jackson L, Dombrink-Kurtzman MA (2006). Patulin. In: Sopers GM, Gomy JR, Yousef AE (eds). Microbiology of fruits and Vegetables. GRC Press Boca Raton Fl. pp 281-311.

Laplace-Buihe C, Kahne K, Hunger W, Trilly Y, Drocourt JL (1993). Application of flow cytometry to rapid microbial analysis in food and drink industries. Biol. Cell., 78:123-128.

Macrae R, Robinson RK, Sedler MJ (1993). Encyclopedia of Food Science, Food Technology and Nutrition Vol 7. Academic Press London UK pp 4344-4349.

Obahiagbon FI (2007). Development of Agronomic Practices for raphia Palms. NIFOR in House Research Review. pp 151-153.

Omamor IB, Asemota AO, Eke CR, Eziashi EI (2007). Fungal contaminants of the oil palm tissue culture. Niger. Institute for Oil Palm Research 2: 534-537.

Otedoh MO (1978). Taxonomic studies in raphia palms – Historical Review. A paper presented at the 14th Annual Conference of the Agricultural Society of Nigeria pp 1-5

Odutayo OI, Oso, RT, Akinyemi, BO Amusa NA (2004). Microbial contaminants of cultured *Hibiscus cannabalis* and *Telfaria occidenttalis* cultured tissue, Afr. J Biotechnol. 3: 301-307.

Okolo EC (2008). Evaluation of raphia hookeri progenies. NIFOR in House Research Review pp 147-148.

Palou L, Usall J, Smilanick JL, Aguilar MJ, Vinas I (2002). Evaluation of food additives and low-toxicity compounds as alternative chemicals for the control of *Penicillium digitatum* and *Penicillium italicum* on citrus fruit, Pest Manag. Sci., 58:459-466.

Palou E, Lopez-Malo A, Barbosa G, Welti J, Davidson P, Swanson B (1998). Effect of oscillatory high hydrostatic pressure treatments on Byssochlamys nivea ascospores suspended in fruit juice concentrates. Lett. Appl. Microbiol., 27:375-378.

Rosenthal A, Silva JL (1979). Alimentos sob Pressao. Emgenharia de Alimentos 14:37

Samson RA, Houbraken J, Varga J, Frisvad JC (2009). Polyphasic taxonomy of the heat resistant ascomycete genus Byssochlamys and its Paecilomyces anamorphic. Personia 22, 2009: 14-27

Splittstoesser DF (1987). Fruits and fruit products. In: Beuchat LR. (ed), Food and beverage mycology, 2nd ed.: 101–128. Van Nostrand Reinhold, NewYork, USA.

Taniwaki MH (1995). Growth and mycotoxin production by fungi under modified atmospheres. PhD thesis, Kensington, NSW: University of New South Wales, Australia.

Tournas V (1994). Heat-Resistant fungi of importance to the food and beverage industry. Crit. Rev. Microbiol., 20:243-263.

Tuley P (1965). How to Tap a raphia Palm. Niger. Field, 30 (3) 120-132.

Udom DS (2000). Investigations on Wine Production from raphia hookeri. Varieties, The Derivable Gross Income in South- Eastern Nigeria. Department Agric. Econ. Ext., University of Calabar, Bull., 12: 1: 4-16.

Waksman SA (1961). The Actinomycetes. Vol. II. Classification, identification and description of genera and species. The Williams & Williams Co. Baltimore.

Xu Z, Yang S (2007). Production of Mycopherolic and Penicillium brevicompactum Immobilized in a Rotating Fibrous-bed Bioreactor Enzyme Microb. Technol., 40: 623-628.

9

Morphology and taxonomy of *Sarcoscypha ololosokwaniensis* sp. nov.: A new Ascomycota species from Serengeti National Park-Tanzania

Donatha Damian Tibuhwa

Department of Molecular Biology and Biotechnology, University of Dar es Salaam,
P.O. Box 35179, Dar es Salaam, Tanzania. E-mail: dtibuhwa@yahoo.co.uk.

Traditional taxonomy emphasizes the morphological features to characterize a taxon. *Sarcoscypha* is a genus in Sarcoscyphaceae family which display wide array of morphological variations. The genus is widespread in northern hemisphere and boreal regions, but also occurs in sub tropical areas and in the southern hemisphere. Both macro and micromorphological features including (ascocarp size, colour, shape, exterior surface of the fruit body, asci size, shape, as well as ascospore size, ends and lipid bodies) were used in a conventional taxonomic analysis of fresh *Sarcoscypha* material collected from southern hemisphere in Tanzania. Results showed that compared with similar species from northern hemisphere, Tanzanian materials were relatively smaller, smooth, vivid sharp red inside a saucers-shaped ascocarp, sessile to substipitate, microscopically unsheathed ascospores with two lipid bodies, distinctive geographical distribution, and unique season of fructification. Furthermore, a dichotomous identification key constructed for the six close similar species proved that Tanzanian material differed from other close species compared. Therefore based on conventional morphological taxonomy Tanzanian material from Serengeti National Park is described for the first time as a new Ascomycota; *Sarcoscypha ololosokwaniensis* sp.nov.

Key words: Ascocarp, *Ascomycota*, *Sarcoscypha*, Serengeti, Tanzania.

INTRODUCTION

Classification of fungi is constantly anguished by contradictions. This is due to the lack of complete knowledge about all the fungal organisms. Ascomycota, also known as sac fungi, is a recently discovered class. In fact, it was once classified within Deuteromycota. Ascomycota is a sister group to basidiomycota. *Sarcoscypha* represents one of the numerous examples of fungal genera in which a sound taxonomy is only achieved on the basis of vital macro- and especially microscopical characters gained from the study of fresh collections ("vital taxonomy"). Harrington (1990) reported the importance of fresh material for species diagnosis, especially for noting ascospore guttulation, which is vital in recognition providing the distinction of species from different bioclimates.

Sarcoscypha Boud. is a genus in a family Sarcoscyphaceae of the ascomycete fungi. It was described in 1885 by Jean Louis Émile Boudier based on Peziza tribe by Fries in 1822 in *Systema mycologicum* (Fries, 1821-1832). Species of *Sarcoscypha* are characterised by a cup-shaped apothecium which is often brightly colored, saprophytic species growing on decaying woody material from various plants such as plants of the rose family, beech, hazel, willow, elm, and in the oak trees (Baral, 2004). According to the fungal dictionary 10th edition of 2008, the genus comprises about 28 species which tentatively are accepted world-wide, about 6 of which occur within Europe and North America. Phylogenetic relationships study in the genus by Harrington (1998) hypothesized that the most recent common ancestor of the genus originated in Europe. Many appear to be endemic to volcanic islands in the

Table 1. Ten species of mushroom described for the first time from Zanzibar-Tanzania by Berkeley in 1885 original described name, and current name.

S/No	Original described name	Present name
1	*Agaricus missionis* Berk.	*Lepiota missionis* (Berk.) Sacc.
2	*Agaricus rhodofephalus* Berk.	*Lepiota rhodofephala* (Berk.) Sacc.
3	*Agaricus vagus* Berk.	*Clitocybe vaga* (Berk.) Sacc.
4	*Hiatula benzonii* (Fr.) Mont.; Berkeley	*Agaricus benzonii* Fr.
5	*Agaricus arethusa* (Berk.) Sacc.	*Omphalina arethusa* (Berk.)
6	*Agaricus obfuscescens* (Berk.) Sacc.	*Pleurotus obfuscescens* (Berk.) Sacc.
7	*Agaricus medius* Fr., Berkeley	*Volvariella media* (Schum.ex Fr.) Singer
8	*Agaricus nicotianus* Berk,	*Agaricus nicotianus* Berk & M.A Curt.
9	*Agaricus alboquadratus* Berk.	*Psilocybe alboquadrata* (Berk.) Sacc.
10	*Agaricus trisulphuratus* Berk.	*Cystoagaricus trisulphuratus* (Berk.) Singer

subtropics. A high number of species (about 60) have ever been combined in the genus *Sarcoscypha*, many of these were often only collected a single time and especially the old descriptions are very inadequate, often re-descriptions of the type material, if any exists, are lacking thus they remain virtually unknown. Many of the taxa have later been found to belong in other genera of the Sarcoscyphaceae (Harrington, 1996; Harrington and Potter, 1997; Spooner, 2002).

Sarcoscypha coccinea (Scop.) Lambotte is known to be used as a medicinal plant by the Oneida Indians (Seaver, 1928), as a table decoration in Scarborough, England by arranging their fruit bodies with moss and leaves (Dickinson and Lucas 1982). The species is also known to be edible (Arora 1986) and a good source of food for rodents in the winter, and slugs in the summer (Brown 1980). Some *Sarcoscypha* species have been also found to posses some bioactive compounds which might be potential in bioremediation (Tortella et al., 2008).

In Tanzania, mushroom forming fungi are poorly collected, sparingly studied and relatively underutilized. The inventory of mushroom in Tanzania was done by Berkeley (1885) who described 10 new species from Zanzibar (summarized in Table1),followed by Hennings (1893) who described several species of Agaricales from western Tanzania.

Pegler (1977) in his Agaric Flora of East Africa book described also exclusively several *Agaricales* from the country. Recently a comprehensive study of edible and poisonous mushrooms of Tanzania was done by Härkönen et al. (1995, 2003) who reported more than one hundred taxa and described several new. More taxonomic studies on specific genus of the basidiomycete include Buyck et al. (2000); Tibuhwa et al. (2008) who all worked on a genus *Cantharellus* Fr. from miombowoodland of Tanzania and Magingo et al. (2004) who studied *Odumensiela* sp. Apart from Härkönen et al. (1995, 2003), who studied a few Ascomycete, the rest of the studies never mentioned the Ascomycete nor

presented any mushrooms in the Serengeti National Park, one among the protected areas of Tanzania with unique complex ecosystem. Serengeti National Park is one of the largest wildlife sanctuaries in the world covering about 14,763 sq km. The park lies in a high plateau between the Ngorongoro highlands and the Kenya/Tanzania border extending North-west almost to Lake Victoria. It is rich in biodiversity ranging from large mammals such as elephants to small microorganism, all making a complex ecosystem. Being a national park, much attention has been paid to study of large organisms viz: the animals (Sinclair and Arcese 1993, Borner et al., 1996); birds (Schmidt, 1982; Fishpool and Evans, 2001), Amphibians and reptiles (Kreulen, 1975), Vegetation (Schmidt, 1975; Belsky, 1987).

Despite the fact that Serengeti National Park is one of the least disturbed and best studied areas in Africa, mushrooms are among the forgotten taxa within the park which has never been studied although they contributes greatly to balancing the ecosystem in terms of nutrient recycling and symbiotic associations. The aim of the present investigation was to characterise the wild *Ascomycota* mushroom, species of *Sarcoscypha* found in Serengeti National Park.

MATERIALS AND METHODS

Study area: Serengeti national park in Tanzania

A two months field trip was made in the park and covered both the dry and wet side of the park (Figure. 1) in two years (2009 to 2010).

Collection, harvesting and preservation of specimens

Mushroom hunting in the park yielded enormous collections among which this striking red mushroom seems to be unique. On the study site, a transect of 5 x 10 meters was fixed using GPS (Global Positioning System- MAGELLAN EXPLORIST BELGIUM) within which mushroom hunting was done randomly throughout the

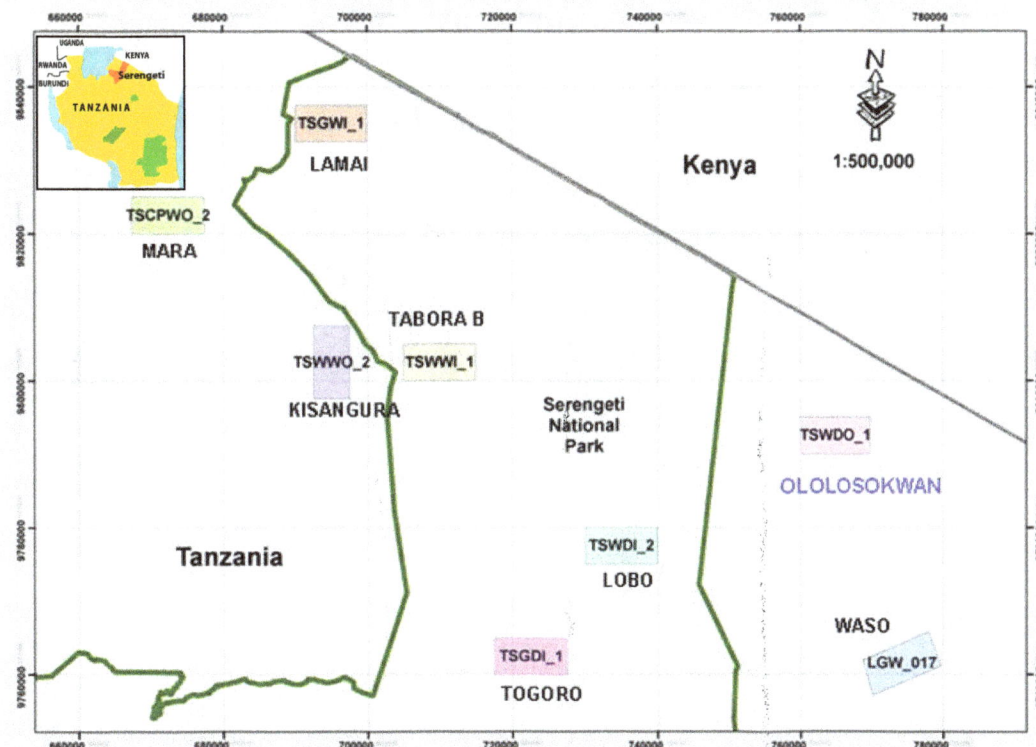

Figure 1. Map of Serengeti National Park, Tanzania showing the study sites, Ololosokwan (in blue color) in the dry side, where the species was collected.

transect. On spotting the mushroom, the GPS reading of the place was noted. Photographs were taken using a digital camera (CYBER-SHOT DSC-W7 JAPAN) and vegetation around were described by a plant taxonomist. The woody substrates lying on the ground were also identified to the generic level and wherever possible to the species level. Ecological parameters of temperature and relative humidity were also noted. Some of the fresh fruiting bodies were harvested then dehydrated using silica gel and kept in air tight plastic bags until further analysis. The type specimens examined come from Tibuhwa's collections that have been deposited in the mycological herbarium of the Uppsala University (UPS) with duplicates at the herbarium of the University of Dar es Salaam (UDSM) in Tanzania.

Macroscopic study

The collected fresh fruit bodies were examined and described in the field. The macromorphology observed in the field included ascocarp colour, size, shape, exterior surface of the fruit body, presence of stipe and how it was attached to the substrates, cup edge curliness, the fruit body fleshiness when fresh and on drying, developmental stages forms, as well as nature of growth. Other field characters such as spore print odor and taste were noted as in Tibuhwa et al. (2008).

Microscopic study

Microscopic observations were made in Ammonia-Congo red solution (TCI DEUTSCHLAND GmbH GERMANY), after a short pretreatment in 10% Ammonium solution. Observations were made at 20, 40 and 100 magnification of a bright field compound microscope Olympus (OLYMPUS BX50 PHASE POL DARKFIELD MICROSCOPE, JAPAN). Microscopic features of the Asci, Ascospore lipid bodies (guttules), mucilaginous envelop in a living spore and paraphyses pictures were taken using a digital camera (CYBER-SHOT DSC-W7 JAPAN) directly mounted on the ocular lens of the microscope, and measured straight using a graduated ocular lens in μm.

Interview for local people's knowledge

Dietary, Culinary, therapeutic, and other ethnomycological utilization of wild mushroom in the area were investigated. Information was collected by face to face interviews for over 150 individuals; detailed results will be presented in a separate study.

Comparative studies and identification of the material studied

Identification of specimens were done using published works on Ascomycetes mushrooms such as Arora (1986), Baral (2004), Don and Dennis (2002), Härkönen et al. (2003), Seaver (1928), Harrington (1996) and Miller and Miller (2006). The materials studied were as follows: Tanzania: Serengeti National Park- Ololosokwan woodland 36M 0761053 9791554: Tibuhwa 1089.2009 (UPS, holotype, isotype, DSM). Other studied material Serengeti-National Park-Loliondo 36M 0771064 9741252: Tibuhwa 1098.2010.

Figure 2. *Sarcoscypha ololosokwaniensis* sp. nov., a and b) General appearance in nature, c) Asci amid the paraphyses d) Ascospores. Scale bar = 11 µm.

RESULTS

Habit and habitat

Fruit bodies of *Sarcoscypha olosokwanii* were found growing singly or in tufts on dead twigs, leaves and fragments of dead wood, usually partly buried on soil and forest litter. The tree associated with them was *Olea europaea* of the family *Oleaceae* although the woodland was also dominated by *Acacia drepanolobium* and *Commiphora africana*. The *Ascomycete* was found fruiting during long rain season of April to June in undisturbed habitat. Average rainfall was 523 mm, Temperature 19 to 21 degree celicius and relative humidity of 32%.

Sarcoscypha ololosokwaniensis Tibuhwa, sp. nov.: MB 519507

Ascocarp 0.5-1.8 cm, sessile or substipitate, smooth, gristly, vivid sharp red inside a saucer which bears the hymenium contrasting a soft paler pellicle which cover the exterior part of the ascocarp.

Holotypus: Tanzania: Serengeti National Park – Ololosokwan woodland 36M 0761053 9791554: Tibuhwa 1089.2009 (UPS, holotype, isotype DSM), April, 2010 on dead twigs, leaves and fragments of dead woods of *Olea europaea* sometime partly buried on soil and forest litter within the vicinity of the same tree species (Figure 2).

Etymology: The *Sarcoscypha* is named as 'ololosokwaniensis' because it was collected from woodland near the Masai village called Ololosokwan under game controlled area of the Serengeti National Park.

Fruiting bodies: 0.5 to 1.8 cm, smooth, gristly, vivid sharp red inside a saucer like which bears the hymenium, contrasting a soft paler pellicle covering exterior part of the saucer and easily removed on holding. The *ascocarp* sessile or substipitate. *Context* very thin and reddish. *Smell* undistinguished. *Ascospores* smooth, hyaline, inamyloid, 28 – 30 x 11 – 13 µm, elliptical with two big oil droplets. *Asci* 8–spored, tubular thinning toward the base, thin walled 120 – 180 x 5 – 9 µm with round end lacking an operculum. *Paraphyses* cylindrical widening toward

Morphology and taxonomy of Sarcoscypha ololosokwaniensis sp. nov.: A new Ascomycota species from Serengeti National...

65

the apex filled irregularly with pigments and droplets.

Scientific classification

S. ololosokwaniensis Tibuhwa, sp. nov. belongs to:

Kingdom: *Fungi*;
Division: *Ascomycota*;
Subdivision: *Pezizomycotina*;
Class: *Pezizomycetes*;
Order: *Pezizales*;
Family: *Sarcoscyphaceae* in the
 Genus: *Sarcoscypha*

DISCUSSION

This species has numerous striking features, both macroscopically and microscopically. Although macroscopically the species of *Sarcoscypha* look outwardly very similar, thus hardly distinguishable, the small size, sharp red colour of the inner part of the saucer ascocarp contrasting the pale-cream pellicle exteriorly instead of hairy to crenulate apothecial margin, sessile or white substipitite, and microscopically, the small unsheathed ascospore with two large lipid bodies as well as asci which are thin walled without operculum are among unifying characters of this taxa. The distribution of *Sarcoscypha* species is known from tropical Asia, United States east of the Rocky Mountains, Central America, Australia and the Caribbean. Members of *Sarcoscypha* have similar appearance with *Peziza* although these two genera are not closely related Landvik et al. (1997).

It is the first time this species in the genus is systematically described from Tanzania. However, Härkönen et al. (2003) presented similar *Sarcoscypha* from Tanzania, but they did not specify the species name probably due to uncertainties of the species identity. Similar red colored "scarlet cup fungus" species of *Sarcoscypha* have been described from different parts of the Northern hemisphere but with restricted distribution. For example, *S. coccinea* (Scop.) Lambotte, is only found in the New World, Central America and the Caribbean, east and Midwest North America, but not in the far west while *Sarcoscypha austriaca* (O. Beck ex Sacc.) Boud and *Sarcoscypha dudleyi* (Peck) Baral are found in eastern regions of the continent (Denison, 1972).

S. olosokwaan sp.nov. is distinguished from other related 'Scarlet Cup Fungi', from north American, *Sarcoscypha coccinea* and *Sarcoscypha occidentalis* (Schwein.) Sacc.; from Hawaii *Sarcoscypha mesocyatha* F.A Harr. and other two *Sarcoscypha austriaca* and *S. dudleyi* in geographical distribution, fruiting season, and macro-micromorphology characters of the fruit body.

While *S. occidentalis* has a vivid differentiated long white stipe (1 to 3 cm), *S. ololosokwaniensis* has none or substipitate. The relatively small size, saucer rather than deep cup and pale pellicle instead of exterior hair, microscopically small size of ascospore, asci, paraphyses and two large lipid bodies also distinguish this species from the closely related *S. coccinea* with numerous small lipid bodies among others. The Hawaiian *S. mesocyatha* differ by having relative large (4.5 cm), shallowly cupulate to flattened discoid ascocarp (Don and Dennis, 2002), while *S. austriaca* and *S. dudleyi* are demarcated by relatively long spores with flattened ends, without full sheath but with small polar caps on either end and typical rounded ascospore in a full sheath respectively (Harrington, 1990). Given the differences in macro-micromorphology, geographic distribution, seasons of fructification, and host differences, this study propose *S. ololosokwaniensis* as a new species (see the key below).

Key to six species of closely similar *Sarcoscypha* species

1. Cup typically < 2 cm across; 2
1. Cup typically > 2 cm across 3

2. Short stem present; spores < 21 μ long; numerous lipid bodies; found east of the Rocky Mountains, North America.............................*S. occidentalis*
2. Short stem absent or present; spores > 21 μ long; two large lipid bodies; found in tropical Africa – Tanzania ....

S. ololosokwaniensis

3. Found in the Pacific Northwest and California; spores usually unsheathed and lacking "polar caps." ....*S. coccinea*
3. Found elsewhere; ascospores with or without sheath 4

4. Found in Hawaii; shallowly cupulate to flattened discoid ascocarp, ascospore unsheathed ....
S. mesocyatha
4. Not found in Hawaii; deep cupulate 5

5. Spores with slightly flattened ends, without full sheath but with a sheath-like covering at each end called "polar caps"........................*S. austriaca*
5. Spores with typical rounded (elliptical) ends; encased by a full sheath..

S. dudleyi.

S. olosokwaan sp. nov. has no cardinal role apart from contributing to the forest ecosystem by degrading complex wood cellulose and lignin since it is a saprophyte. Its small size, insubstantial fruiting, tough texture would daunt most to collect them for food. Nevertheless, the information for the species edibility was also impaired by the fact that the Masai people who live around Ololosokwan village, by their culture never eat mushroom at all.

Conclusions

The new species described in the present study named *S. ololosokwaniensis* sp.nov. is well distinguishable from all known species in this genus macro-micromophologically. With the description of this new species it is obvious that "Vital taxonomy" remains a strong tool in delimiting *Sarcoscypha* species using living material. *Sarcoscypha* species are well known for several applications including culinary use, and ability to produce interesting enzymes with potential uses in biotechnological processes like bioremediation, biodegradation biopulping and detoxification of recalcitrant substances since they have some bioactive compounds. This study thus, recommends a thorough investigation on the possible bioactive compound found in this new described *Sarcoscypha* species for biotechnological applications.

ACKNOWLEDGEMENTS

The author is grateful to the Association for Strengthening Agricultural Research in Eastern and Central Africa project that sponsored the field work, Dr. Mligo Cosmas who helped in identifying the associated plant species as well as the Loliondo Municipal council for providing armed guard during mushroom gathering in the park.

REFERENCES

Arora D (1986). Mushrooms Demystified: a Comprehensive Guide to the Fleshy Fungi Berkeley, CA: Ten Speed Press, p. 836.

Baral HO (2004). Host specificity, plant communities. The European and North-American species of Sarcoscypha. http://www.gbimycology.de/HostedSites/Baral/Sarcoscypha.htm. Retrieved 2010-08-22.

Berkeley MJ (1885). Notices of some fungi collected in Zanzibar in Berkeley R E, Ann. Mag. Nat. History, p. 384.

Belsky A (1987). Revegetation of natural and human-caused disturbances in the Serengeti National Park, Tanzania. Vegetation, pp. 51-60.

Borner M, Fitzgibbon C, Borner M, Caro T, Lindsay W, Collins D, Bristow M (1996). Dog Jabs to Save Lions. BBC Wildlife, p. 61.

Buyck B, Eyssartier G, Kivaisi A (2000). Addition to the inventory of the genus Cantharellus (Basidiomycotina, Cantharellaceae) in Tanzania. Nova Hedwigia, 71: 491–502.

Brown RP (1980). "Observations on Sarcoscypha coccinea and Disciotis venosa in North Wales during 1978–1979". Bull. Brit. Mycol. Soc., 14 (2): 130–135.

Denison WC (1972). Central American Pezizales. IV. The genera Sarcoscypha, Pithya, and Nanoscypha. Mycologia, 64(3): 609–623.

Dickinson C, Lucas J (1982). VNR Color Dictionary of Mushrooms. Van Nostrand Reinhold, pp. 20–21.

Don EH, Dennis ED (2002). Mushrooms of Hawai`I, An identification guide. Ten Speed Press 212p.

Fishpool L, Evans M (2001). Important Bird Areas for Africa and Associated Islands. Priority Sites for Conservation. BirdLife International, Cambridge, UK.

Fries EA (1822). Systema Mycologicum. (Lundae), 2(1): 78.

Härkönen M, Niemelä T, Mwasumbi L (1995). Edible Mushrooms of Tanzania. Karstenia, p. 92.

Härkönen M, Niemelä T, Mwasumbi L (2003). Tanzanian Mushrooms: Edible, Harmful and other Fungi. Norrlinia, p. 200.

Harrington FA (1990). Sarcoscypha in North America (Pezizales, Sarcoscyphaceae). Mycotaxonomy, 38: 417–458.

Harrington FA (1996). Systematic studies of Sarcoscypha (Ascomycetes, Pezizales). Ph.D. Dissertation, L. H. Baily Hortorium, Cornell University, p. 223.

Harrington FA, Potter D (1997). Phylogenetic relationship within Sarcoscypha based upon nucleotide sequences of the internal transcribed spacer of nuclear ribosomal DNA. Mycologia, 89: 258–267.

Harrington FA (1998). Relationships among Sarcoscypha species: Evidence from molecular and morphological characters. Mycologia 90(2): 235–243.

Hennings P (1893). Fungi Africani II Engl. Botanot. Jahrbuch., 17:1–42.

Landvik S, Egger KN, Schumer T (1997). Toward sub ordinal classification of the Pezizales Ascomycota: Pyhlogenetic analysis of SSU rDNA sequences. Nordic J. Bot.. 403–418.

Kreulen D (1975). Amphibians and reptiles of the Serengeti National Park Tanzania. Bulletin de la Societe Zoologique de France, pp. 673–674.

Magingo FS, Oriyo NM, Kivaisi AK, Danell E (2004). Cultivation of Oudemensiella Tanzanica nom. prov. on agric solid wastes in Tanzanaia. Mycologia, 96(2): 197–204.

Miller HR, Miller OK (2006). North American Mushrooms: A Field Guide to Edible and Inedible Fungi. Guilford, CN: Falcon Guide, p. 536.

Pegler DN (1977). A Preliminary Agarics Flora of East Africa. Kew Bull., p. 615.

Schmidt W (1975). The vegetation of the Northeastern Serengeti National Park, Tanzania. Phytocoenolgia, pp. 30–82.

Schmidt D (1982). The Birds of the Serengeti National Park, Tanzania. BOU Check-list No. 5, SRI Publication No. 225. British Ornithologists' Union, London.

Seaver FJ (1928). The North American Cup-Fungi (Operculates). New York: Self published, pp. 191–192.

Sinclair A, Arcese P (1993). Serengeti II: Research, Management and Conservation of an Ecosystem, p. 152.

Spooner BM (2002). The Larger Cup Fungi in Britain – part 4. Sarcoscyphaceae and Sarcosomataceae. Field Mycol., 3: 9-14.

Tibuhwa DD, Buyck B, Kivaisi AK, Tibell, L (2008). Cantharellus fistulosus sp. nov. from Tanzania. J. Mycol., 129–135.

Tortella GR, Rubilar O, Gianfreda L, Valenzuela E, Diez MC (2008). Enzymatic characterization of Chilean native wood-rotting fungi for potential use in the bioremediation of polluted environments with chlorophenols. World J. Microbiol. Biotechnol., I24: 2805–2818.

10

Identification of *Candida glabrata* and *Candida parapsilosis* strains by polymerase chain reaction assay using RPS0 gene fragment

Emira Noumi[1,2]*, Mejdi Snoussi[1,3], Maria del Pilar Vercher[2], Eulogio Valentin[2], Lucas Del Castillo[2] and Amina Bakhrouf[1]

[1]Laboratoire d'Analyse, Traitement et Valorisation des Polluants de l'Environnement et des Produits, Département de Microbiologie, Faculté de Pharmacie, Monastir, Tunisia.
[2]Departamento de Microbiología y Ecología, Facultad de Farmacia, Universidad de Valencia, Burjassot, Valencia, Spain.
[3]Laboratoire de Traitement et de Recyclage des Eaux. Centre de Recherches et des Technologies des eaux, Technopôle de Borj-Cédria, BP 901, 2050 Hammam-Lif, Tunisia.

Two *Candida* species were identified by the amplification of the RPS0 gene intron fragment. For this, two pairs of primers were used in PCR analysis performed with genomic DNA of clinical isolates of Candida. The primers designed are highly specific for their respective species and produce amplicons of the expected sizes and fail to amplify any DNA fragment from the other species tested. For *Candida glabrata*, the size of the amplicon was 406 pb and 150 bp for *C. parapsilosis*. The designed primers were able to amplify all *C. glabrata* isolates. One of three *C. parapsilosis* strains was confirmed as *C. orthopsilosis*, when we used the designed oligonucleotides. The used primers cannot amplify the other *Candida* species such as *C. albicans*. These results indicate that sequences of intron genes can be useful to specifically identify Candida strains by PCR. This molecular identification will be considered as an early identification of *Candida* species responsible for all candidiasis.

Key words: *Candida glabrata, C. parapsilosis, C. orthopsilosis*, PCR- identification, RPS0 intron.

INTRODUCTION

Candidiasis constitutes the majority of fungal infections with an increased incidence (Baquero et al., 2002). *Candida albicans* is the most common pathogenic *Candida* species. Last year, a shift in the spectrum of *Candida* species has been observed, with an increase of other non *Candida albicans* species (Viscoli and Castagnola, 1999) such as *Candida glabrata* and *Candida parapsilosis* (Malani et al., 2001).

C. parapsilosis has emerged as an important nosocomial pathogen. Due to its variable genetic composition, two new species named *Candida metapsilosis* and *Candida orthopsilosis*, have been recently identified replacing the existing designation of *C.*

parapsilosis groups II and III (Tavanti et al., 2005). The identification of the two new species is currently performed with the aid of DNA-based techniques (Tavanti et al., 2005). DNA-based techniques of *C. parapsilosis* strain collections revealed that *C. orthopsilosis* and C. metapsilosis constitute 10% of all infections previously attributed to *C. parapsilosis* (Gomez-Lopez, 2008; Lochart, 2008). These three separate species, *C. parapsilosis*, *C. orthopsilosis* and *C. metapsilosis* have been identified using multigenic sequence analysis and internal transcribed spacer sequencing (Tavanti et al., 2005).

For all Candida strains and in the absence of pathogenic signs or symptoms, the diagnosis of candidiasis is usually based on the isolation and identification of Candida by conventional morphological and carbohydrates assimilation tests which take many days (Warren, 1995; Velegraki et al., 1999). Furthermore,

*Corresponding author. E-mail : emira_noumi@yahoo.fr.

Table 1. Origin and designation of strains of *C. glabrata* and *C. parapsilosis* tested for identification by *RPS0* intron amplification.

Species	Strain	Origin
C. glabrata	E$_{69}$	Vaginal
C. glabrata	P$_2$	Vaginal
C. glabrata	P$_5$	Vaginal
C. glabrata	P$_8$	Vaginal
C. glabrata	P$_9$	Vaginal
C. glabrata	P$_{11}$	Vaginal
C. glabrata	P$_{12}$	Vaginal
C. glabrata	P$_{13}$	Vaginal
C. glabrata	P$_{14'}$	Vaginal
C. glabrata	P$_{15'}$	Vaginal
C. glabrata	P$_{17}$	Vaginal
C. glabrata	P$_{19'}$	Vaginal
C. glabrata	P$_{21}$	Vaginal
C. glabrata	P$_{25}$	Vaginal
C. glabrata	P$_{26}$	Otitis pus
C. glabrata	P$_{26'}$	Otitis pus
C. glabrata	E$_{68}$	Urine
C. glabrata	8	Oral
C. glabrata	15$_T$	Oral
C. parapsilosis	H$_{11}$	Vaginal
C. parapsilosis	H$_{12}$	Vaginal
C. parapsilosis ATCC 22019	R$_2$	Type strain
C. orthopsilosis J981226	R	Type strain

clinical yeast isolates are sometimes misidentified when automated biochemical systems are used (Dooley et al., 1994). Thus, rapid and accurate identification methods of pathogenic fungi at the species level would prove very helpful in clinical terms. Thus, it is necessary to develop a simple and rapid system that can identify the majority of *Candida* species. The conventional antibody detection tests for the direct detection of *Candida* species antigens has been shown to have potential as an early diagnostic test (Lemieux et al., 1990). Also, CHROMagar Candida is another conventional method of the identification. The sensitivity of CHROMagar identification of Candida was 66.7% to 100%, and specificity was 95.7% to 100% (Willinger and Manafi, 1999).

In recent years, several molecular biology-based methods have been developed to diagnose Candida infections and for specific identification of Candida to the species level (Xiang et al., 2007). PCR methods are particularly promising because of their simplicity, specificity and sensitivity. One of some sensitive and specific method to rapidly and simultaneously identify the most common pathogenic Candida yeast species is based on the use of primers targeted to the yeast RPS0 gene (intron or exon) to obtain a DNA fragment specific for each yeast species by PCR assay. The RPS0 gene codes for a protein which is a component of the

translational machinery and is extremely conserved among species (Baquero et al., 2002, Garcia et al., 2010). Its homology extends to the whole DNA coding sequence, allowing the design of degenerate primers of this gene for amplification purposes, even if the sequence is unknown. However, more yeast species and almost all the fungal species, contain one or more introns that completely differ in size and sequence and enable the design of specific primers for the identification of the species.

In this study, we used the oligonucleotide primers deduced from the *C. glabrata* and *C. parapsilosis* RPS0 intron to specifically amplify, by PCR assay, a DNA fragment in these two species. The RPS0 gene is the same gene Ca YST1 amplified in *C. albicans* strains with difference in intron sequence.

MATERIALS AND METHODS

Clinical strains, media and growth conditions

A total of twenty two strains (19 *C. glabrata* and three *C. parapsilosis*) were subject of this study. Two *C. glabrata* isolates (8 and 15T) were collected from patients admitted to the dental hospital of Monastir (Tunisia) and suffering from denture stomatitis. The other 17 *C. glabrata* isolates were obtained from women attending the service of genecology of Farhat Hached hospital (Sousse, Tunisia). Fourteen of these *C. glabrata* strains were vaginal isolates, two of them were isolated from otitis pus and one strain was isolated from urine. Two *C. parapsilosis* strains (H11 and H12) were isolated from oral cavity of patients; the other strain was *C. parapsilosis* ATCC 22019 reference strain (R2) (Table 1).

The *C. orthopsilosis* J981226 strain was kindly provided by Dr. Odds (University of Aberdeen, Scotland) and served as a positive control. All samples were cultured on Sabouraud chloramphenicol agar (Bio-rad, France) for 48 h at 30°C. All clinical isolates were identified by standard microbiological methods: macroscopic test of culture on Sabouraud chloramphenicol agar, microscopic test and carbohydrates assimilation test by using the ID 32 C system (bio-Mérieux, Marcy l'Étoile, France).

Biochemical identification of Candida strains

All the strains have been identified by assimilation tests using the ID32 C strips (bio-Mérieux) according to the manufacturer's specification. Identification is produced using identification software (Gutierrez et al., 1994). Strains were stored at 4°C on Sabouraud dextrose broth (Bio-rad, France) supplemented with glycerol at 10% (v/v).

Genotypic identification of *Candida* spp.

Preparation of DNA

For DNA extraction, yeasts were routinely grown on Sabouraud dextrose agar plates at 28°C for 24 h to 48 h. A single colony was then grown overnight on YPD broth (1% yeast extract, 2% peptone, 2% dextrose) at 28°C, with shaking at 200 rpm. DNA was extracted from cultures by adapting the method described previously to yeast (Del Castillo et al., 1995) as described by Garcia et al. (2010). DNA concentrations and A260/A280 ratios were determined by means of a "Gene Quant Spectrophotometer" (Pharmacia). An A260/A280

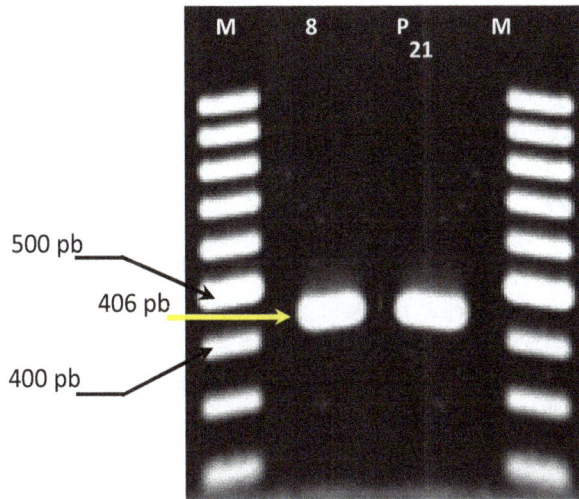

Figure 1. Representative agarose gel electrophoresis (1% agarose) showing the amplification of products obtained for oral (strain 8) and vaginal (strain P21) *C. glabrata* isolates. M: 100-pb molecular weight marker (Fermentas).

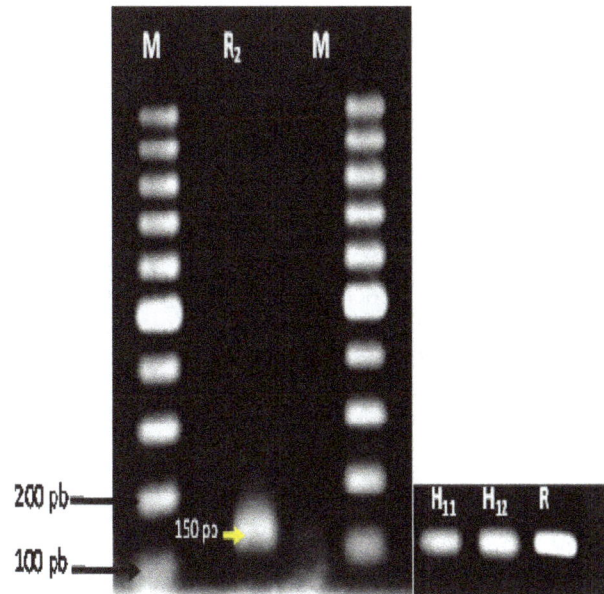

Figure 2. Representative agarose gel electrophoresis (1% agarose) showing the amplification of products obtained from *C. parapsilosis* ATCC 22019 (R2) and *C. orthopsilosis* (H11, H12 and J981226 (R)) isolates. M: 100-pb molecular weight marker.

ratio of 1.8 to 2.1 was considered acceptable.

Primers, conditions of PCR amplification and agarose gel electrophoresis

The sequences of synthetic oligonucleotides used as primers were CG1 (5'-acatatgtttgctgaaaaggc-3') and CG2 (5'-actttttcttagtgttcaggacttc-3') for *C. glabrata*, CP1 (5'-agggattgccaatatgccca-3') and CP2 (5'-gtgacattgtgtagatccttgg-3') for *C. parapsilosis* (Garcia et al., 2010). The primers used for the identification of *C. orthopsilosis* strains were generously provided by Dr. Lucas Del Castillo (University of Valencia, unpublished data). CO1 (5'-tttcaatatgcctagagccacattgtgaatac-3') and CO2 (5'-gcattagttagtatcgtcttttattaaata-3') for *C. orthopsilosis* (M.P. Vercher, unpublished results).

The amplification was performed in an automated thermocycler in a final volume of 25 µl containing 2.5 µl of 10x buffer, 1 µl of 50 mmol of $MgCl_2$, 2.5 U of Eco Taq polymerase (MBI fermentas), 2.5 µl of dNTP (2.5 mmol) (Sigma, St. Louis, MO, USA), and optimum concentrations of each primer (4 µmol). One microlitre of DNA suspension (30 to 50 ng) was amplified in a PCR thermal cycler (PTC-150 MinicyclerTM) by using 1 cycle at 95°C for 5 min and then 35 cycles as follows: 30 s of denaturation at 95°C, 30 s of annealing at 56°C, and 90 s of primer extension at 72°C. At the final cycle, an additional 10 min of incubation at 72°C was added for complete polymerization.

The resultant fragments of amplified DNA were analyzed by electrophoresis through 1% agarose gels, run in Tris-acetate-EDTA buffer (TAE) for 1 h at 90 volts. A 100-bp ladder (Fermentas) was used as a size marker. Gels were stained in a solution of ethidium bromide (10 µg/ml) and photographed by Gel printer plus and image analysis software ScionImage (TDI. S.A. Madrid, Spain).

RESULTS

Identification of Candida isolates based on carbohydrates assimilation using ID 32 C system identified our isolates as *C. glabrata* (19 strains) and *C. parapsilosis* (3 strains). Molecular identification of this two species has been carried out by PCR amplification of RPS0 gene intron fragment. For *C. glabrata*, primers CG1 and CG2 (Garcia et al., 2010) produce a specific amplicon of 406 pb (Figure 1). This amplicon was sequenced and presented a 100% identity with RPS0 intron (Figure 1). We obtained 100% of positive PCR products from 19 strains of *C. glabrata* (an amplicon size of 406 bp). Since the intron sequences are poorly conserved among microorganism strains, we decided to use this 406 bp amplicon, which we termed CgRPS0-INT, to identify *C. glabrata* from different sources. Our results showed that these primers can specifically identify only *C. glabrata* strains isolated from the oral cavity and vaginal site.

The second strain object of this study was *C. parapsilosis*. The size of the amplicon was 150 pb (Figure 1). Amplification with specific primers CP1 and CP2 (Garcia et al., 2010) only identify as *C. parapsilosis* the control strain ATCC 20019, shown by an amplicon of 150 bp on the gel. No amplicon was obtained for the two oral isolates (H11 and H12). These strains were further identified as *C. orthopsilosis* using specific primers recently designed at the Department of Microbiology of the Valencia University, Spain (Vercher, 2009).

For the two Candida species, only one amplicon was obtained (406 pb for *C. glabrata* and 150 pb for *C. parapsilosis*). As shown in Figure 2, the 406 bp fragments were amplified in all 19 *C. glabrata* isolates tested, confirming the high sensitivity of the method.

DISCUSSION

PCR approaches for the identification of *Candida* species are important in both epidemiological and taxonomic studies. Genes containing intron sequences could prove useful to design specific primers for the identification of yeast strains at the species level. The present study extends the work of Baquero et al. (2002) and Garcia et al. (2010) that used RPS0 intron based primers to identify different *Candida* species by designing a set of primers for the identification of yeast species of relevant clinical interest. These primers are based on the RPS0 gene and are mainly derived from intron sequences.

According to the previous results, PCR primers based on RPS0 introns are an important tool for the identification of yeasts and fungi of clinical and environmental interest. For the few yeast species with no RPS0 intron, primers from the less conserved regions of the gene can be designed. Alternatively, another gene intron could be eventually used for these few species. These data indicate that sequences of intron genes can be useful to specifically identify Candida strains by PCR.

The discriminatory power of the described test (Garcia et al., 2009) is strongly supported by the fact that, two strains of *C. parapsilosis* were confirmed as non *C. parapsilosis* by this molecular identification, whereas they were identified as *C. parapsilosis* with the Api ID 32 C system. The confirmation of the identity of these two strains as *C. parapsilosis* group III or *C. orthopsilosis* by PCR amplification with primers specific for *C. orthopsilosis* RPS0 intron.

Conclusion

Molecular approaches may have interest in epidemiological and taxonomic studies and will be considered as an early identification of *Candida* species responsible for all candidiasis, in this work, we showed the value of RPS0 gene intron for identification of pathogen yeast species and we concluded that, this molecular identification will be considered as an early identification of *Candida* species responsible of all candidiasis.

ACKNOWLEDGMENT

The experimental work was partially supported by grant BFU2006-08684/BFU from the Spanish Ministry of Science and Technology. We gratefully acknowledge Mrs Hajer Hentati (Dental Hospital of Monastir, Tunisia) and Mrs Fatma Saghrouni (Service de parasitologie, Hospital Farhat Hached de Sousse, Tunisie) for their help in the collection of *Candida* strains.

REFERENCES

Baquero C, Montero M, Sentandreu R, Valentin E (2002). Identification of Candida albicans by polymerase chain reaction amplification of a CaYST1 gene intron fragment. Rev. Iberoam. Micol., 19: 80-83.

Del Castillo Agudo L, Gavidia I, Pérez-Bermúdez P, Segura J (1995). BioTechniques 18: 766-768.

Dooley DP, Beckius ML, Jeffrey BS (1994). Misidentification of clinical yeast isolates by using the updated Vitek Yeast Biochemical Card. J. Clin. Microbiol., 32: 2889-2892.

Garcia Martínez JM, Valentín Gómez E, Peman J, Cantón E, Gómez Garcia Martínez M, del Castillo Agudo L (2010). Identification of pathogenic yeast species by polymerase chain reaction amplification of the RPS0 gene intron fragment. J. Appl. Microbiol., 108(6): 1917-1927.

Gomez-Lopez A, Alastruey-Izquierdo A, Rodriguez D (2008). Prevalence and susceptibility profile of Candida metapsilosis and Candida orthopsilosis: results from population-based surveillance of candidemia in Spain. Antimicrob. Agents. Chemother, 52: 1506–9.

Gutierrez J, Martin E, Lozano C, Coronilla J, Nogales C (1994). Evaluation of the ATB 32 C, Automicrobien system and API 20 C using clinical yeast isolates. Ann. Bio. Clin., 50: 443-446.

Lemieux C, St-Germain G, Vincelette J, Kaufman L, de Repentigny L (1990). Collaborative evaluation of antigen detection by a commercial latex agglutination test and enzyme immunoassay in the diagnosis of invasive candidiasis. J. Clin. Microbiol., 28(2): 249-53.

Lochart SR, Messer SA, Pfaller MA (2008). Geographic distribution and antifungal susceptibility of the newly described species Candida orthopsilosis and Candida metapsilosis, in comparison to the closely-related species Candida parapsilosis. J. Clin. Microbiol. 46: 2659–64.

Malani PN, Bradley SF, Little RS, Kauffman CA (2001). Trends in species causing fungaemia in a tertiary care medical centre over 12 years. Mycoses., 44(11-12): 446-9.

Tavanti A, Davidson AD, Gow NA, Maiden MC, Odds FC (2005) Candida orthopsilosis and Candida metapsilosis spp. nov. to replace Candida parapsilosis groups II and III. J. Clin. Microbiol., 43: 284–292.

Velegraki A, Kambouris ME, Skiniotis G, Savala M, Mitroussia-Ziouva A, Legakis NJ (1999) Identification of medically significant fungal genera by polymerase chain reaction followed by restriction enzyme analysis. FEMS Immunol. Med. Microbiol., 23: 303-312.

Viscoli C, Castagnola E (1999) . Epidemiology and therapy of mycotic infections in immunocompromised host with special regard to the role of lipid formulations of amphotericin B. Recenti. Prog. Med., 90(10): 545-57.

Vercher M. P. (2009). Diagnostico molecular diferencial del grupo psilosis. Graduation thesis. Universidad de Valencia, Valencia, Spain.

Willinger B, Manafi M (1999) . Evaluation of CHROMagar Candida for rapid screening of clinical specimens for Candida species. Mycoses, .42(1-2): 61-5.

Xiang H, Xiong L, Liu X, Tu Z (2007). Rapid simultaneous detection and identification of six species Candida using polymerase chain reaction and reverse line hybridization assay. J. Microbiol. Meth., 69: 282-287.

11

Evidence of antagonistic interactions between rhizosphere and mycorrhizal fungi associated with *Dendrocalamus strictus* (Bamboo)

Rohit Sharma[1,2]*, Ram .C. Rajak[1] and Akhilesh .K. Pandey[1]

[1]Department of Biological Sciences, Mycological Research Laboratory, R. D. University, Jabalpur- 482 001, Madhya Pradesh, India.
[2]Microbial Culture Collection, Affiliated to National Centre for Cell Science, University of Pune, Ganeshkhind, Pune- 411 007, Maharashtra, India.

The paper deals with interactions of some microfungal strains isolated from rhizosphere soils from three different sites with ectomycorrhizal fungus *Cantharellus tropicalis* mycelium grown *in vitro* on agar plates. The rhizospheric fungi were isolated from 3 different sites of bamboo forest and grown against *Cantharellus*. The cross inoculation method showed that *C. tropicalis* was highly active against some fungi, thus resulting in different types and strength of interactions. Overgrowth was the most common interaction (45%), followed by inhibition at distance (29%), intermingling (17%) and contact inhibition (13%). The competitive strength of the ectomycorrhizal fungus was high and only affected by some fast growing sterile mycelia, an unidentified fungus and *Trichoderma viride*.

Key words: *Cantharellus,* ectomycorrhiza, mycorrhizal systems, biological control, soil micro fungi.

INTRODUCTION

The microbiota of forest soils is dominated by ectomycorrhizal (ECM) and saprotrophic decomposer fungi involved respectively in supply of nutrients to trees and decomposition of woody plant litter. Saprotrophic basidiomycetes are also abundant in bamboo forests (Sharma, 2008) degrading cellulose, lignin and ligno-cellulose. Ectomycorrhizal fungal mycelia are ubiquitous in forest soils and associate with host trees to fulfill various ecological functions. Each ectomycorrhizal fungus with its special physiology can use either in-organic nutrients or utilize organic sources. In addition to increasing absorptive surface area of root systems, ECM fungi provide an increased surface area for interactions with other microorganisms, thus translocating products of photosynthesis to soil. These interactions may be inhibitory or stimulatory, some are clearly competitive, others mutualistic. An understanding of interactions between ECM and saprotrophic organisms is important given their central roles in biogeochemical cycling in ecosystems of both managed and natural forests. However,

saprotrophs obtain their C from decaying organic matter while ECM fungi obtain most of their C directly from their host plants (Leake and Johnson, 2004). Antagonistic interactions between rhizosphere microorganisms and mycorrhizal fungi have an important role in functions of mycorrhizal systems (Stark and Kytöviita, 2005). Moreover, exudation and re-absorption of fluid droplets at ECM hyphal tips helps in conditioning the hyphal environment in the vicinity of tips (Sun et al., 1999).

Mycorrhizal fungi also modify the interactions of plants with other soil organisms, both pathogens (nematodes and fungi) and mutualists (nitrogen-fixing bacteria). Pathogenic fungi, may invade roots and mycorrhizal fungi can alter host response to these pathogens. *Laccaria bicolor* prevented the spread of *Fusarium oxysporum* in Douglas-fir roots as a result of flavanoid wall infusions (Fitter and Garbaye, 1994). Wu et al. (2003) explored interactions between saprotrophic microbes and ECM fungi using a protein-tannin complex as N source by red pine (*Pinus resinosa*). Olsson (1999) studied the role of fatty acids to determine the distribution and interactions of mycorrhizal fungi in soil. Mycorrhizal fungi colonize feeder roots and thereby interact with root pathogens that parasitize them. In a natural ecosystem where uptake of phosphorus is low,

---

*Corresponding author. E-mail: rsmushroom@gmail.com.

mycorrhizal fungi protect root system from endemic pathogens such as *Fusarium* spp. Mycorrhizal fungi may reduce the incidence and severity of root diseases (Whipps, 2004). Over the last 30 years, there has been an increasing interest in potential role that ECM fungi can play in control of plant diseases. It is possible to exploit these interactions to improve mycorrhizal function (Finlay, 2004) and restrict pathogenic organisms in the form of biological control.

There have been a few laboratory studies of interactions between pure cultures of representatives of both ECM and saprotrophs fungi in axenic microcosm system. In the same way no remarkable studies of competitive interactions between mycorrhizal *Cantharellus tropicalis* Rahi, Rajak and Pandey and saprotrophic fungi in soils have been done. The objectives of the present study were to examine whether interactions occur between species of different fungal group from bamboo forest. Interactions between ectomycorrhizal fungi and rhizospheric soil microfungi were studied *in vitro*, providing us an insight into the ecology of ectomycorrhizal fungi associated with *Dendrocalamus*.

MATERIALS AND METHODS

Rhizosphere soil samples of *D. strictus* were collected from three sites of bamboo forests in the districts of Balaghat (site 1), Lamta (site 2), and Nainpur (site 3), district Balaghat, Madhya Pradesh, India. With the help of a trowel the samples were collected at a depth of 5 – 10 cm along with root bits into sterile polythene bags. The composite soil samples were immediately brought to the laboratory and stored in refrigerator at 4±2°C for further analysis in order to determine the soil type and nutritional status.

The cross inoculation methods used by Baar and Stanton (2000) and Vaidya (2005) was followed with modifications. MMN (modified Melin Norkrans) medium with agar (15 gl^{-1}) was used for growth of all fungi. Many ECM grow well on MMN media. Also, soil micro fungi grow well on MMN medium. For inoculation, mycelial plugs of 9 mm diam. were cut from edge of ECM mycelia and transferred to MMN agar. Pair-wise combinations were made by plating mycelial plugs of ECM and soil micro fungi on opposite corners of plate, about 70 – 80 mm away from each other. Each pair-wise combination was replicated three times. Agar plates were incubated at 28±2°C. Radial growth towards other mycelium was determined by measuring colony radius. The experiment was terminated when radial growth of ECM fungi reached other mycelium.

RESULTS

The cross inoculation method has shown that *C. tropicalis* was highly active against some rhizosphere soil microorganism. Pair-wise combinations of ECM *C. tropicalis* and soil micro fungi from sites (site 1 - Figures 1, 2; site 2 - Figures 3 and 4; site 3 - Figures 5 and 6) differed between species. This resulted not only in different types of interactions between fungi, but also in differences in strength of interactions. Overgrowth was the most common interaction (45%), followed by inhibition at a distance (29%) intermingling (17%) and contact inhibition (13%). Details of observation are given in Table 1.

Overgrowth was observed when *C. tropicalis* interacted with *A. flavus Link ex Fr.*. Inhibition at a distance was observed for combinations between *C. tropicalis* and either *Aspergillus niger* Link ex Fr. or *Emericella* sp. Isolated from site 1 (Figures 1c and 2a); *Aspergillus* sp. 2 isolated from site 2, however, the mycelia of *Aspergillus* sp. 2 were larger and less suppressed by *C. tropicalis* (Figures 3a); and similar results were observed for *Aspergillus* sp. 1 isolated from site 3 (Figure 5a).

Sterile mycelium from site 3 overgrew *C. tropicalis* (Figures 6b and 6c). Both species of *Trichoderma* spp. (site 1 and 2) restricted the growth of chanterelle. However, when chanterelle was grown in combination with *A. niger* Van Tiegh, a clear zone with no hyphae of either fungi was formed (Figures 1a, 5c and 5d).

DISCUSSION

The inhibition of soil micro fungi, mostly at a distance by *C. tropicalis*, suggests that this fungus prevented invasion by potential competitors. This type of defense mechanism has been reported for other ectomycorrhizal fungi (Baar and Stanton, 2000). Herein, the inhibition of soil micro fungi by *C. tropicalis* might be caused by production of secondary metabolites. However, antibiotics produced by ECM species (that is, *Amanita, Boletus* and *Cenococcum* spp.) have been reported in earlier studies (Santoro and Casida, 1962).

The overgrowth of *C. tropicalis* by relatively fast-growing sterile mycelial fungus, unidentified fungus and *Trichoderma viride* Pers. ex Fr. was remarkable. The results of an earlier study by Shaw et al. (1995) showed growth suppression of *Rhizopogon roseolus* by several saprotrophic basidiomycetes. Furthermore, the growth of *Suillus granulatus* (L.:Fr.) Rouss, was inhibited by rhizoplane fungi of *Pinus halepensis* (Girlanda et al., 1995). Baar and Stanton (2000) have attributed low investment of N in mycelial biomass to reduced competition for some ECM fungi. Hardly any sporocarps of saprotrophic basidiomycetes occur in bamboo forest but species of *Ramaria, Clavaria* and *Clitocybe* have been found growing near the bamboo plants, but could not be isolated. They can be studied for their competitiveness with chanterelle. In previous studies, *Clitocybe marginella* Harmaja inhibited the growth of *Cenococcum geophilum* and *L. bicolor* (Baar and Stanton, 2000). In a similar microcosm experiment in which mycelium of *Suillus bovinus* (L.:Fr.) Rouss mycorrhizal with *Pinus sylvestris* were grown alone and in interaction with *Phanerochaete velutina* (Leake et al., 2001), the effect of mycorrhiza on growth of saprotroph was limited.

In the present study, the ectomycorrhizal fungi suppressed the soil micro fungi in maximum number of the pair wise comparisons indicating that *Cantharellus* mycelia has higher competitiveness than soil micro fungi. This may be attributed to several alkaloids, terpenes,

Figure 1. Dual culture interaction between *C. tropicalis* (*Ct*) and soil microfungi (site 1). a. *Ct-Aspergillus niger*, b. *Ct-A. terreus*, c. *Ct-A. flavus*, d. *Ct-Fusarium* sp.1, e. *Ct-Curvularia* sp., f. enlarged zone of *Ct-Alternaria* sp. interaction.

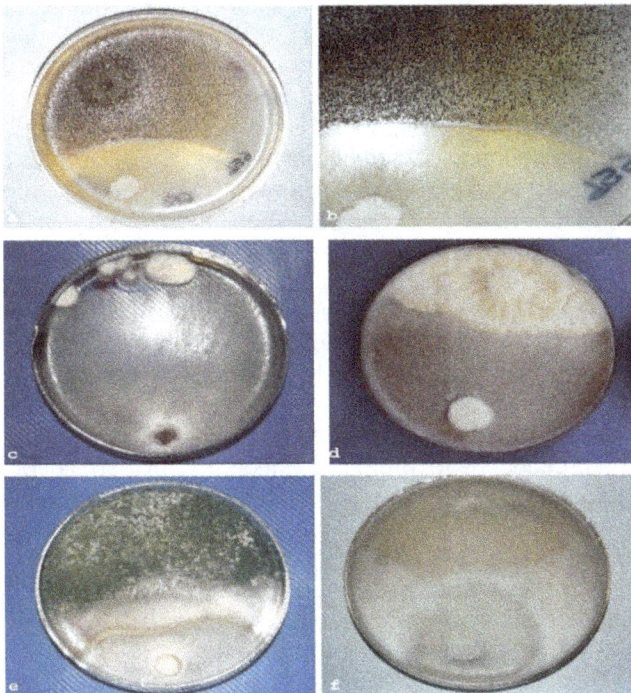

Figure 3. Dual culture interaction between *C. tropicalis* (*Ct*) and soil microfungi (site 2- Lamta). a. Ct-*A.niger*, b. *Fusarium* sp.1, c. Ct-*Alternaria* sp., d. enlarged zone of Ct-*Alternaria* sp. interaction, e. Ct-*Curvularia*, f. enlarged zone of Ct-*Curvularia* sp. interaction.

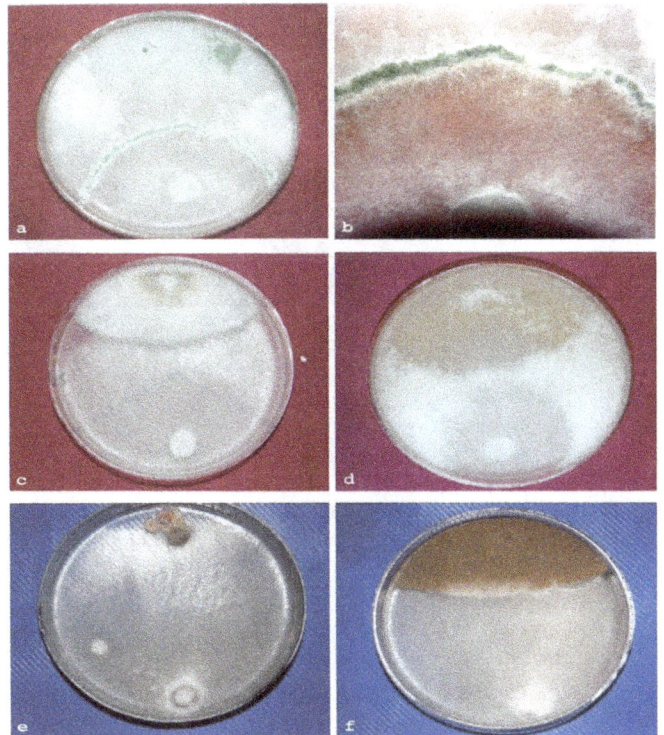

Figure 2. Dual culture interaction between *C. tropicalis* (*Ct*) and soil microfungi (site 1- Balaghat). a. *Ct-Emericella* sp., b. enlarged zone of *Ct-Emericella* interaction, c. *Ct-Penicillium* sp.1, d. *Ct-Fusarium* sp.2, e. *Ct-Trichoderma viride*, f. *Ct*-Unidentified fungus.

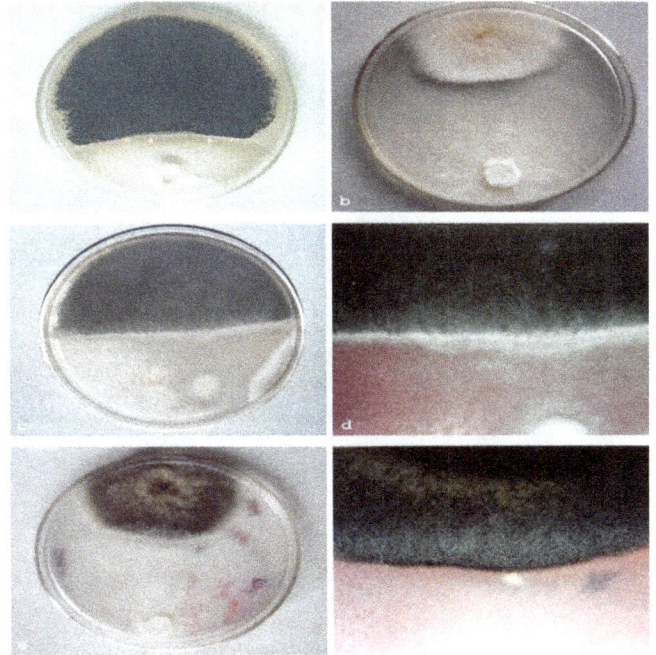

Figure 4. Dual culture interaction between *C. tropicalis* (*Ct*) and soil microfungi (site 2- Lamta). a. *Ct-Trichoderma* sp., b. enlarged zone of *Ct-Trichoderma* sp. interaction, c. *Ct-Fusarium* sp.1, d. *Ct*-Unidentified fungus, e. *Ct-Penicillium* sp.3, f. *Ct-Aspergillus* sp.2.

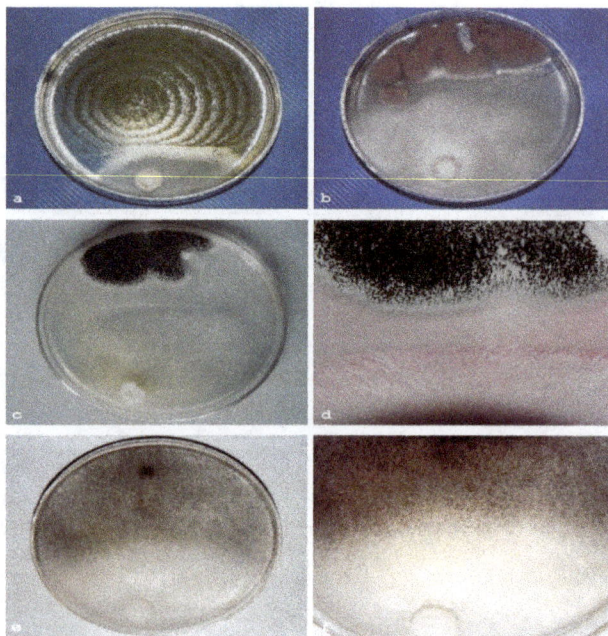

Figure 5. Dual culture interaction between *C. tropicalis* (*Ct*) and soil microfungi (site 3- Nainpur). a. *Ct-Aspergillus* sp.1, b. *Penicillium* sp.1, c. *Ct-A. niger*, d. enlarged zone of interaction, e. *Ct-Mucor* sp., f. enlarged zone of Ct-*A. flavus* interaction.

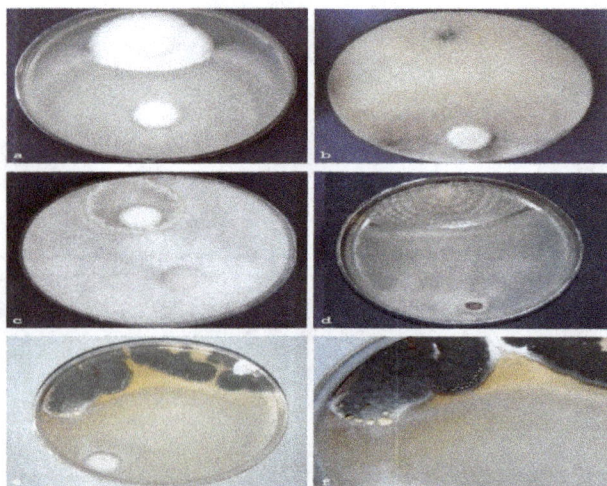

Figure 6. Dual culture interaction between *C. tropicalis* (*Ct*) and soil microfungi (site 3 - Nainpur). a. *Ct-Fusarium* sp.1, b. Sterile mycelium, c. *Ct*-Unidentified fungus, d. *Ct-Emericella* sp., e. *Ct-A. flavus*, f. Enlarged zone of *Ct-A. flavus* interaction.

polysaccharides produced by *Cantharellus* mycelia. Different strategies were observed for soil micro fungi such as inhibitor at a distance (29%), contact inhibition (13%), intermingling (17%) and overgrowth (45%). *Penicillium* sp., a known mycotoxins producer was hardly combative against the ECM fungus (Table 1, Figures 2c, 4e and 5b). Low competitiveness of some of the soil

micro fungi *viz., Curvularia, Alternaria, Mucor* and *Fusarium* may suggest that these species occupy different niches or are weak organisms when competing with *Cantharellus*. Moreover there are seldom reports of any root disease in *Dendrocalamus*.

ECM fungi have been shown to have inhibitory effects on root pathogenic fungi but their interactions with saprophytic fungi have received surprisingly little attention (Johansson et al, 2004). There has been reports of strong inhibition of root pathogens like *Phytophthora cinnamomi* Rands, *Pythium debaryanum* Hesseltine and *P. sylvaticum in vitro* by ectomycorrhizal fungi. Stark and Kytöviita (2005) provided evidence of antagonistic interactions between rhizosphere microorganisms and mycorrhizal fungi associated with birch (*Betula pubescens* Ehrh.). Isolates of *Laccaria* sp. protected young seedlings of *Picea abies* (L.) Karst. and *Pseudostuga menziesii* from *F. oxysporum* (Sampangiramaiah and Perrin, 1990). Natarajan and Govindaswamy (1990) have tested *Amanita muscaria, Laccaria laccata, L. fraterna,* and *Suillus brevipes* against six root pathogens *viz., Armillaria mellea* (Vahl in Fl. Dan. ex Fr.), *Cylindrocladium parvum* Anderson, *C. scoparium* Morg., *F. oxysporum* Schlecht., *F. solani* (Mart.) App. and Wollenw and *Rhizoctonia solani* Kuehn *S. brevipes* inhibited all root pathogens tested. In another study, *Tricholoma* sp., *Paxillus involutus* and *Hebeloma cylindrosporum* inhibited growth of *Cylindrocladium floridanum* in Petri dishes, while *L. bicolor* was inhibited and completely covered by *C. floridanum* (Morin et al., 1999). Growth (in paired culture) and colony forming units (in the rhizosphere of *Pinus banksiana* Lamb. seedlings) of *F. oxysporum* was reduced significantly by *L. laccata*. When grown in co-culture, Werner and Zadworny (2003) observed suppression of *Mucor hiemalis* by *L. laccata*. They also studied interactions between the *L. laccata* and soil fungus *Trichoderma virens* in co-culture and in the rhizosphere of *P. sylvestris* seedlings growing *in vitro*, where growth of *T. virens* was inhibited in co-culture (Werner et al., 2002). Antifungal and antibacterial action of ECM fungi *Pisolithus* and *Scleroderma in vitro* have been *tested* against 8 fungi and 6 bacteria and showed higher activity against all fungi except three *Aspergillus* spp. (Vaidya et al., 2005).

While our knowledge is currently limited, it seems that interactions have profound effects on mycorhizosphere processes. The ability to redistribute nutrients between compartments of forest floor is a fundamental activity of many saprotrophic and mycorrhizal fungi. More extensive research is warranted to enhance our knowledge on interactions within fungal community and exploring potential for manipulating ectomycorrhizosphere environment for biotechnological purposes (Bruns and Bidartondo, 2002; Cairney and Meharg, 2002). The intensity of interactions between different soil fungi and ECM fungus *C. tropicalis* highlights the potential importance of interactions on functioning of these microorganisms in forest ecosystems.

Table 1. Results of the fungal interactions and estimated average size (% of Petri dish covered) of the mycelia of *C. tropicalis* (*Ct*) with soil micro fungi at the harvest time of the three sites studied (site 1 - Balaghat; site 2 - Lamta; site 3 - Nainpur).

S/No.	Interactions	Site 1[†]		Site 2[†]		Site 3[†]	
		Fungal interaction	% PD covered	Fungal interaction	% PD covered	Fungal interaction	% PD covered
1	*Ct-A. alternata*	-	-	O (s)	47 - 53	-	-
2	*Ct-A. flavus*	O (e)	81 - 19	-	-	O (e)	68 - 32
3	*Ct-A. niger*	HD (e)	52 - 48	HD (s)	32 - 68	HD (e)	58 - 42
4	*Ct-Aspergillus* sp.1	CH (e)	58 - 42	-	-	CH (s)	29 - 71
5	*Ct-Aspergillus* sp.2	-	-	CH (e)	62 - 38	-	-
6	*Ct-Curvularia* sp.	O (e)	55 - 45	O (e)	56 - 44	-	-
7	*Ct-Emericella* sp.	HD (s)	39 - 61	-	-	HD (s)	67 - 33
8	*Ct-Fusarium* sp.1	O (e)	56 - 44	O (e)	66 - 34	O (e)	61 - 39
9	*Ct-Fusarium* sp.2	O (e)	51 - 49	O (e)	62 - 38	-	-
10	*Ct-Mucor* sp.	-	-	-	-	M (s)	44 - 56
11	*Ct-Penicillium* sp.1	CH (e)	83 - 17	-	-	-	-
12	*Ct-Penicillium* sp.2	-	-	-	-	HD (e)	60 - 40
13	*Ct-Penicillium* sp.3	-	-	HD (e)	80 - 20	-	-
14	*Ct-T. viride*	O (s)	28 - 72	-	-	-	-
15	*Ct-Trichoderma* sp.	-	-	O (s)	46 - 54	-	-
16	*Ct-Sterile mycelium*	-	-	-	-	M (s)	22 - 78
17	*Ct-Unidentified fungus*	M (e)	49 - 51	M (e)	48 - 52	M (e)	61 - 39

* Interactions distinguished were: contact inhibition (CH), inhibition at distance (HD), intermingling (M), and overgrowth (O). † Letters in brackets indicate which fungus exerted a specific interaction effect upon its opponent: ECM fungus (e), soil micro fungus (s). Dash (-) indicates that particular soil micro fungus was not isolated from that site.

ACKNOWLEDGEMENTS

The authors thank the Department of Biotechnology, New Delhi, India for financial assistance as research project (No: BT/PR3916/PID/20/153/2003) and Junior Research Fellowship to Rohit Sharma. Authors also thank Head of the Department of Biological Sciences, R. D. University, Jabalpur, India for laboratory facilities.

REFERENCES

Baar J, Stanton NL (2000). Ectomycorrhizal fungi challenged by saprotrophic basidiomycetes and soil micro fungi under different ammonium regimes *in vitro*. Mycol. Res., 104:691-697.

Bruns TD, Bidartondo MI (2002). Molecular windows into the below-ground interactions of ectomycorrhizal fungi. Mycologist 16:47-50.

Cairney JWG, Meharg AA (2002). Interactions between ectomycorrhizal fungi and soil saprotrophs: implications for decomposition of organic matter in soils and degradation of organic pollutants in the rhizosphere. Can. J. Bot., 80:803–809.

Finlay RD (2004). Mycorrhizal fungi and their multifunctional role. Mycologist 18:91-96.

Fitter AH, Garbaye J (1994). Interactions between mycorrhizal fungi and other soil organisms. Plants and Soil 159:123-132.

Girlanda M, Varese GC, Luppi-Mosca AM (1995). *In vitro* interactions between saprotrophic microfungi and ectomycorrhizal symbionts. Allionia 33:81-86.

Johansson J, Paul L, Finlay RD (2004). Microbial interactions in the mycorrhizosphere and their significance for sustainable agriculture. FEMS Microbiol. Ecol., 48 (1): 13.

Leake JR, Johnson D (2004). Networks of power and influence: the role of mycorrhizal mycelium in controlling plant communities and agro ecosystem functioning. Can. J. Bot., 82:1016-1045.

Leake JR, Donnelly DP, Saunders EM, Boddy L, Read DJ (2001). Rates and quantities of carbon flux to ectomycorrhizal mycelium following ^{14}C pulse labeling of *Pinus sylvestris* L. seedlings: effects of litter patches and interaction with a wood-decomposer fungus. Tree Physiology, 21:71-82.

Morin C, Samson J, Dessureault M (1999). Protection of black spruce seedlings against *Cylindrocladium* root rot with ectomycorrhizal fungi. Can. J. Bot., 77:169–174.

Natarajan K, Govindaswamy V (1990). Antagonism of ectomycorrhizal fungi to some common root pathogens. In: Current trends in mycorrhizal research- The proceedings of the national conference on mycorrhiza (eds. Mukerji KG, Chamola BP, Singh J). Harayana Agricultural University, Hissar, India. pp. 98-99.

Olsson PA (1993). Signature fatty acids provide tools for determination of the distribution and interactions of mycorrhizal fungi in soil. FEMS Microbiol. Ecol., 29:303-310.

Sampangiramaiah K, Perrin R (1990). Interactions between isolates of ectomycorrhizal *Laccaria* spp. and root rot fungi of conifers. In: Current trends in mycorrhizal research- The proceedings of the national conference on mycorrhiza (eds. Mukerji KG, Chamola BP, Singh J). Harayana Agricultural University, Hissar, India, pp. 124-125.

Santoro T, Casida LE Jr (1962). Elaboration of antibiotics by *Boletus luteus* and certain other mycorrhizal fungi. Can. J. Microbiol., 8:43-48.

Sharma R (2008). Studies on ectomycorrhizal mushrooms of MP and Chhattisgarh, Ph.D. thesis, Rani Durgavati University, Jabalpur, India.

Shaw TM, Dighton J, Sanders FE (1995). Interactions between ectomycorrhizal and saprotrophic fungi on agar and in association with seedlings of lodgepole pine (*Pinus contorta*). Mycol. Res., 99: 159 - 165.

Stark S, Kytöviita MM (2005). Evidence of antagonistic interactions between rhizosphere microorganisms and mycorrhizal fungi associated with birch (*Betula pubescens*). Acta Oecologica, 28:149-155.

Sun YP, Unestam T, Lucas SD, Johanson KJ, Kenne L, Finlay R (1999). Exudation-reabsorption in a mycorrhizal fungus, the dynamic interface for interaction with soil and soil microorganisms. Mycorrhiza 9:137-144.

Vaidya GS, Shrestha K, Wallander H (2005). Antagonistic study of ectomycorrhizal fungi isolated from Baluwa forest (Central Nepal) against with pathogenic fungi and bacteria. Scientific World 3:49-52.

Werner A, Zadworny M (2003). *In vitro* evidence of mycoparasitism of the ectomycorrhizal fungus *Laccaria laccata* against *Mucor hiemalis* in the rhizosphere of *Pinus sylvestris*. Mycorrhiza 13:41-47.

Werner A, Zadworny M, Idzikowska K (2002). Interaction between *Laccaria laccata* and *Trichoderma virens* in co-culture and in the rhizosphere of *Pinus sylvestris* grown *in vitro*. Mycorrhiza 12:139-145.

Whipps JM (2004). Prospects and limitations for mycorrhizas in biocontrol of root pathogens. Can. J. Bot., 82:1198-1227.

Wu T, Sharda JN, Koide RT (2003). Exploring interactions between saprotrophic microbes and ectomycorrhizal fungi using a protein-tannin complex as an N source by red pine (*Pinus resinosa*). Special Issue: Functional genomics of plant-pathogen interactions. New Phytol., 159:131-139.

12

The research of infection process and biological characteristics of *Rhizoctonia solani* AG-1 IB on soybean

Aiping Zheng* and Yanran Wang

Rice Research Institute, Sichuan Agricultural University, Wenjiang 611130, China.

The isolate *Rhizoctonia solani* AG-1 IB collected from the diseased leaves of soybean were identified by the method of internal transcribed spacer sequence analysis. In this study, we focus on the biological characteristics and infection process of AG-1IB. Morphological, nucleus, chromosome, and infection process were observed. Typical infectious structures as infection cushion and appressorium were observed during the infection process.

Key words: *Rhizoctonia solani* AG-1 IB, soybean, biological characteristics, infection process.

INTRODUCTION

Rhizoctonia solani, a basidiomycete fungus, was divided into 12 anastomosis groups (AG1-11 and BI) according to hyphal anastomosis behavior, cultural morphology, host range, pathogenicity and so on (Ogoshi, 1987). Among these groups, isolates of AG-1 have been recovered from many hosts (Huang et al., 2003). Furthermore, AG-1 has been subdivided into three subgroups designated as AG-1IA, AG-1IB and AG-1IC (Ogoshi, 1987; Sneh et al., 1991). The common symptoms of *Rhizoctonia* disease are referred to as damping-off, sheath blight, sheath spot, leaf blight and rot (Duan et al., 2008; Grosch et al., 2003; Takeshi et al., 1998; Yang et al., 2005). *R. solani* AG-1 IB is a widely existing fungus with great harm to many plants.

In Xishuangbanna District of China, *R. solani* isolates collected from the diseased leaves of Chinese cabbage, mint and lettuce were identified to belong to anastomosis group AG-1 IB (Sneh et al., 1991). Web-blight disease of European pear in Okayama prefecture is caused by *R. solani* AG-1, especially IB (Kuramae et al., 2003). However, these studies are only in the level of identified pathogens were *R. solani* AG-1 IB. *R. solani* can cause damping off, root rot, and hypocotyl lesions on soybeans in the United States as well as web blight in the southern United States (Yang, 1999) and *R. solani* rot is a major disease of soybean (Sinclair and Dhingra, 1975; Yang, 1999). To facilitate breeding of resistant cultivars, it is important to understand the infection process. So we focused on biological characteristics and infection process.

MATERIALS AND METHODS

Pathogen isolation

R. solani isolates were collected from the diseased leaves of soybean. Specimens were rinsed gently in tap water, then cut into small pieces (2 to 5 mm), washed 3 times in sterile distilled water and blotted dry on sterile paper towels. Pieces were placed on 5.0% water agar (WA) and incubated at 28°C for 1 or 2 days. Emerging hyphal tips were transferred to plates of potato dextrose agar (PDA, 200 g of potato, 20 g of dextrose and 20 g of agar) and pure cultures were transferred to PDA slants for storage at 4°C until use (Zhou and Yang, 1998).

Pathogen identification

Mycelium for DNA extraction was grown by inoculating 50 ml PDB in 250-ml conical flasks with mycelia fragments. Cultures were incubated on an orbital shaker (28°C) for 2 to 3 days, depending on the growth rate of the isolate. The culture products were washed twice with sterile distilled water, then dried with filter paper and stored at -20°C for use. The culture products were then ground in liquid nitrogen and total genomic DNA extracted from isolates followed the method of CTAB (Wu et al., 2009). PCR were

*Corresponding author. E-mail: aipingzh@yahoo.cn.

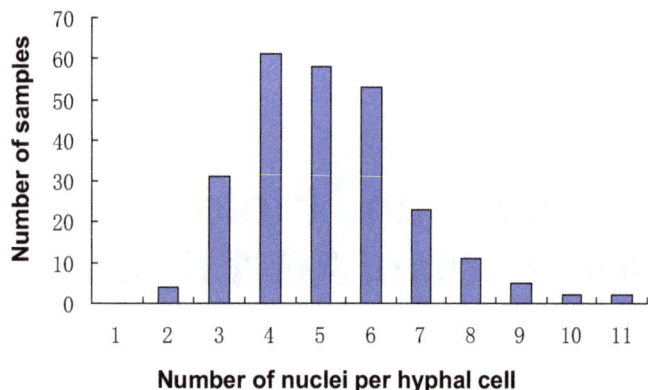

Figure 1. Statistics of nuclei number.

performed in a final mixture of 50 µl containing 100 ng template DNA, 5 µl 10 × reaction buffer (with 1.5 mmol/L Mg^{2+}), 0.2 mmol/L dNTP mixture, 0.2 µmol/L of each primer (ITS1-F : 5'-CTTGGTCATTTAGAGGAAGTAA-3'; ITS4: 5'-TCCTCCGCTT ATTGATAGC-3') (Chen et al., 2010; Takeshi et al., 1998), and 1 Unit of Taq DNA polymerase with the thermal profile: 2 min at 95°C, initial denaturation cycle; 30 s at 94°C, 30 s at 55°C, 60 s at 72°C, 30 cycles; 5 min at 72°C, 1 cycle. PCR products were sent to Premier Scientific Partner for sequence analysis, and sequences were compared on NCBI (http://www.ncbi.nlm.nih.gov/).

Morphological characterization

A 6-mm mycelial disc from a 3-day-old PDA culture of IB was placed in the middle of PDA Petri dishes and the Petri dishes were incubated at 28°C. The diameter of colony was measured and colony morphology was observed every 24 h. Sterile slides were inserted into the plates of PDA which were inoculated. The Petri dishes were incubated at 28°C until mycelium covered half of the slides. A drop of lactophenol cotton blue was added on slide, mycelial were examined microscopically 5 min later.

Nucleus and chromosome observations

Slides which were covered with mycelium were fixed in 1:3 glacial acetic acid: ethyl alcohol for 3 to 5 h, and transferred to 75 and 95% ethyl alcohol 30 min respectively. Then slides were kept in 4 mg/ml lysozyme 1 to 3 h (28°C). Phenol magenta was added on slides after slides were rinsed with distilled water. DAPI was used to count the number of nucleus, by adding stain on slides which were covered with mycelium (Coleman et al., 1981).

Infection process observations

A 6-mm mycelial disc from a 2-day-old PDA culture of IB was placed on leaves of soybean, cultured in a high humidity environment at 28°C. Culturing started at 4 h, and was observed and sampled each hour until 12 h. Then it was further observed and sampled at 16, 20, 24, 36, 48, 72, 96, and 130 h. Half of the samples were fixed and discolored in 1:3 glacial acetic acid: ethyl alcohol for 24 h. Trypan blue was added on samples to observed infectious structures by optical microscope (Xie, 2008). The other samples were fixed in 2.5% glutaraldehyde in 0.1 M phosphate buffer, pH 6.2, overnight.

Subsequently, the samples were rinsed with phosphate buffer and gradually dehydrated in a series of ascending concentrations of ethyl alcohol. This sample was replaced with isoamyl acetate and critical point drying, then coated with gold particles and observed by scanning electron microscope (Takuya et al., 2008).

RESULTS

Pathogen identification

The internal transcribed spacer sequence was amplified using PCR. Agarose gel showing internal transcribed spacer sequence length is about 750 bp, which is in accord with except length (Figure 1). Sequences were compared with standard on NCBI, the coincidence rate is 100%. So pathogen was identified to belong to anastomosis group AG-1 IB.

Morphological characteristics

The vegetative mycelium of R. solani AG-1 IB is colorless when young, but become brown as they grow (Figure 2A and B). Aerial mycelium is undeveloped. Mycelium forms a white group, and gradually becomes a brown sclerotium, which are scattered on the plate (Figure 2C and D). The surface of sclerotia will appear with some drops in a high humidity environment (Figure 2E). The hyphae often branch lass than a 90° angles (Figure 3). The mycelium consists of hyphae partitioned into individual cells by a septum, and hyphal fusion can be seen normally (Figure 4). The growth rate of mycelium is 1.11 mm/h.

Nucleus and chromosome characteristics

We census the number of nuclei per cell by the use of DAPI (Figure 5). Mycelia possess more than 2 and less than 11 nuclei, and usually possess between 4 and 6 nuclei per hyphal cell.

Infection process observations

The infection process of R. solani AG-1 IB on soybean includes pre-infection, penetration of epidermal cell, spreading in the codex, colonization and showing of symptoms. Hyphae begin to grow to leaves at 4 h after inoculation. 6 h later, tips of hyphae begin to swell (Figure 6). Infection cushion and appressorium are formed at 8 to 10 h after inoculation (Figure 7A and B), and then penetrate host. A layer of mycelium tile in the leaves can be observed at 12 h (Figure 8). Either the hyphae at the base of infection or the infection hyphae developing from the appressorium penetrated host cuticle directly or through stomata (Figure 9A, B and C) hyphae, spread

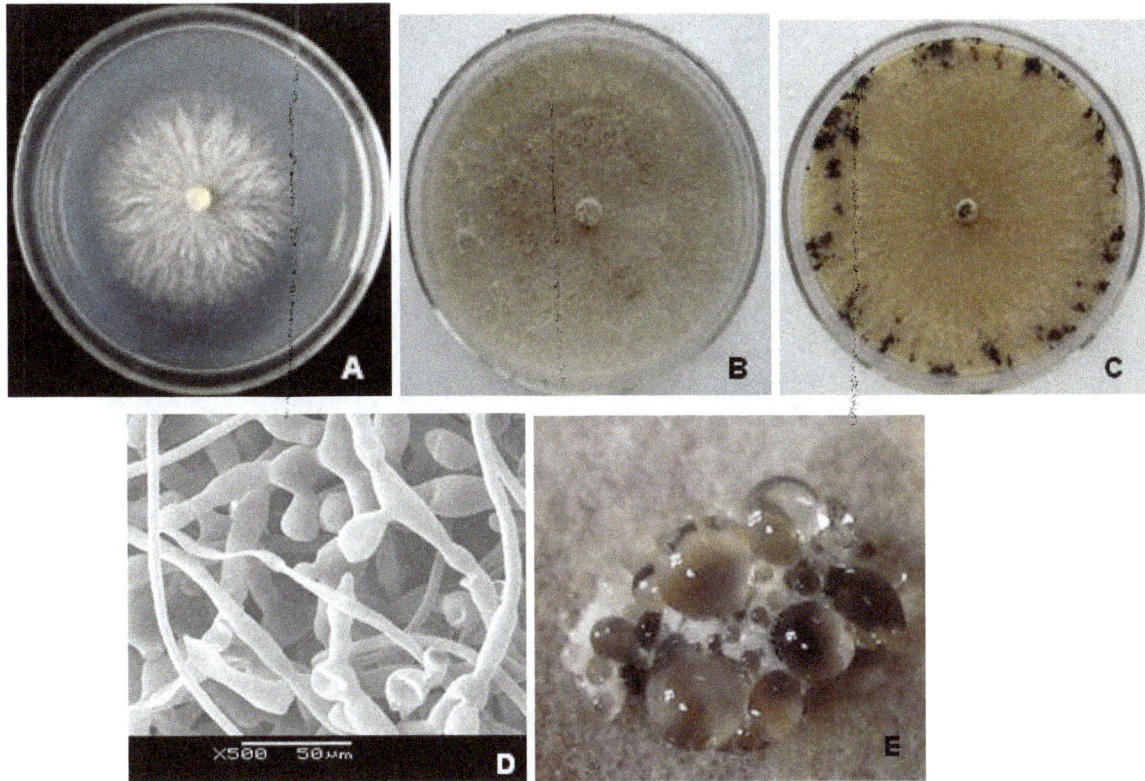

Figure 2. A) Young vegetative mycelium (48 h); B) Matured vegetative mycelium(96 h); C) Sclerotium scattered on the plate (7 days); D) The surface of sclerotia; E) Appearance of much drops on the surface of sclerotia.

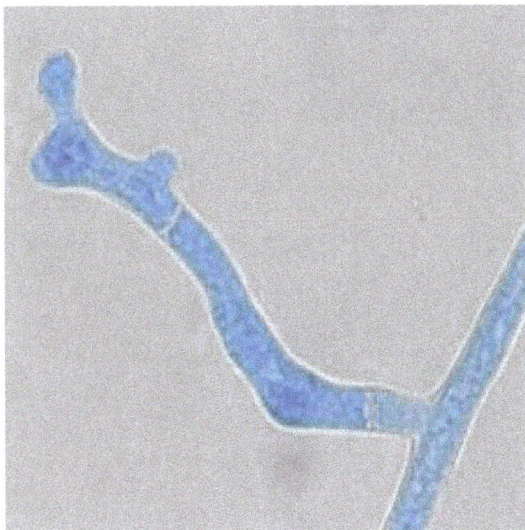

Figure 3. Branch lass than a 90° angles and septums.

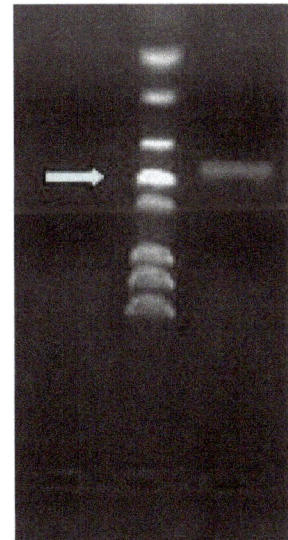

750 bp

Figure 4. Agarose gel showing PCR amplified product.

thought plant cells (Figure 10), and leaves show symptoms 24 h after inoculation. As the disease progresses, the fungus causes lesions on leaves. Small oval or circular, greenish-gray spots appear on leaves. The spots soon enlarge, with irregular dark brown margins and bleached to grayish white centers or light green centers (Figure 11). On high humidity and high temperature conditions, diseased organization will rot. Four days later, white sclerotia appear and become brown after maturity (Figure 12).

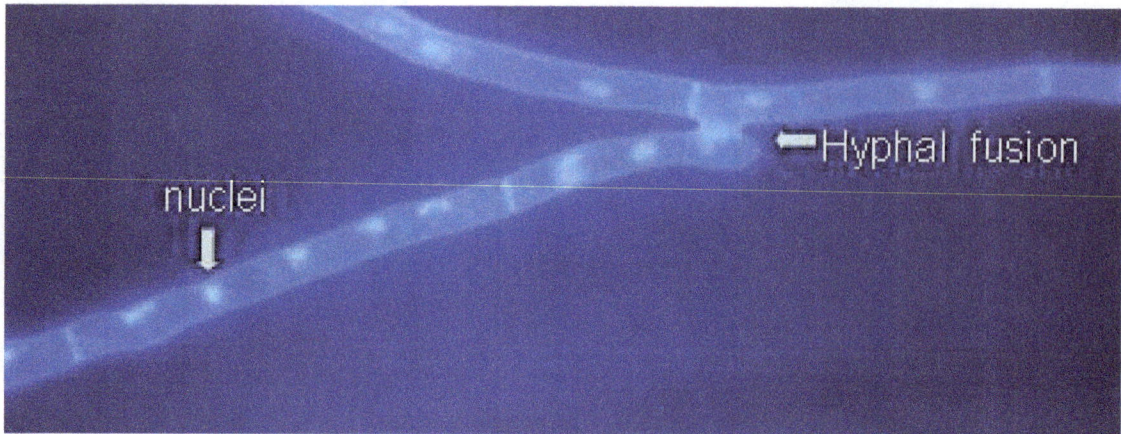

Figure 5. Hyphal fusion and nuclei.

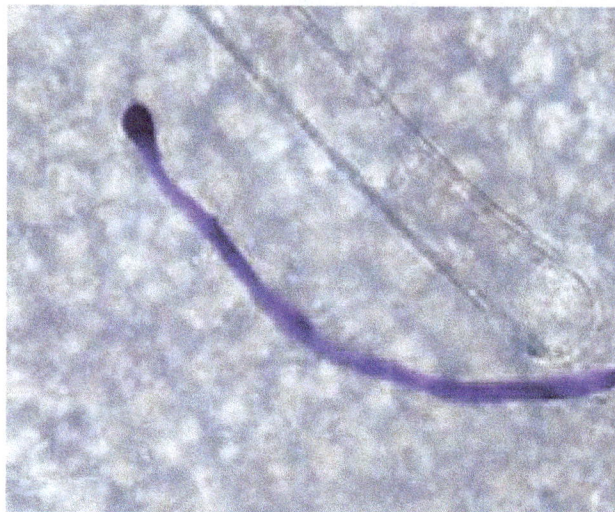

Figure 6. Tips of hyphae begin to swell.

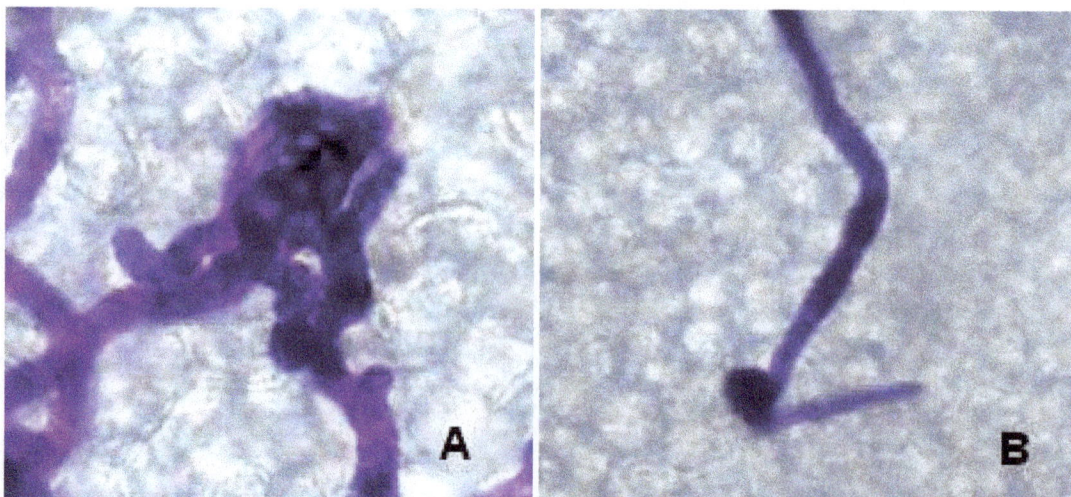

Figure 7. A) Infection cushion; B) Appressorium.

Figure 8. A layer of mycelium tile in the leaves.

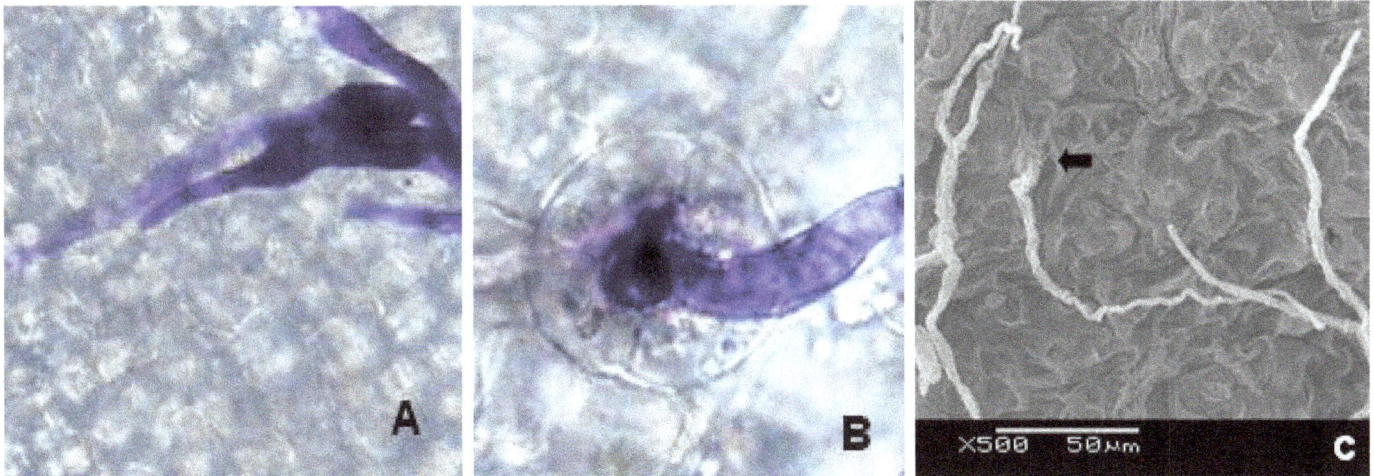

Figure 9. A) Hyphae penetrated directly; B) Hyphae penetrated through stomata (optical microscope); C) Hyphae penetrated through stomata (scanning electron microscope).

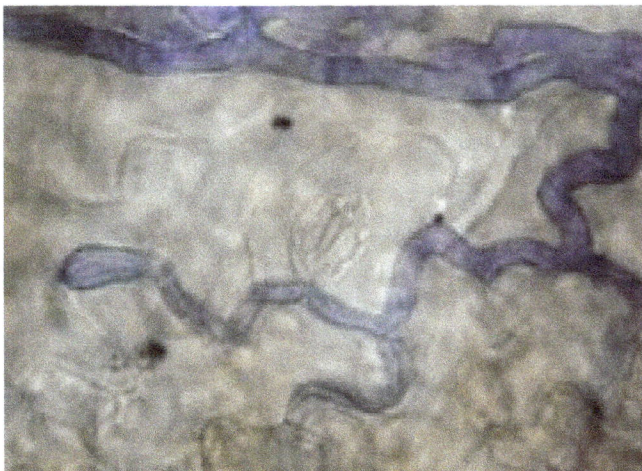

Figure 10. Hyphae spread thought plant cells.

DISCUSSION

The process of infection includes mycelial growth on plant surface before infection, the formation of infectious structures, penetration, expansion of pathogen within the plant tissue and the appearance of symptom. Among these steps, penetration is the most important cause whether the plant disease takes place or not is depended on the penetration's chances of success. So, the formation and progress of infectious structures play a vital role in the occurrence of plant disease. Infection cushion and appressorium are *R. solani*'s infectious structures. Although these two structures have been deeply considered in the process of *R. solani* AG-1 IA infection of rice; we first observed infection cushion and appressorium in the process of *R. solani* AG-1 IB infection of soybean. Comparing these processes, we can find that infection

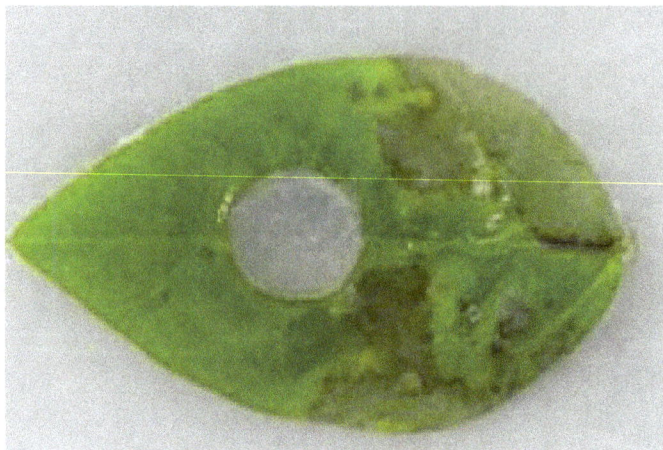

Figure 11. Symptom on leaf.

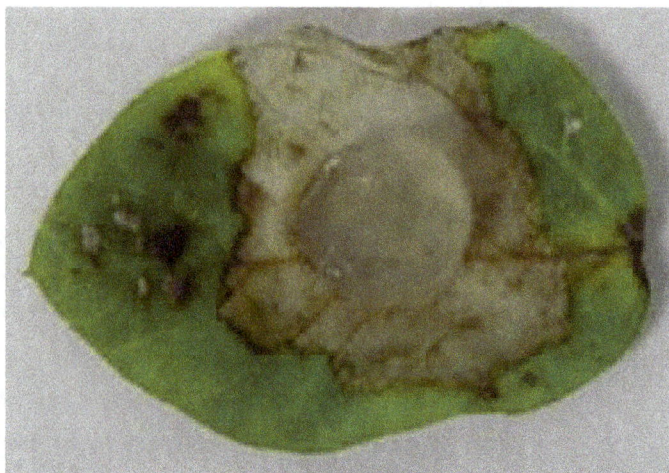

Figure 12. Sclerotia on leaf.

cushion on soybean leaf is sparser than that on rice leaf (Chen et al., 2000; Yang et al., 2008; Liu and Xiao, 1999; Tao and Tan, 1995; Zheng and Wu, 2007). The invasion of *R. solani* has diverse ways. The hyphae at the base of the appressorium or the infection hyphae developing from the appressorium can penetrate host cuticle directly. Also, some hyphae penetrate host cuticle through stomata or intercellular space indirectly. We observed both of these ways in our study. The infection process is a progress of interaction between plants and pathogens. In this study, we focused on the side of pathogens, and we hope to concentrate on plants in future to have a comprehensive understanding of *R. solani* AG-1 IB's infection process.

REFERENCES

Chen J, Tan C, Cao Z (2000). On Penetration Process of Sheath Blight Pathogen in Maize. J. Shenyang Agric. Univ., 31: 503-566.

Chen T, Zhang Z, Chai RY, Wan JY, Mao XQ, Qiu HP, Du XF, Jiang H, Wan LA, Wan YL, Sun G (2010). Genetic Diversity and Pathogenicity Variation of Different Rhizoctonia solani Isolatesin Rice from Zhe jiang Province, China. Chinese J. Rice Sci., 24:67-72.

Coleman AW, Maguire MJ, Coleman JR (1981). Mithramycin-and 4'-6-Diamidino-2-phenylindole (DAPI)-DNA staining for fluorescence microspectrophotometric measurement of DNA in nuclei, plastids and virus particles. J. Histochem. Cytochem., 29: 959-968.

Duan C, Yang G, Ni Z (2008). Occurrence of Foliar Rot of Chinese Cabbage, Mint and Lettuce Caused by Rhizoctonia solani AG-1 IB in China. J. Yunnan Agricul. Univ., 23: 422-425.

Yang GH, Conner RL, Chen YY, Chen JY, Wang YG (2008). Frequence and Pathogenicity Distribution of *Rhizoctona* spp. Causing Sheath Blight on Rice and Banded Leaf Disease on Maize in Yunnan, China. J. Plant Pathol., 90: 387-392.

Grosch R, Kofoet A, Elad Y (2003). Characteristics of Rhizoctonia solani isolates associated with bottom rot of lettuce. Turkey.

Huang J, Zhou E, QI P (2003). Identification of 'Rhizoctonia spp. Isolated from Thirteen Crops in Guangzhou Region in China. J. South China Agric. Univ. (Natural Science Edition), 24: 24-27.

Kuramae EE, Buzeto AL, Ciampi MB (2003). Identification of Rhizoctonia solani AG 1-IB in lettuce,AG 4 HG -I in tomato and melon.and AG 4 HG -III in Broccoli and Spinach,in Brazil. Eur. J. Plant Pathol., 109: 391-395.

Liu X, Xiao J (1999). Histopathological Studies on Infection Process of Wheat Sheath Blight Rhizoctonia Cerealis. Myeosystema, 18: 288-293.

Ogoshi A (1987). Ecology and pathogenicity of anastomosis and intra specific groups of *Rhizoctonia solani* Kühn. Ann. Rev. Phytupathol., 25: 125-143.

Sinclair JB, Dhingra OD (1975). An annotated bibliography of soybean diseases 1882-1974. INTSOY Series, No. 7. University of Illinois, Urbana-Champaign.

Sneh B, Burpee B, Ogoshi A (1991). Identification of *Rhizoctonia* species. APS Press Inc., St Paul, p. 133.

Takeshi T, Hideo N, Koji K (1998). Genetic identification of web-blight fungus (Rhizoctonia solani AG-1) obtained from European pear using RFLP of rDNA-ITS and RAPD analyses. Res. Bull. Fac. Agric. Gifu Univ., 63: 1-9.

Takuya S, Mitsuyo Hi, Makoto S (2008). Efficient Dye Decolorization and Production of Dye DecolorizingEnzymes by the Basidiomycete Thanatephorus cucumeris Dec 1 in a Liquid and Solid Hybrid Culture. J. Biosci. Bioengr., 106: 481-497.

Tao J, Tan F (1995). Studies on the Infection by Rhizoctonia solani on Maize. Acta Phytopathologica Sinica. 25: 253-257.

Wu F, Huang DY, Huang XL, Zhou X, Cheng WJ (2009). Comparing Study on several Methods for DNA Extraction from endophytic fungi. Chinese Agricultural Science Bulletin. 25:62-64.

Xie J (2008). Genetic diversity of rice sheath blight fungus and genes up-regulated during early stage of infection to rice leaves. Huanzhong Agricultural University.

YANG GH, CHEN HR, NAITO S (2005). Occurrence of foilar rot of pak choy and Chinese mustard caused by Rhizoctonia solani AG-1 IB in China. J. Gen. plant pathol., 71:377-379.

Yang XB (1999). Rhizoctonia damping-off and root rot. Pages 45-46 in: Compendium of Soybean Diseases, 4th ed. G. L. Hartman, J. B. Sinclair, and J. C. Rupe, eds. American Phytopathological Society, St. Paul, MN. ZHENG L, WU X (2007). Advances on Infection Structures of Plant Pathogenic Fungi. J. Nanjing Forestry University (Natural Sci. Edition).31: 90-94.

Zhou E, Yang M (1998). A rapid and simple technique for the isolation of Rhizoctonia solani from diseased plant tissues. J. South China Agric. Univ. 19:125-126.

13

Occurrence and identification of yeast species isolated from Egyptian Karish cheese

Neveen S. M. Soliman and Salwa A. Aly*

Department of Food Hygiene, Faculty of Veterinary Medicine, Cairo University, Egypt.

This study aims at identifying the diversity and abundance of yeast associated with Egyptian Karish cheese, employing comparison between conventional laboratory techniques and API20 kits techniques in yeast identification. A total of one hundred samples (fifty each) of Egyptian raw and pasteurized Karish cheese milk were randomly collected from farmers and markets in Cairo and Giza Districts. The occurrence of yeast in raw and pasteurized Karish cheese milk were 100 and 38% with a mean value of 7 ± 1.1 l and 1 ± 0.31 $\log_{10}$ cfu g^{-1}, respectively. Yeast strains isolated from both raw and pasteurized karish cheese samples were identified and characterized using both conventional methods and API 20 C AUX as a commercial identification system. The most prevalent isolates belonged to *Trichosporon cutaneum* (25%), *Candida catenulata* (23%), *Yarrowia lipolytica* (13%), *Debaryomyces hansenii* (13%), *Kluyveromyces lactis* (6%), *Geotrichum candidum* (7%), *Candida zeylanoides* (5%), *Candida lambica* (3%), *Candida albicans* (2%), *Cryptococcus formans* (1%), *Rhodotorula glabrata* (1%) and *Saccharomyces cerevisiae* (1%). There was no significant difference ($P \geq 0.05$) between the conventional method and the API20 kits test. However, the results of this study reveal that API 20 kits are simple, highly useful and are commercially available kits that considerably shorten the time required for the identification of yeast in cheese.

Key words: Yeast identification, Karish cheese, API 20 kits.

INTRODUCTION

Karish cheese is one of the most popular local types of fresh soft cheese in Egypt. The increasing demand by Egyptian consumers is mainly attributed to its high protein content and low price (Osman et al., 2010). Karish cheese is traditionally made from skim cow or buffalo's milk which is extracted directly into special earthenware pots known as (shalia) and kept undisturbed in a suitable place to allow the fat to rise to the surface forming a cream layer. Then the cream layer is removed and the curd is poured onto a mat which is tied and hung with its contents to allow the drainage of the whey. This process of squeezing takes two or three days until the desired texture of the cheese is obtained. Finally, the cheese is cut into suitable pieces and salted cheese is left for a few hours in the mat till whey no longer drains out, then it is ready to be consumed as fresh soft cheese (Ojokoh, 1998).

This traditional method affords many opportunities for microbial contamination. It is generally made from raw milk often of poor bacteriological quality and produced under unsatisfactory conditions. Also, this product is sold uncovered without a container, thus the risk of contamination is very high. Therefore, it can be considered as a good medium for the growth of different types of spoilage and pathogenic microorganisms (Yousef, 2007; Dawood et al., 2009).

It is widely recognized that yeasts can be an important component of the microflora of many cheese varieties because of the low pH, low moisture content, high salt concentration and refrigerated storage of these products (Devoyod, 2008). Nevertheless, yeasts play a dual role depending on the cheese. In fact, in some cheese types they make a positive contribution to the development of flavor and texture during the stage of maturation, while in other varieties, yeasts can be regarded as spoilage organisms. Yeast spoilage is recognized as a problem primarily in fermented milk and cheese (Brocklehurst and Lund, 1985; Fleet, 1990).

The sources of these yeast infections are located along the whole chain of production from the farm to the final product. The main defects caused by spoilage yeasts are fruity, bitter or yeasty off-flavors, gas production, discolorration changes and texture. Assessment of cheese spoilage by yeasts is complicated by subjective judgments on whether yeast activity during maturation is detrimental or beneficial to product quality. The main mechanisms by which yeast growth influences the final quality of cheese are: fermentation of lactose, utilization of lactic acid and lipolytic and proteolytic activities (Tudor and Board, 2010). Particularly, over-ripening during maturation could be interpreted as spoilage. In fact, continued lactose fermentation could lead to increased acidity, gassiness and fruity flavors, while continued hydrolysis of protein and fat could contribute to bitter and rancid flavors as well as a softening of product texture. However, the use of yeasts in dairy industry could determine potential advantages including production of flavor components and acceleration of ripening by means of its lipolytic and proteolytic properties, fermentation of lactose, assimilation of lactate and positive interactions with the primary starter cultures (Rohm et al., 2010).

Yeasts in some cheese types can periodically cause both economic and public health problems. Yeasts themselves are not commonly the cause of defects in cheese unless they ferment lactose. In this case, they can grow rapidly and produce a characteristic yeasty or fruity flavor and obvious gas (Dennis and Buhagiar, 2007; Dillon and Board, 2008; Daryaei et al., 2010). There are numerous references concerning the significance of the presence of yeasts in dairy products, where they may contribute positively to the characteristic taste and flavor development during the stage of maturation or, on the contrary, may lead to product spoilage (Ebrahim, 2008; Mahmoud, 2009).

In Egypt, the information about the involvement of Karish cheese in human illness and economic losses are unknown. Therefore, this study was designed to cover the following items: (1) Enumeration of the yeast populations in raw and pasteurized Karish cheese milk samples, and (2) Isolation and identification of the yeast species, using both conventional method and commercial identification system API 20 Aux test.

MATERIALS AND METHODS

Collection of samples

One hundred (fifty each) raw and pasteurized Karish cheese milk samples were obtained from retail farmers and supermarkets in Cairo and Giza governorates. Each cheese sample was represented by one whole cheese (500 g). All samples were transported to the laboratory under refrigeration and analyzed on arrival for both sensory and chemical examination as well as isolation and identification of yeast.

Sensory examination

Karish cheese samples were scored using a score card for flavor

(50 points), body and texture (35 points), and appearance and color (15 points). The scores were averaged by five well trained panelists of stuff members from the Food hygiene department, Faculty of veterinary medicine, Cairo University, according to the methodologies of Nelson and Trout (1981).

Chemical analysis

All Karish cheese samples were chemically examined for pH using a pH meter (model SA 720). Moisture and salt content were applied according to AOAC (2003).

Isolation of yeast according to the method of Van der Walt and Yarrow (2009)

Ten grams of the product were taken from the inner part of the cheese, diluted in 90 ml of sterile solution of 2% (w/v) sodium citrate (Sigma, St. Louis, MO, USA) and homogenized in a Stomacher (PBI, Milan Italy) for 30 s. For all samples, ten fold serial dilutions were prepared in a sterile solution of 2% (w/v) sodium citrate and the numbers of yeasts were determined by surface plating on yeast potato dextrose agar (PDA) (Microbiol, Cagliari, Italy) with chloramphenicol (0.01%) (Microbiol), after incubation at 25°C for 5 days. All samples were prepared and analyzed in duplicate. Yeast colonies were sorted on the basis of their morphology (smoothness of surface, regularity of border, consistency, color, etc.), streaked to single colonies on yeast potato dextrose agar media (1% yeast extract, 2% dextrose, 2% peptone and 1.5% agar), incubated for 5 days at 25°C and checked for purity. Counts for each individual type of colony were made in order to estimate the relative occurrence of the various yeasts present in the samples. Yeast species counts were calculated as number of colony forming units per gram of sample and were reported as $\log_{10}$ cfu g^{-1}.

Identification of the isolated strains

The isolates were identified using the conventional tests and were checked using the API 20 Aux kits (bioMerieux, Rome, Italy).

API 20 C method according to Dolan and Woodward (2007)

As recommended by the manufacturer, each isolate was subcultured prior to testing, to ensure viability and purity. Yeast inoculum suspensions were prepared from 48 h cultures grown on sabouraud dextrose agar plates at 30°C. Yeast cells were suspended in 2 ml of RapID Yeast Plus Inoculation Fluid to achieve a turbidity which completely obliterate the black lines of the inoculation card supplied with the kits. Each yeast suspension was dispensed into a RapID Yeast Plus panel, and the panels were then incubated for 4 h at 30°C. Immediately after the incubation time, RapID Yeast Plus Reagents A and B were added to the designated cavities and color reactions were evaluated by following the manufacturer's directions. A six-digit microcode was derived and compared to the codes in the RapID Yeast Plus Code Compendium for the identification of the isolate. All microcodes were also sent to the manufacturer for confirmation. Molten (50°C) API basal medium ampoules were inoculated with yeast colonies, and the suspension was standardized to a density below 1+ (lines can be clearly distinguished) on a Wickerham card. Each cupule was inoculated, and the trays were incubated for 72 h at 30°C. Cupules showing turbidity significantly heavier than that of the negative control cupule (0 cupule) were considered positive. Identification was made by generating a microcode and using the API 20C Analytical Profile

Table 1. Statistical analytical results of raw and pasteurized Karish cheese milk samples based on sensory examination (50 samples each).

Cheese samples	Organoleptic scores			
	Flavo (50)	Body and texture (35)	Appearance and color (15)	Total scores (100)
Raw milk	49±0.01	34±0.08	15±0.0.06	98
Pasteurized milk	44±0.01	34±0.04	14±0.05	92

Table 2. The mean chemical composition of the examined Karish cheese samples.

Cheese samples	Moisture (%)	Salt (%)	pH
Raw milk	77.4 ±0.049	1.28±0.03	4.16±0.8
Pasteurized milk	60.0±0.08	1.27±0.03	4.20±0.01

Index or Voice Response System (for profiles not found in the index).

Conventional method according to Kurtzman and Fell (2000)

Tubes were read after 24 and 48 h and again after 10 days for evidence of gas production, which indicated fermentation of the carbohydrate substrate.

Analysis of data according to Ott (2009)

The results are presented as mean and standard errors. The analysis of variance (ANOVA) test was conducted to test the possible significance (P≥0.05) among mean values of sensory, chemical and yeast count using Fishers Least Significance Difference (LSD).

RESULTS AND DISCUSSION

The data illustrated in Table 1 show the total score of raw Karish cheese milk in comparison with pasteurized milk samples. There was a significant difference (P>0.05) between raw and pasteurized Karish cheese samples. Practically, similar findings were reported by Yousef et al. (2001), Hamam (2005) and El -Batawy (2009). The flavor of raw Karish cheese milk had the highest total score compared to pasteurized cheese samples. This may be due to the natural flora initially present in raw milk which participates in flavor production (Brocklehurst and Lund, 1985).

As shown in Table 2, the mean chemical composition of raw and pasteurized milk cheese samples had a mean moisture content of 77.4 ± 0.04, while the pasteurized milk samples had 60.0 ± 0.08. There was no significant difference (P>0.05) between raw and pasteurized samples. The moisture content of pasteurized cheese samples was lower than raw cheese samples. This may be attributed to the effect of heat treatment on the capacity of the cheese protein to hold water (Shabatai, 2010). However, almost similar findings were reported by Abd El- Salam et al. (2003) and Ghosh et al. (2006). Higher

results were recorded by Kanka et al. (2007), Schaffer et al. (2008) and El-Batawy (2009). The mean salt content and pH value in both raw and pasteurized Karish cheese samples were 1.28±0.03 and 1.27±0.03, and 4.16±0.8 and 4.20±0.01, respectively. There was no significant difference (P>0.05) between pH and salt content in both raw and pasteurized cheese samples. Nearly similar findings were obtained by Abd El-Salam et al. (2003), Lalaguna (2008), and Omer and Elshirbiny (2003).

The composition of most cheese falls within certain compositional ranges. Moisture and salt content are considered the most important compositional factors (Fox and McSweeney, 2009). The higher moisture means more potential for off flavors, which result in many soluble breakdown products of acids, sugars, proteins and lipids (Ceylan et al., 2003; El-Sharoud et al., 2009). There was no regulation concerning the addition of sodium chloride to this product and the amount of salt added depended on the cheese markers themselves. The apparent variation among the chemical examinations of examined raw and pasteurized cheese samples was due to the fact that this type of cheese relied upon individual dairies. Thus, there is no standard production process and thus there is variation in the composition and properties of the milk processed (Turkoglu et al., 2007).

Data depicted in Table 3 revealed that the highest total yeast count was found in raw Karish cheese milk samples (100%), followed by 38% in pasteurized Karish cheese milk samples with a mean log of 7.89±1.1 and 1.00±0.31, respectively. There was a significant difference (P>0.05) between raw and heat treated Karish cheese milk samples, although nearly similar findings were reported by Ahmed et al. (2008), Devoyod (2008), Salwa et al. (2008) and Qing et al. (2010). Higher results were obtained by Kaldes et al. (2006), while lower results were obtained by Brocklehurst and Lund (1985), Elkohly (2001) and Said et al. (2009). The Egyptian Standards (2005) specify that the total yeast count does not exceed 10 cfu g^{-1} detected in the cheese. The International Commission on Microbiological Specifications for Foods

Table 3. Statistical analysis based on total yeast count of the examined Karish cheese samples (50 samples each).

Cheese samples	Total no.	No. of positive sample	Percentage	Mean log.
Raw milk	50	50	100	$7.89^a \pm 1.1$
Pasteurized milk	50	19	38	$1.00^b \pm 0.31$

Table 4. Incidence of the isolated yeast strains using conventional method and APi 20 C Aux.

Isolated yeast species	API20 Aux	Conventional method
	Percentage of correctly identified strains	Percentage of correctly identified strains
Trichosporon cutaneoum	25	25
Candida catenulata	23	23
Yarrowia lipolytica	13	13
Debaryomyces hansenii	13	13
Kluyveromyces lactis	6	6
Geotricum candidum	7	6
Candida zeylanoides	5	5
Candida lambica	3	3
Candida albicans	2	2
Cryptococcus formans	1	1
Saccharomyces cerevisiae	1	1
Rhodotorula glabrata	1	1

(2005) has classified cheese as a high risk potential hazard. A high yeast count often indicates neglected hygienic measures during production and handling, contamination of raw material, unsatisfactory sanitation, or unsuitable time and temperature during storage and/or production. It may also refer to the suitable pH of cheese for yeast growth as well their wide distribution in the environment (Aponte et al., 2010).

From the obtained results, it is obvious that most of the examined raw milk of Karish cheese samples failed to conform to the Egyptian standard (2005) as they exceeded the accepted level. The Egyptian standards for Karish cheese have proposed a limit for the total yeast count to be less than 10 g^{-1}. The high incidence in the examined samples may be attributed to poor sanitation during preparation and or storage of the product. There are numerous sources of yeast contamination. These include the use of contaminated milk, the observed dirty premises and utensils used as observed during sample collection, the use of bare hands in preparing the products (personal communication with the handlers), equipment, through persons taking part in manufacturing and handling of the product, improperly cleaned servers and debris falling into uncovered raw karish cheese milk. The high total yeast counts have resulted from inadequate processing (Aly et al., 2010). Yeast spoilage constitutes a major economic loss in the cheese industry through developing undesirable changes, such as slimness, red color and yeasty flavor (Sarais et al., 2009). This may be due to their capacity to produce lipolytic and proteolytic enzymes (Fleet and Mian, 2009; Tornadijo et al., 2010).

A comparison of a number of correctly identified organisms by both methods is presented in Table 4. The API 20 C system correctly identified about 100% of the isolates compared to 99% by the conventional method. The API system identified all of the members of the yeast genera. The conventional method correctly identified all organisms except for four isolates. The identification of isolated yeast species revealed the presence of *Trichosporon cutaneum* (25%), *Candida catenulate* (23%), *Yarrowia lipolytica* (13%), *Debaryomyces hansenii* (13%), *Kluyveromyces lactis* (6%), *Geotrichum candidum* (7%), *Candida zeylanoides* (5%), *Candida lambica* (3%), *Candida albicans* (2%), *Cryptococcus formans* (1%), *Rhodotorula glabrata* (1%) and *Saccharomyces cervisiae* (1%).

It was reported that *Trichosporon* spp. caused formation of a surface film on the cottage cheese leading to spoilage (Nichol and Harden, 2006; Welthagen and Viljoen, 2009). Presence of this species in high concentrations could indicate poor hygiene and ineffective cleaning procedures (Seiler and Busse, 2009; Viljoen et al., 2010). *Y. lipolytica* resulted in a browning spoilage of cheese (Vorbeck and Cone, 2009; Westall and Filtenborg, 2010), while *C. zeylanoides* was isolated from feta cheese, but it was not possible to determine whether spoilage was associated with this species or not (Eklund et al., 2005; Diriye et al., 2007; Rohm et al., 2010).

The source of the isolation of T. *cutaneum* varies considerably, although many of them are of human and animal origin (Kreger-van Rij, 2009). The presence of this species in high concentration could indicate poor hygiene and ineffective cleaning procedures, and show the need for improved sanitization procedures. *Y. lipolytica and C. zeylanoides* have also been isolated from spoiled cottage cheese. *Y. lipolytica* in high concentration resulted in an unwanted texture of feta-cheese due to the degradation of fat through production of lipolytic and proteolytic enzymes (Westall and Filtenborg, 2010).

Candida spp. are the most common cause of fungal infection in immune compromised persons known as candidiasis. Candidiasis is caused by infection of species within the genus *Candida,* predominantly with *C. albicans. Candida* species are ubiquitous fungi that represent the most common fungal pathogens that affect humans. Oropharyngeal colonization is found in 30 to 55% of healthy young adults, and *Candida* species may be detected in 40 to 65% of normal fecal florae. *C. albicans* can infect all areas of the skin as well as the mucous membranes. Infections by *C. albicans*, especially the ones found in the mucous membranes, are contagious (Hidalgo, 2011).

At present, the identification of yeast is generally performed by biochemical methods. However, many research findings have demonstrated that this method is not only complicated and labor intensive, but also time-consuming. This study confirmed the results on the identification of yeasts by biochemical and API 20 kit methods and found that there was no significant difference (P>0.05) between the conventional method and API 20. Therefore, it may be a useful tool for the biochemical identification of yeast isolates. They have almost the same sensitivity, but API 20 is easier, faster and rapid. However, almost similar findings were reported by Donald et al. (2009). On the other hand, Kitch and Badawi (2008) found significant differences between the conventional method and API 20 kits test. The data obtained in this study revealed that 'API 20 kits' was highly simple, useful, commercially available and considerably shortened the time required for identification of yeast in cheese. Conclusively, there is a need for continuous monitoring of Egyptian Karish cheese by educating producers, distributors and retailers on good sanitary practices during processing and sale of the product and the possible danger of contaminated Karish cheese.

REFERENCES

Abd El-Salam OG, Meyers SP, Nichols RA (2003). Extracellular proteinases of yeasts and yeast-like fungi. Appl. Microbiol., 16(12): 1370–1374.

Ahmed D, Bowman PI, Ahern DG (2008). Evaluation of the Uni-Yeast-Tek Kit for the identification of medically important yeasts. J. Clin. Microbiol., 2: 354-358.

Aly MM, Al-Seeni MN, Qusti SY, El-Sawi NM (2010). Mineral content and microbiological examination of some white cheese in Jeddah,

Saudi Arabia during summer 2008. Food Chem. Toxicol., 48(11): 3031-3034.

Aponte M, Pepe O, Blaiotta G (2010). Short communication: Identification and technological characterization of yeast strains isolated from samples of cottage cheese. J. Dairy Sci., 93(6): 2358-2361.

AOAC Association of Official Analytical Chemists (2003). Official analytical chemists. 15th Ed., Inc. Arlington, Virginia, USA.

Kaldes N, Barnett JA, Payne RW, Yarrow D (2006) Yeasts: Characteristics and Identification. (2nd Ed.), Cambridge Univ. Press, Cambridge.

Brocklehurst TF, Lund BM (1985). Microbiological changes in cottage cheese varieties during storage at 7 C. Food Microbiol., 2(3): 207–233.

Ceylan ZG, Turkoglu H, Dayisoylu KS (2003): The microbiological and chemical quality of sikma cheese produced in Turkey. Pakistan J. Nutr., 2: 95-97.

Daryaei H, Coventry J, Versteeg C, Sherkat F (2010). Combined pH and high hydrostatic pressure effects on Lactococcus starter cultures and Candida spoilage yeasts in a fermented milk test system during cold storage. Food Microbiol., 27(8): 1051-1056.

Dawood M, Abdou R, Salem G (2009): Effect of irradiation of skim milk on the quality of karish cheese. Egyptian J. Dairy Sci., 32(4): 500-509.

Dennis C, Buhagiar R (2007). Yeast spoilage of fresh and processed fruits and vegetables. In: Skinner, F.A., Passmore, S.M., Davenport, R.R. Eds.., Biology and Activities of Yeasts. Academic Press, London, pp. 123–133.

Devoyod JJ (2008). Yeasts in cheese making. In: Spencer, F.J., Spencer, D.M. Eds.., Yeast Technology. Springer-Verlag, Berlin, pp. 228–245.

Diriye FU, Scozzetti G, Martini A (2007). Incidence of yeast isolated from dairy foods. International J. Food Microbiol., 39(1): 27–37.

Dillon VM, Board RG (2008). Yeasts associated with cheese. J. Appl. Bacteriol., 71(2): 93–108.

Dolan CT, Woodward MR (2007). Identification procedures: yeast and yeast-like organisms, Cryptococcus species using API 20 kits in the diagnostic laboratory. Am. J. Clin. Pathol., 65(3): 591-559.

Donald C, Edwige D, Zubair H, Pierre R (2009). Evaluation of the Auxacolor System for Biochemical Identification of Medically Important Yeasts, J. Clin. Microbiol., 36(12): 3726–3727.

Ebrahim M (2008). Occurrence of Staphylococcus and Enteropathogens in soft cheese commercialized in the city of Rio de Janeiro, Brazil. J. Appl. Microbiol., 92 (6): 1172-1177.

El-Batawy A (2009). Biodiversity of bacterial ecosystems in traditional Egyptian Domiati cheese. Appl. Environ. Microbiol. 73 (4) 1248-1255.

Elkohly G (2001). Biodiversity of yeast species in traditional Egyptian karish cheese. Appl. Environ. Microbiol. 73(4): 1248-1255.

Egyptian Standard (2005). Karish cheese Es. 1185/01, Egyptian Organization for Standardization and Quality Control, Ministry of Industry, Cairo, Egypt.

El-Sharoud W, Belloch C, Peris D, Querol A (2009). Molecular identification of yeasts associated with traditional Egyptian dairy products. J. Food Sci., 74(7): 341-346.

Eklund N, Spinelli J, Miyanchi D, Grominger H (2005). Characteristics of yeasts isolated from Pacific crab meat. Appl. Microbiol., 13: 985-990.

Fleet GH (1990). Yeasts in dairy products: a review. J. Appl. Bacteriol., 68(1): 199–211.

Fleet GH, Mian MA (2009). The occurrence and growth of yeasts in dairy products. Int. J. Food Microbiol., 41(2): 145–155

Fox PF, McSweeney PL (2009): Advanced Dairy Chemistry, Vol.1, Proteins.Kluwer Academic/Plenum Publishers, and New York.

Ghosh A, Nafisa AE, Seham IF, Orsi F (2006). Effect of pasteurization and storage condition on the microbial, chemical and organoleptic quality of soft cheese. J. Dairy Sci., 94(1): 177-190.

Hamam K (2005). The microbiological and chemical quality of orgu c heese produced in Turkey. Food Microbiol 72 (2) 92-94.

Hidalgo H (2011). Candidiases. Interna. Med. J.88 (1) 52-58. International Committee on Microbiological Specifications for Foods, ICMSF, (2005). Microorganisms in food, their significance and methods of enumeration. 2nd Ed. Academic press, London.

Kitch J, Badawi RM (2008). Survival of micro entrapped bifido bacteria during storage of white cheese and their effect on cheese quality. Monofia J. Agric. Res. 24(2): 493-513.

Kanka M, Kamaly K, Zedan N, Zaghlol A (2007). Acceleration of ripening of Domiati cheese by accelase and lipozyme enzyme. Egyptian J. Dairy Sci., 35(1): 75-90.

Kreger van Rij NJW (2009). The Yeasts: A Taxonomic Study. , Elsevier, Amsterdam.

Kurtzman CP, Fell JW (2000). The Yeasts: A Taxonomic Study. , Elsevier, Amsterdam.

Lalaguna S (2008). Microbiological and compositional status of white soft cheeses. J. Food Technol., 49(1): 115-119.

Mahmoud I (2009). Safety of traditional food: regulations and technical aspects. Scienza Tecnica – Lattiero- Casearia, 55 (3): 159-165.

*Marth K, Steel A (2005). Chemical and bacteriological characteristics of Pichtogalo Chanion cheese and mesophilic starter cultures for its production. J. Food Protect., 61(2): 688-692.

Nelson JA, Trout GM (1981): Judging of dairy products, 4th Ed. INC Westport, Academic Press, pp. 345-567.

Nichol AW, Harden TJ (2006). Enzymic browning in mould ripened cheeses. Austr. J. Dairy Technol., 48(1): 71–73.

Ojokoh J (1998). Comparison of RapID Yeast Plus System with API 20C System for Identification of Common, New, and Emerging Yeast Pathogens J. Clin. Microbiol. 36(4): 883-886.

Omer D, Elshirbiny H (2003). Identification of yeasts in karish cheese. J. Food Control, 55 (3): 977- 981.

Osman O, Ozturk I, Bayram O, Kesmen Z, Yilmaz MT (2010), Characterization of cheese Spoiling Yeasts and their Inhibition by Some Spices. Egyptian J. Dairy Sci., 75 (2): 637–640.

Ott L (2009). An Introduction to Statistical Methods and Data Analysis. 2nd edition, PWS publisher, Boston, MA, USA.

Qing M, Bai M, Sun Z, Zhang H, Sun T (2010). Identification and biodiversity of yeasts from Qula in Tibet and milk cake in Yunnan of China. Wei Sheng Wu Xue Bao. 50(9): 1141-1146.

Rohm H, Lechner F, Brauer M (2010). Diversity of yeasts in selected dairy products. J. Applied Bacteriol., 72 (3): 370–376.

Said F, Białasiewicz D, Kłosiński M (2009). Contamination of dairy products by yeasts-like fungi. J. Dairy Technol., 73 (2): 655-658.

Salwa A Aly, Morgan SD, Moawad AA, Metwally B (2008). Effect of moisture, salt content and pH on the microbiological quality of traditional Egyptian Domiati cheese. Assiut Vet. Med. J., (2007) 53(115): 68-81.

Sarais I, Piussi D, Aquili V, Stecchini ML (2009). The behavior of yeast populations in stracchino cheese packaged under various conditions. J. Food Protect. 59(4): 541–544.

Seiler H, Busse M (2009). The yeast of cheese brines. Intern. J. Food Microbiol., 41(3): 289–304.

Shabatai Y (2010). Isolation and characterization of a lipolytic bacterium capable of growing in a low-water content oil–water emulsion in cheese. Appl. Environ. Microbiol., 57(9): 1740–1745.

Schaffer M, Adams ED, Cooper BH (2008). Evaluation of a modified Wickerham medium for identifying medically important yeasts in cottage cheese. Am. J. Med. Technol., 40(3): 377- 388.

Tornadijo ME, Sarmento RM, Carballo J (2010). Study of yeasts during the ripening of Armada cheeses from raw goat's milk. Lait, 78(2): 647 659.

Turkoglu H, Ceylan Z, Dayisoylu K (2007). The microbiological and chemical quality of orgu cheese produced in Turkey. Pak. J. Nutr. 2(1): 92-94.

Tudor DA, Board RG (2010). Food spoilage yeasts. In: The Yeast. Yeasts Technology (Rose, A.H., Harrison, J.S. Eds). 2nd ed. Academic Press, London, 5: 436–451.

Van der Walt and Yarrow, D., (2009). Methods for the isolation, manteinance, classification and identification of yeasts. In: Kreger-van Rij, N.J.W. Ed.., The Yeasts: A Taxonomic Study. 3rd edn. Elsevier, Amsterdam, pp. 45–104.

Viljoen BC, Dykes GA, Callis M, von Holy A (2010). Yeasts associated with cheese packaging. Int. J. Food Microbiol., 42 (1) 53–62.

Vorbeck ML, Cone JF (2009). Characteristics of an intracellular proteinase system of a *Trichosporon* species isolated from trappist-type cheese. Applied Microbiol., 11: 23–27.

Welthagen JJ, Viljoen BC (2009). Yeast profile in Gouda cheese during processing and ripening. International J. Food Microbiol., 41(2): 185–194.

Westall S, Filtenborg O (2010). Yeast occurrence in Danish feta cheese. Food Microbiol., 15(1): 215–222.

Yousef A (2007). Effect of technological factors on the manufacture of karish cheese. Egyptian J. Dairy Sci., 31(1): 161-165.

Yousef H, Sobieh M, Nagedan K (2001). Microbial status of Domiati cheese, at El-Gassiem area, Saudi Arabia. 8th Sci. Cong., Fac. Vet. Med., Assiut Univ., pp. 91-97.

14

Gasteroid mycobiota of Rio Grande do Sul, Brazil: Boletales

Vagner G. Cortez[1]*, Iuri G. Baseia[2] and Rosa Mara B. Silveira[3]

[1]Universidade Federal do Paraná, Rua Pioneiro 2153, Jardim Dallas, CEP 85950-000, Palotina, PR, Brazil.
[2]Departamento de Botânica, Universidade Federal do Rio Grande do Sul,
Av. Bento Gonçalves 9500, CEP 91501-900, Porto Alegre, RS, Brazil.
[3]Departamento de Botânica, Ecologia, Universidade Federal do Rio Grande do Norte, e Zoologia, CEP 59072-970, Natal, RN, Brazil.

Boletales is an order of the subclass Phallomycetidae, which comprises of a wide variety of morphological types of macrofungi, including the boletes and earthballs. In this paper, the gasteroid members of the Boletales from Rio Grande do Sul State, in southern Brazil, were revised. Specimens were collected during the years 2006 to 2009, analyzed macro and microscopically and the collections are preserved at the herbarium ICN. The following taxa were recorded: *Rhizopogon roseolus* (*Rhizopogonaceae*), *Calostoma zanchianum*, *Pisolithus arhizus*, *Scleroderma albidum*, *Scleroderma bovista*, *Scleroderma citrinum*, *Scleroderma dictyosporum*, *Scleroderma fuscum*, *Scleroderma laeve* and *Scleroderma verrucosum* (*Sclerodermataceae*). *Scleroderma dictyosporum* and *S. laeve* are reported for the first time from Brazil. A key for the identification of the species of *Scleroderma* is provided and colour photographs and line drawings of the basidiospores are presented for all taxa studied.

Key words: Basidiomycota, calostomataceae, ectomycorrhizal fungi, *Eucalyptus*, *Pinus*, pisolithaceae, taxonomy.

INTRODUCTION

The order Boletales E. J. Gilbert comprises a wide grouping of macrofungi of several morphological typologies, such as boletoid (poroid and lamellate), gasteroid, secotioid, agaricoid, corticioid, merulioid, hydroid, and polyporoid (Binder and Hibbett, 2006). The phylogeny of this group has been investigated in recent years by a number of authors (Høiland, 1987; Kretzer and Bruns, 1999; Jarosch, 2001; Binder and Bresinsky, 2002), and the current trend in their classification is the recognition of six well-supported lineages within Boletales, considered at the subordinal level: Boletineae, Coniophorineae, Paxillineae, Sclerodermatineae, Suillineae, and Tapinellineae (Binder and Hibbett, 2006). Although

some taxa are saprotrophs or even mycoparasites, the mycorrhizal association with a large number of plant families is a noteworthy feature in the group, and genera such as *Boletus* Dill. Ex. Fr., *Rhizopogon* Fr. and Nordholm, *Scleroderma* Pers. and *Suillus* Gray, play an important role in natural forest ecosystems and forestry (Cairney and Chambers, 1999; Futai et al., 2008).

In Brazil, the knowledge of the *Boletales* is almost limited to surveys of boletoid members (Putzke et al., 1994; Watling and Meijer, 1997; Giachini et al., 2000), and little information on the other representatives is available. In a recently published Brazilian checklist, Neves and Capelari (2007) reported 20 genera and 70 species (seven *Rhizopogon* species) belonging to the *Boletales sensu* Kirk et al. (2001), thus not including the Sclerodermataceae. The gasteroid genera of *Boletales* are currently classified in to the *Diplocystaceae Kreisel*

*Corresponding Author. E-mail: cortezvg@yahoo.com.br.

(Astraeaceae Zeller ex. Jülich.), Rhizopogonaceae Gäum. and C.W. Dodge, and *Sclerodermataceae* Corda (*Calostomataceae* E. Fisch., Pisolithaceae Ulbrich), and other unknown families in Brazil. *Scleroderma* is the largest and best known genus, with approximately 30 species, 13 of which are known from Brazil (Guzmán, 1970; Baseia and Milanez, 2000; Giachini et al., 2000; Meijer, 2006; Gurgel et al., 2008). A number of *Pisolithus*, *Rhizopogon*, and *Scleroderma* species were reported from Rio Grande do Sul State by Rick (1961), Guzmán (1970), Sobestiansky (2005) and Cortez et al. (2008a). This study aimed to provide information on the diversity of the gasteroid members of the Boletales in Rio Grande do Sul State, southern Brazil, and comprises partial results of a comprehensive survey of the gasteroid fungi of the state (Cortez et al., 2008a, b, 2009, 2010, 2011a, b; Sulzbacher et al., 2010).

MATERIALS AND METHODS

Fresh material was collected from March 2006 to 2009 during gasteromycete surveys in Rio Grande do Sul State, southern Brazil. Collected specimens are preserved in the ICN herbarium (Instituto de Biociências, Universidade Federal do Rio Grande do Sul). Macroscopical analysis comprised the study of external and internal features of the peridium (color, texture, thickness) and the gleba (color, consistency) and also peristome and rhizomorphs. Microscopical study was done with free-hand sections of the basidiomata mounted on slides in 5% KOH (potassium hydroxide) or 1% Congo Red. Measurements and line drawings were made with a light microscope equipped with a camera lucida; 25 measurements for each microstructure were considered. In all basidiospore measurements, the diameter of the complex ornamentation is included. Morphology and taxonomy followed BY Kirk et al. (2008) and for identification of *Scleroderma* spp., Guzmán's (1970) monograph and the revised key by Sims et al. (1995) were followed. Colour terminology is according to Kornerup and Wanscher (1978).

RESULTS AND DISCUSSION

Rhizopogonaceae Gäum. and C.W. Dodge

Rhizopogon roseolus (Corda) Th. Fr., Svensk Bot. Tidskrift 1: 282, 1909. Figures 1a and 2a. Basidiomata 14 to 44 mm diam., 19 to 40 mm high, subglobose to broad pyriform, with basal rhizomorphs attached to soil and roots. Peridium 1 to 1.5 mm thick, white (5A1), reddish blond (5C3) to brownish orange (5C6) with pastel red (9A4) spots where handled. Gleba loculate, with a slightly gelatinous consistency but very hard when dried, white (5A1) when young, becoming greyish yellow (5C5) to yellowish brown (5D5) at maturity. Basidiospores (7–) 8.5–10 x (3.5–) 4.2–5 µm, ellipsoid, with a slightly truncate apex, guttulate, walls smooth and thickened, hyaline to pale greenish. Basidia 21–29.5 x 5–9.2 µm, ventricose to sublageniform, bearing six sterigmata, hyaline. Peridium composed of prostrate to interwoven

hyphae, 3.5 to 7.5 µm diameter, yellowish brown, walls thin but encrusted with abundant brown crystals. Hyphae of the trama 2.5 to 5 µm diameter, hyaline, smooth and thin-walled, simple-septate, gelatinized and interwoven.

Examined specimens: BRAZIL. Rio Grande do Sul State. Itaara, 10/V/2006, *V.G. Cortez 043/06* (ICN); 27/IV/2007, *V.G. Cortez 069/07* (ICN). Santa Maria, 15/V/2007, *V.G. Cortez 096/07* (ICN); 16/V/2008, *V.G. Cortez 099/08* (ICN); 18/VI/2008, *V.G. Cortez 118/08* (ICN).

Distribution: widespread in *Pinus* spp. plantations worldwide. Brazil: São Paulo (Baseia and Milanez, 2002), Paraná (Meijer, 2006) and Rio Grande do Sul.

Discussion: *Rhizopogon* is an ectomycorrhizal genus introduced in Brazil through seedlings of exotic North American *Pinus* spp. (Baseia and Milanez, 2002). The basidiomata grow semi-hypogeously around *Pinus* trees, associated with their roots, especially in autumn. The specimens from a *Pinus* plantation reported by Sobestiansky (2005) in Rio Grande do Sul possibly belong to this species, but that material was not included in this study. Giachini et al. (2000) reported the following species from the neighboring state of Santa Catarina: *Rhizopogon fuscorubens* A.H. Sm., *Rhizopogon nigrescens* Coker and Couch, *Rhizopogon rubescens* Tul., *Rhizopogon vulgaris* (Vittad.) M. Lange, and *Rhizopogon zelleri* A.H. Sm. *Rhizopogon roseolus* is a new record from Rio Grande do Sul.

Sclerodermataceae Corda

(i) *Calostoma zanchianum* (Rick) Baseia and Calonge, Mycotaxon 95: 114, 2006.
≡ *Mitremyces zanchianus* Rick, Iheringia, Série Botânica 9: 456, 1961.

Examined specimen: Brazil. Rio Grande do Sul State. São João do Polêsine (formerly Cachoeira do Sul), III.1943, *R. Zanchi* (PACA 19673, holotype).

Discussion: This is the only known *Calostoma* from southern Brazil. The type and only preserved specimen was described by Rick (1961) as *Mitremyces zanchianus*. Descriptions, illustrations (including SEM images of the basidiospores) and a discussion on this species are found in Baseia et al. (2006, 2007).

(ii) *Pisolithus arhizus* (Pers.) Rauschert, Zeitschrift für Pilzkunde 25: 50, 1959.
= *Pisolithus tinctorius* (Pers.) Coker and Couch, Gasterom. East. USA and Canada: 170, 1928 Figures 1b and 2b. Basidiomata globose, subglobose to hemispheric, 40 to 64 mm diamater, 44 to 129 mm high, with a

Figure 1. Basidiomata of gasteroid *Boletales*. (a) *Rhizopogon roseolus*. (b) *Pisolithus arhizus*. (c) *Scleroderma albidum*. (d) *S. bovista*. (e) *S. citrinum*. (f) *S. fuscum*. (g) *S. laeve*. (h) *S. verrucosum*. Scale bar = 20 mm. Photos: (a-c, e-f, h) by Vagner G. Cortez, (d) by M.A. Sulzbacher, (g) by M.A. Reck.

distinct rhizomorphic base, commonly forming pseudo-stipe in larger and older specimens. Peridium thin (<0.5 mm thick), smooth, greyish yellow (4C5) olive brown (4E8) when young to olive brown (4F8) when mature;

dehiscence irregular, exposing the mature gleba and peridioles. Gleba composed of irregular shaped peridioles, 2 to 4.5 mm diam., which mature from the base toward the top of the basidioma, mature gleba pulverulent, olive brown (4E8). Basidiospores 7.5 to 11 µm diam., globose, yellowish brown in KOH, echinate, with spines 1 to 1.7 µm long.

Examined specimens: Brazil. Rio Grande do Sul State. Capão do Leão, 06/V/1996, *C. Rodrigues 355* (PEL 15419); Porto Alegre, 18/VI/1965 (SP 91481); Rio Grande, 25/V/1992, V.L.N. Susin (HURG 3732); 25/VI/1990, V.L.N. Susin et al. (HURG 3530); 27/IX/1993, C.M. Abreu and M.S. Farias (HURG 3008); 07/VI/2008, V.G. Cortez 108/08 (ICN); Santa Maria, 23/III/2007, V.G. Cortez 025/07 (ICN); 24/III/2008, V.G. Cortez 066/08 (ICN); Viamão, 13/IX/1967, O. Dieffenbach (ICN 5818).

Distribution: cosmopolitan in *Eucalyptus* plantations worldwide. Brazil: known records from the States of Espirito Santo (Vinha, 1988), São Paulo (Bononi et al., 1981), and Rio Grande do Sul (Guerrero and Homrich, 1999).

Discussion: *P. arhizus* (often referred as *P. tinctorius*) is recognized by its globose, echinate basidiospores, which are larger than those of *P. microcarpus* [5–7 µm diameter. sensu Cunningham (1942)]. This is the best known species of the genus and has been intensively studied due to its importance in forestry as an ectomycorrhizal partner of eucalypts and other cultivated trees (Cairney and Chambers, 1997, 1999). In Rio Grande do Sul, specimens were collected in eucalypt and pine (*Pinus elliottii* Engelm.) plantations and near native pink trumpet trees [*Tabebuia heptaphylla* (Vell.) Vell.].

(iii) *Scleroderma albidum* Pat. and Trab. emend. Guzmán, Darwiniana 16: 295, 1970.
  Figures 1c and 2c. Basidiomata 14 to 24 mm diameter, 17 to 40 mm high, depressed globose to pyriform, with a small rhizomorphic base. Peridium <1 mm thick when fresh, leathery, surface partially smooth with scattered, small and thin scales, greyish yellow (4C5) to brownish orange (5C5). Gleba olive brown (4F8), compact to pulverulent at maturity. Basidiospores 12 to 17 µm diam., globose, brown, densely echinate and not reticulate, the spines are pointed and slightly curved. Peridium formed by an external layer composed of orange brown, clavate hyphae, 13–21 x 5–9 µm, and an internal layer formed by interwoven, yellowish to hyaline hyphae, 4–7.5 µm diameter, not clamped.

Examined specimens: Brazil. Rio Grande do Sul State. Capitão, 30/V/2007, V.G. Cortez 117/07 (ICN); Minas do Leão, 26/V/2008, V.G. Cortez 093/08 (ICN); 18/VI/2008, V.G. Cortez 121/08 (ICN); Rio Grande, 23/V/1997, A.C.S.

Campos (HURG 1669); Santa Cruz do Sul, 19/05/2008, M.A. Sulzbacher (ICN); 03/VI/2007, M.A. Sulzbacher and V.G. Cortez 122/07 (ICN); 29/VII/2007, V.G. Cortez 149/07 (ICN); Santa Maria, 03/VI/2006, V.G. Cortez 053/06 (ICN, UFRN-Fungos), G. Coelho and V.G. Cortez 011/08 (ICN); 14/III/2008, V.G. Cortez 058/08 (ICN), São Leopoldo, 07/V/1960 (SP 61721); Viamão, 22/V/2004, V.G. Cortez 024/04 (ICN).

Distribution: common in southern hemisphere, Asia and North America, but infrequent in Europe (Guzmán, 1970). In Brazil it is known from the states of Pernambuco (Gurgel et al., 2008), Minas Gerais, Rio de Janeiro, São Paulo (Guzmán, 1970), Santa Catarina (Giachini et al.,2000) and Rio Grande do Sul (Cortez et al., 2008a).

Discussion: *Scleroderma albidum* belongs to *Scleroderma* section *Aculeatispora* because of the spiny and non-reticulate basidiospores and absence of clamp connections in the hyphae (Guzmán, 1970; Sims et al., 1995). This is one of the most common species of the genus in Rio Grande do Sul, apparently associated with several trees, although more frequently collected under *Eucalyptus* spp., where it sometimes grows sub-hypogeously.

(iv) *Scleroderma bovista* Fr., Syst. Mycol. 3: 48, 1829. Figures 1d and 2d. Basidiomata 13–24 mm diam., 10–16 mm high, depressed subglobose to subglobose, with a short rhizomorphic base (<7 mm high). Peridium <1 mm thick when dried, fleshy and soft when fresh, surface partially smooth towards the base and cracked on top, reddish blond (4C3) to brownish orange (4C4). Gleba pulverulent when mature, olive (3E5). Basidiospores 12–14.4 µm diam., globose, orange brown to dark brown, reticulate-echinate, the reticulation not continuous, and the spines are pointed to slightly curved. Peridium formed by yellowish to hyaline, prostrate to more or less interwoven, clamped hyphae.

Examined specimens: Brazil. Rio Grande do Sul State. Santa Maria, 03/II/2009, M.A. Sulzbacher and V.G. Cortez 002/09 (ICN).

Distribution: known from all continents, mostly distributed in the Americas (Guzmán, 1970). Brazil: known from the states of Pernambuco, São Paulo (Gurgel et al., 2008), Santa Catarina (Giachini et al., 2000), and Rio Grande do Sul.

Discussion: *S. bovista* has been differently interpreted by many authors and the taxonomic problems have been discussed in detail by Guzmán (1970). The *S. bovista* material reported from Brazil by Guzmán (1970) was probably collected by J. Rick in Rio Grande do Sul.

(v) *Scleroderma citrinum* Pers., Syn. Meth. Fung. 153,

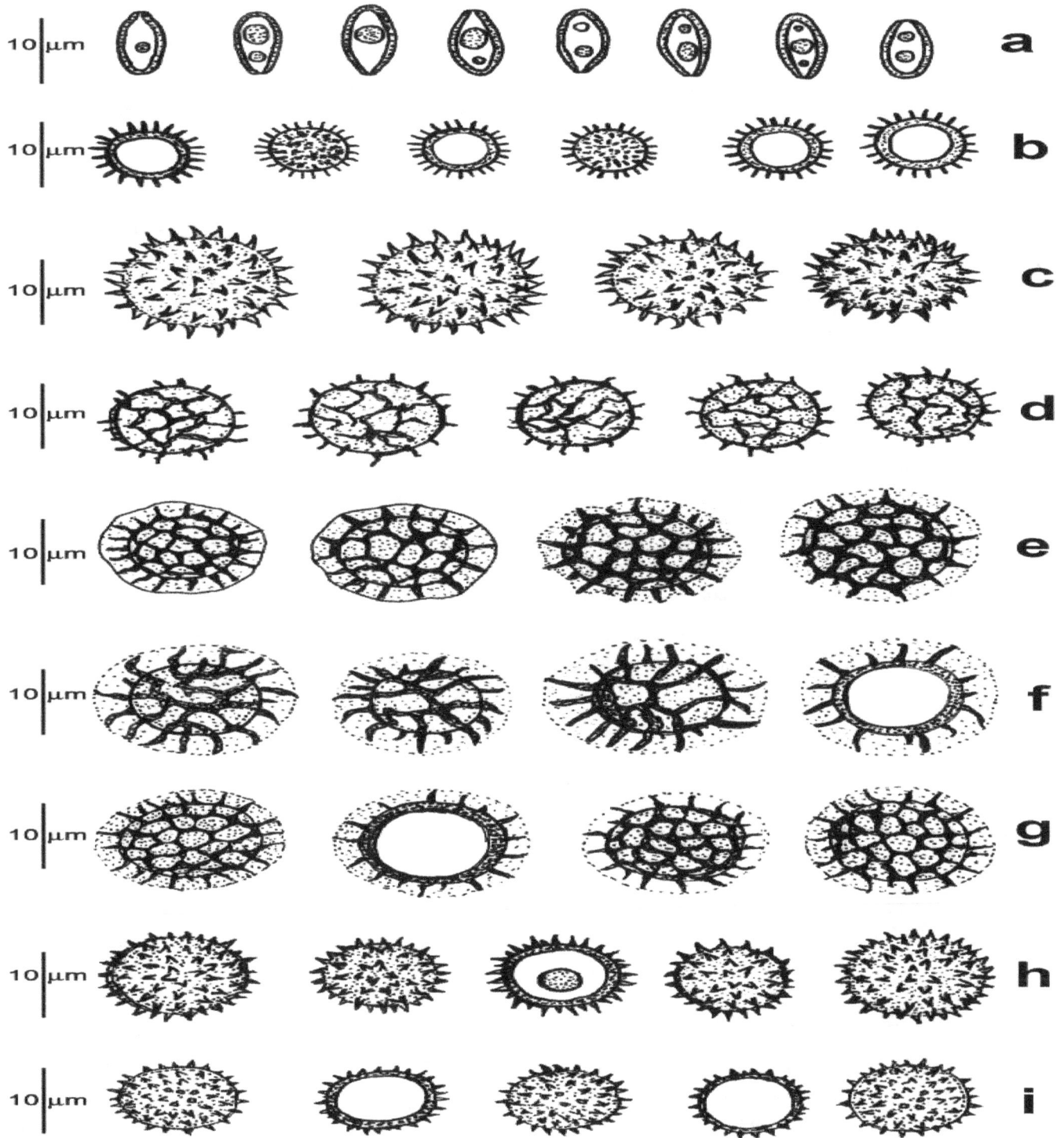

Figure 2. Basidiospores of gasteroid *Boletales*. (a) *Rhizopogon roseolus*. (b) *Pisolithus arrhizus*. (c) *Scleroderma albidum*. d) *S. bovista*. (e) *S. citrinum*. (f) *S. dictyosporum*. (g) *S. fuscum*. (h) *S. laeve*. (i) *S. verrucosum*. All line drawings by Vagner G. Cortez.

1801. Figures 1e and 2e. Basidiomata 31–72 mm diameter, 24–52 mm high, depressed globose to subglobose, with a small to often large and fasciculate rhizomorphic base, not forming a pseudostipe. Peridium <2 mm thick when dried, fleshy when fresh, drying very hard, surface scaly and cracked in small yellowish brown (5E6) scales on a greyish yellow (3C5) background. Gleba olive brown (4F8) to black, compact to pulverulent. Basidiospores 12.5–16 µm diameter, globose, dark brown, reticulate, with spines <2 µm long. Peridium composed externally by fascicles of erect, orange brown, cystidioid terminal hyphae, arising from a pseudoparenchymatic internal layer composed of hyaline to pale yellow, smooth-walled, clamped hyphae.

Examined specimens: Brazil. Rio Grande do Sul State. Santa Maria, 22/I/2008, G. Coelho and V.G. Cortez 010/08 (ICN); 08/V/2008, V.G. Cortez 075/08 (ICN); São Francisco de Paula, 08/IV/2006, V.G. Cortez 015/06 (ICN); 21/IV/2007, V.G. Cortez 051/07 (ICN).

Distribution: widespread, mostly growing under *Pinus* spp. plantations (Guzmán, 1970). Brazil: known from the states of Paraíba (Gurgel et al., 2008), São Paulo (Bononi et al., 1981), Paraná (Meijer, 2006), Santa Catarina (Giachini et al., 2000), and Rio Grande do Sul (Sobestiansky, 2005).

Discussion: This species is among the most common macrofungi in *Pinus* plantations in southern Brazil. It grows abundantly, solitary or in small groups among the needles. This earthball has been tested for use as an ectomycorrhizal for pines with some good results (Chen et al., 2006). The scaly peridium makes it macroscopically similar to *S. verrucosum*, but this species has echinate and non-reticulate basidiospores in contrast to the reticulate basidiospores of *S. citrinum* (Pegler et al., 1995).

(vi) *Scleroderma dictyosporum* Pat., Bull. Soc. Mycol. Fr. 12: 135, 1896 Figure 2f. Basidiomata 9–15 mm diameter, 11–24 mm high, globose to subglobose, with a small to usually well-developed rhizomorphic base, dehiscence not observed. Peridium <1 mm thickness and leathery when dried, surface partially smooth with scattered small scales at the top, light brown (6D4) to brown (6E4). Gleba when mature olive brown (4F7), fairly pulverulent. Basidiospores 12–16 µm diameter, globose, dark brown, strongly reticulate, with a continuous ornamentation 2–4 µm thick. Peridium composed externally of collapsed, prostrate hyphae, reddish brown with encrusted walls, and internally of interwoven, hyaline to yellowish, smooth-walled, clamped hyphae.

Examined specimens: Brazil. Rio Grande do Sul State. Santa Maria, 03/V/2007, V.G. Cortez 071/07 (ICN); 13/VI/2007, V.G. Cortez 124/07 (ICN).

Distribution: known from tropical areas of Africa, Asia and Americas (Guzmán, 1970). Brazil: only known from Rio Grande do Sul.

Discussion: According to Guzmán (1970), *S. dictyosporum* exhibits basidiospores 8.8–13.6 µm, with fully reticulate ornamentation measuring 1.9–3.5 µm long. Our specimens have basidiospores slightly larger, but display the same ornamentation pattern and other macroscopic features as those described by Guzmán (1970). It is similar to *S. bovista*, but the ornamentation of the latter is smaller (<2 µm) and the reticulation is not continuous (Guzmán 1970). This species is distributed across dry regions of Africa, Asia and America, as well the subtropical zone. The Brazilian specimens were found growing near the base of *Acacia caven* (Molina) Molina, a native species from southern South America. In Africa, *S. dictyosporum* has also been found as ectomycorrhizal partner of other acacias, such as *A. holosericea* and *A. mangium* (Founoune et al., 2002; Duponnois et al., 2005; Sanon et al., 2009). This is the first report of this species from Brazil.

(vii) *Scleroderma fuscum* (Corda) E. Fisch., Nat. Pflanz. 1: 336, 1900 Figures 1f and 2g. Basidiomata 26–32 mm diameter, 19–24 mm high, globose to subglobose, with a small but fasciculate rhizomorphic base, dehiscence not observed. Peridium <1.5 mm thick and leathery when dried, surface squamulose, cracking into small yellowish brown (5E8) scales on a greyish yellow (3C4) background. Gleba mature olive brown (4F8) pulverulent. Basidiospores 12–17.5 µm diameter, globose, dark brown, with a reticulate ornamentation, <2.5 µm diam. Peridium composed externally of prostrate orange brown hyphae and internally of interwoven, hyaline to yellowish hyphae, with smooth walls, clamps present.

Examined specimens: Brazil. Rio Grande do Sul State. Porto Alegre, 03/V/1990, *J. Pereira* (ICN ex HASU 1305). Santa Maria, 15/V/2007, V.G. Cortez 083/07 and 088/07 (ICN).

Distribution: widespread, associated with *Pinus* plantations, especially in South America (Guzmán, 1970). Brazil: known from the states of Santa Catarina (Giachini et al., 2000) and Rio Grande do Sul.

Discussion: This species forms an ectomycorrhizal association with *Pinus* spp. and for this reason it has been reported from several regions, including South America. Although *S. fuscum* has been reported from Brazil by Giachini et al. (2000) in the state of Santa Catarina, this is the first record from Rio Grande do Sul.

(viii) *Scleroderma laeve* Lloyd emend Guzmán, Darwiniana 16: 301, 1970 Figures 1g and 2g. Basidiomata 32–58 mm diameter, 32–65 mm high, depressed subglobose to subglobose, finally stellate at maturity, with a distinct and abundant rhizomorphic base but not forming a pseudostipe. Peridium <1.5 mm dried, color greyish yellow (4B4) at the base, light brown (5D5) to a yellowish brown (5E6) at the top, surface scaly at top of peridium, smooth towards the base, dehiscence stellate. Gleba compact to pulverulent, color yellowish brown (5F8) at mature stage. Basidiospores 9.2–14 µm diameter, globose, echinate, acute spines up to 1.5 µm long, color yellowish brown. Peridium with an external layer formed by yellowish, fairly prostrate and subparallel hyphae, 2.5–7 µm diam., internal layer composed of interwoven, hyaline hyphae, 6.5–12 µm diameter, clamp connections absent.

Examined specimen: Brazil. Rio Grande do Sul State: Porto Alegre, 22/IX/2006, V.G. Cortez 067/06 (ICN); 19/VI/2009, P.S. Silva and M.A. Reck (ICN).

Distribution: North America, Africa (Guzmán, 1970), Australia (Malajczuk et al., 1982), Japan (Kasuya et al., 2002) and Brazil (Rio Grande do Sul).

Discussion: *S. laeve* is similar to *S. albidum* (both members of the *Scleroderma* sect. *Aculeatispora*) but they differ in the scaly peridium and smaller basidiospores of the former (Guzmán, 1970). This is an ectomycorrhizal partner of *Eucalyptus* and possibly of some interest in forestry since it is a non-specific associate (Malajczuk et al., 1982). This is the first record of this species from Brazil.

(ix) *Scleroderma verrucosum* (Bull.) Pers., Syn. Meth. Fung.: 154, 1801 Figures 1h and 2i. Basidiomata (5–)14–20 mm diam., (4–)10–14 mm high, globose to depressed subglobose, with a small rhizomorphic base, dehiscence through a wide irregular to substellate pore. Peridium < 0.5 mm thick and leathery when dried, yellowish white at the base to pastel yellow (2A4) at the top, surface covered by small and thin brown (6E8) scales. Gleba greyish brown (5E3) when mature, compact. Basidiospores 10–12.5 µm diameter, globose, yellowish brown, echinate, spines <1.5 µm long. Peridium composed externally of a layer of interwoven, reddish brown hyphae, and internally by interwoven, hyaline hyphae, with smooth and slightly thickened walls, clamp connections absent.

Examined specimens: Brazil. Rio Grande do Sul State. Santa Maria, 15/V/2007, V.G. Cortez 093/07 (ICN); 03/II/2009, M.A. Sulzbacher & V.G. Cortez 017/09 (ICN).

Distribution: Cosmopolitan (Guzmán, 1970). Brazil: known from the states of Bahia, Rio de Janeiro (Guzmán, 1970), Paraná (Meijer, 2006), and Rio Grande do Sul (Sobestiansky, 2005).

Discussion: Guzmán (1970) described this as a poorly understood species and according to his concept, it is diagnosed by the small, echinate basidiospores and thin peridium (<0.5 mm). This species is widely distributed over every continent, and reported from several regions of Brazil. Wright and Albertó (2006) considered it a poisonous mushroom.

Key for the identification of *Scleroderma* from Rio Grande do Sul:

1. *Basidiospores echinate*..2
    Basidiospores reticulate 4

2. Peridium smooth to squamulose, basidiospores 12–17 µm .... *S. albidum*
    Peridium with small to large scales, basidiospores smaller 3

3. Peridium <0.5 mm thick, scales small, dehiscence irregular...*S. verrucosum*
    Peridium <1.5 mm thick, scales large, dehiscence stellate .... *S. laeve*

4. Basidiospore ornamentation >2 µm long ....
.... *S. dictyosporum*
    Basidiospore ornamentation <2 µm long.....................5

5. Peridium smooth to subtly scaly............................ 6
    Peridium conspicuously scaly, basidiospores 12.5–16 µm .... *S. citrinum*

6. Peridium pale-colored, basidiospores 12–14.4 µm .... *S. bovista*
    Peridium dark-colored, basidiospores 12–17.5 µm .... *S. fuscum*

Doubtful or excluded records

Astraeus hygrometricus (Pers.) Morgan – Reported by Rick (1961) as *Geaster hygrometricum*, but no specimens were gathered or preserved by the author. Rick (1961, p. 468) wrote: "quamquam in RGS (Rio Grande do Sul) nondum inventus". Two specimens preserved at the PACA herbarium are from Germany (Berlin, 26/IX/1930, *B. Hennig*, PACA 12225) and Portugal (J. Rick, PACA 15965). However, its occurrence in southern Brazil can be expected because it is known from Argentina (Nouhra and Dominguez de Toledo, 1998) and Brazil (Baseia and Galvão, 2003; Phosri et al., 2007; Vinha, 1988).

Calostoma cinnabarinum Desvaux: This taxon was not reported by Rick (1961), but one specimen of *C. cinnabarinum* is preserved in the PACA herbarium (*C.G. Lloyd*, PACA 14331). However, the material reveals no information on location and date, except that it was collected by Lloyd, probably in the USA. Baseia et al.

(2007) reported it in Brazil from the states of São Paulo and Pernambuco.

ACKNOWLEDGEMENTS

The authors thank Prof. Gastón Guzmán (INECOL, México) for kindly providing relevant literature. Gilberto Coelho, Mateus A. Reck and Marcelo A. Sulzbacher are thanked for providing specimens and photographs. CNPq is thanked for financial support.

REFERENCES

Baseia IG, Cortez VG, Calonge FD (2006). Rick's species revision: *Mitremyces zanchianus* versus *Calostoma zanchianum*. Mycotaxon, 95: 113-116.

Baseia IG, Galvão TCO (2003). Some interesting Gasteromycetes (*Basidiomycota*) from areas of the Northeastern Brazil. Acta Bot. Bras., 16: 1-8.

Baseia IG, Milanez AI (2000). First record of *Scleroderma polyrhizum* Pers. (Gasteromycetes) from Brazil. Acta Bot. Bras., 14: 181-184.

Baseia IG, Milanez AI (2002). *Rhizopogon* (Gasteromycetes): hypogeous fungi in exotic forests from the State of São Paulo, Brazil. Acta Bot. Bras., 16: 55-60.

Baseia IG, Silva BDB, Leite AG, Maia LC (2007). O gênero *Calostoma* (*Boletales*, *Agaricomycetidae*) em áreas de cerrado e semi-árido no Brasil. Acta Bot. Bras., 21: 277-280.

Binder M, Bresinsky A (2002). Derivation of a polymorphic lineage of Gasteromycetes from boletoid ancestors. Mycologia, 94: 85-98.

Binder M, Hibbett DS (2006). Molecular systematics and biological diversification of *Boletales*. Mycologia, 98: 971-981.

Bononi VLR, Trufem SFB, Grandi RAP (1981). Fungos macroscópicos do Parque Estadual das Fontes do Ipiranga depositados no Herbário do instituto de Botânica. Rickia, 9: 37-53.

Cairney JWG, Chambers SM (1997). Interactions between *Pisolithus tinctorius* and its hosts: a review of current knowledge. Mycorrhiza, 7: 117-131.

Cairney JWG, Chambers SM (1999). Ectomycorrhizal fungi: key genera in profile. Heidelberg: Springer.

Chen YL, Kang LH, Malajczuk N, Dell B (2006). Selecting ectomycorrhizal fungi for inoculating plantations in south China: effect of *Scleroderma* on colonization and growth of exotic *Eucalyptus globulus*, *E. urophylla*, *Pinus elliottii*, and *P. radiata*. Mycorrhiza, 16: 251-259.

Cortez VG, Baseia IG, Silveira RMB (2008a). Gasteromicetos (*Basidiomycota*) no Parque Estadual de Itapuã, Viamão, Rio Grande do Sul, Brasil. Rev. Bras. Bioci., 6: 291-299.

Cortez VG, Baseia IG, Guerrero RT, Silveira RMB (2008b). Two sequestrate cortinarioid fungi from Rio Grande do Sul, Brazil. Hoehnea, 35: 513-518.

Cortez VG, Baseia IG, Silveira RMB (2009). Gasteroid mycobiota of Rio Grande do Sul, Brazil: *Tulostomataceae*. Mycotaxon, 108: 365-384.

Cortez VG, Baseia IG, Silveira RMB (2010). Gasteroid mycobiota of Rio Grande do Sul, Brazil: *Arachnion* and *Disciseda* (*Lycoperdaceae*). Acta Biol. Par., 39: 19-27.

Cortez VG, Baseia IG, Silveira RMB (2011a). Gasteroid mycobiota of Rio Grande do Sul, Brazil: *Lysuraceae* (*Basidiomycota*). Acta Scient., Biol. Sci., 33: 87-92.

Cortez VG, Sulzbacher MA, Baseia IG, Antoniolli ZI, Silveira RMB (2011b).
New records of *Hysterangium* (*Basidiomycota*) in *Eucalyptus* plantations of south Brazil. R. Bras Bioci.: in press.

Cunningham GH (1942). The Gasteromycetes of Australia and New Zealand. Dunedin: John McIndoe.

Duponnois R, Founoune H, Masse D, Pontanier R (2005). Inoculation of

Acacia holosericea with ectomycorrhizal fungi in a semiarid site in Senegal: growth response and influences on the mycorrhizal soil infectivity after 2 years plantation. For. Ecol. Manag., 207: 351-362.

Founoune H, Duponnois R, Bâ AM (2002). Ectomycorrhization of *Acacia mangium* Willd. and *Acacia holosericea* A. Cunn. ex G. Don in Senegal. Impact on plant growth, populations of indigenous symbiotic microorganisms and plant parasitic nematodes. J. Arid Environ., 50: 325-332.

Futai K, Taniguchi T, Kataoka R (2008). Ectomycorrhizae and their importance in forest ecosystems. In: Z.A. Siddiqui, M.S. Akhtar and K. Futai (eds.) Mycorrhizae: Sustainable Agriculture and Forestry. Berlin: Springer, pp. 241-285.

Giachini AJ, Oliveira VL, Castellano MA, Trappe JM (2000). Ectomycorrhizal fungi in *Eucalyptus* and *Pinus* plantations in Southern Brazil. Mycologia, 92: 1166-1177.

Guerrero RT, Homrich MH (1999). Fungos Macroscópicos Comuns no Rio Grande do Sul – Guia para Identificação. 2ª ed. Porto Alegre: Ed. UFRGS.

Gurgel FE, Silva BDB, Baseia IG (2008). New records of *Scleroderma* from Northeastern Brazil. Mycotaxon, 105: 399-405.

Guzmán G (1970). Monografia del género *Scleroderma*. Darwiniana, 16: 233-407.

Høiland K (1987). A new approach to the phylogeny of the order *Boletales* (*Basidiomycotina*). Nordic J. Bot., 7: 705-718.

Jarosch M (2001). Zur Molekularen Systematik der *Boletales*: *Coniophorineae*, *Paxillineae* und *Suillineae*. Bibliotheca Mycologica p. 191. Berlin: J. Cramer.

Kasuya T, Guzmán G, Ramirez-Guillén F, Kato T (2002). *Scleroderma laeve* (Gasteromycetes, *Sclerodermatales*), new to Japan. Mycoscience, 43: 475-476.

Kirk PM, Cannon PF, David JC, Stalpers JA (2001). Ainsworth & Bisby's Dictionary of the Fungi. 9th ed. Wallingford: CABI Publ.

Kirk PM, Cannon PF, Minter DW, Stalpers JA (2008). Dictionary of the Fungi. 10th ed. Wallingford: CABI Publ.

Kornerup A, Wanscher JE (1978). Methuen Handbook of Colour. 3rd ed. London: Eyre Methuen.

Kretzer AM, Bruns TD (1999). Use of atp6 in fungal phylogenetics: an example from the *Boletales*. Mol. Phylogen. Evol., 13: 483-492.

Malajczuk N, Molina R, Trappe JM (1982). Ectomycorrhiza formation in *Eucalyptus* I. Pure culture synthesis, host specificity and mycorrhizal compatibility with *Pinus radiata*. N. Phytol., 91: 467-482.

Meijer AAR (2006) Macromycete survey from Brazilian State of Paraná. Bol. Mus. Bot. Mun., Curitiba 68: 1-59.

Neves MA, Capelari M (2007). A preliminary checklist of the *Boletales* from Brazil and notes on *Boletales* specimens at the Instituto de Botânica (SP) Herbarium, São Paulo, SP, Brazil. Sitientibus, Ser. Ciên. Biol., 7: 163-169.

Nouhra ER, Dominguez de Toledo L (1998). The first record of *Astraeus hygrometricus* from Argentina. Mycologist, 12: 112-113.

Pegler DN, Læssøe T, Spooner BM (1995). British puffballs, earthstars and stinkhorns. London: Royal Botanic Gardens, Kew.

Phosri C, Martín MP, Sihanonth P, Whalley AJS, Watling R (2007). Molecular study of the genus *Astraeus*. Mycol. Res., 111: 275-286.

Putzke J, Pereira AB, Maria L (1994). Os fungos da família *Boletaceae* conhecidos do Rio Grande do Sul (Fungi, *Basidiomycota*). Cad. Pesq., Sér. Bot., 6: 75-100.

Rick J (1961). *Basidiomycetes* Eubasidii in Rio Grande do Sul - Brasilia. 6. Iheringia, Sér. Bot., 9: 451-479.

Sanon KB, Bâ AM, Delaruelle C, Duponnois R, Martin F (2009). Morphological and molecular analyses in *Scleroderma* species associated with some Caesalpinioid legumes, *Dipterocarpaceae* and *Phyllanthaceae* trees in southern Burkina Faso. Mycorrhiza, 19: 571-584.

Sims KP, Watling R, Jeffries P (1995). A revised key to the genus *Scleroderma*. Mycotaxon, 56: 403-420.

Sobestiansky G (2005). Contribution to a macromycete survey of the States of Rio Grande do Sul and Santa Catarina in Brazil. Braz. Arch. Biol. Technol., 48: 437-457.

Sulzbacher MA, Cortez VG, Coelho G, Jacques RJ, Antoniolli ZI (2010). *Chondrogaster pachysporus* in a *Eucalyptus* plantation of Southern

Brazil. Mycotaxon 113: 377-384.

Vinha PC (1988). Fungos macroscópicos do estado do Espírito Santo depositados no herbário central da Universidade Federal do Espírito Santo, Brasil. Hoehnea, 15: 57-64.

Watling R, Meijer AAR (1997). Macromycetes of the state of Paraná, Brazil 5. Poroid and lamellate boletes. Edinb. J. Bot., 54: 231-251.

Wright JE, Albertó E (2006). Guía de los Hongos de la Región Pampeana. II: Hongos in laminillas. Buenos Aires: L.O.L.A.

15

Two new schifffnerulaceous fungi from Kerala, India

V. B. Hosagoudar[1], J. Thomas[1] and D. K. Agarwal[2]*

[1]Tropical Botanic Garden and Research Institute, Palode - 695 562, Thiruvananthapuram, Kerala, India.
[2]Plant pathology Division, IARI, New Delhi 110 012, India.

Two new species namely, *Sarcinella loranthacearum* and *Schiffnerula meliosmatis* were collected from Peppara and Neyyar Wildlife sanctuaries and Silent Valley National park are described and illustrated in detail.

Key words: Black mildews, new species, Kerala, India.

INTRODUCTION

Black mildews are the group of fungi belonging to several orders of the fungi. Of these, the present work gives an account of two new species of the genus *Schiffnerula*, and of which one is with teleomorph on the member of Sabiaceae and the other with anamorph on Loranthaceae. These are the obligate biotrophs having high degree of host specificity.

METHODOLOGY

Infected plant parts were selected in the field, field notes were made regarding their pathogenicity, nature of colonies, nature of infection and the collection locality. For each collection, a separate field number was given. In the field, each infected plant was collected separately in polythene bags along with the host twig (preferably with the reproductive parts to facilitate the identity of the corresponding host). These infected plant parts were pressed neatly and dried in-between blotting papers. After ensuring their dryness, they were used for microscopic study. Scrapes were taken directly from the infected host and mounted in 10% KOH solution. After 30 min, KOH was replaced by Lactophenol. Both the mountants work well as clearing agents and made the septa visible for taking measurements.

To study the entire colony in its natural condition, a drop of high quality natural colored or well transparent nail polish was applied to the selected colonies and carefully thinned with the help of a fine brush without disturbing the colonies. Colonies with hyper parasites showing a woolly nature were avoided. The treated colonies along with their host plants were kept in dust free chamber for half an hour. When the nail polish on the colonies dried fully, a thin, colorless or slightly apple rose colored (depending upon the colour tint in the nail polish) film or flip was formed with the colonies firmly

embedded in it. In case of soft host parts, the flip was lifted off with a slight pressure on the opposite side of the leaves and just below the colonies. In case of hard host parts, the flip was eased off with the help of a razor or scalpel. A drop of DPX was spread on a clean slide and the flip was spread properly on it. One or two more drops of DPX were added additionally on the flip and a clean cover glass was placed over it. By gently pressuring on the cover glass, excessive amount of DPX was removed after drying. Care was taken to avoid air bubbles. These slides were labeled and placed in a dust free chamber for one to two days for drying. These permanent slides were then used for further studies. For innate fungi, sections were made and stained in cotton blue. After the study of each collection, part of the material was retained in the regional herbarium, Tropical Botanic Garden, Thiruvananthapuram (TBGT) and part of it was deposited in the Herbarium Cryptogamae Indiae Orientalis (HCIO), IARI, and New Delhi.

TAXONOMY

Sarcinella loranthacearum sp. nov. (Figure 1)

Coloniae epiphyllae, densae, ad 2 mm diameter, confluentes. Hyphae brunnneae, rectae vel subrectae, alternate vel opposite acuteque vel laxe ramosae, arte reticulatae, cellulae 16-29 × 4-7 µm. Appressoria alternata, unicellularis, ovata vel globosa, integra, 9-12 × 7-10 µm. Conidiophorae micronemataw, mononematae, simplices, rectae, leniter brunnneae, exoriens hyphis lateralis, glabra, 9-15 × 7-10 µm; cellulae conidiogenae integratae, plerumque terminalis, monoblasticae, determinatae, cylindraceae; conidia simplices, solitaria, sicca, acrogena, globosa, glabra, brunnnea vel nigra, constrictus ad septatus, celluae 5-8, sarciniformes, 24-29 µm diam.Conidia *Questieriella* visa, recta vel

*Corresponding author. E-mail:vbhosagoudar@rediffmail.com.

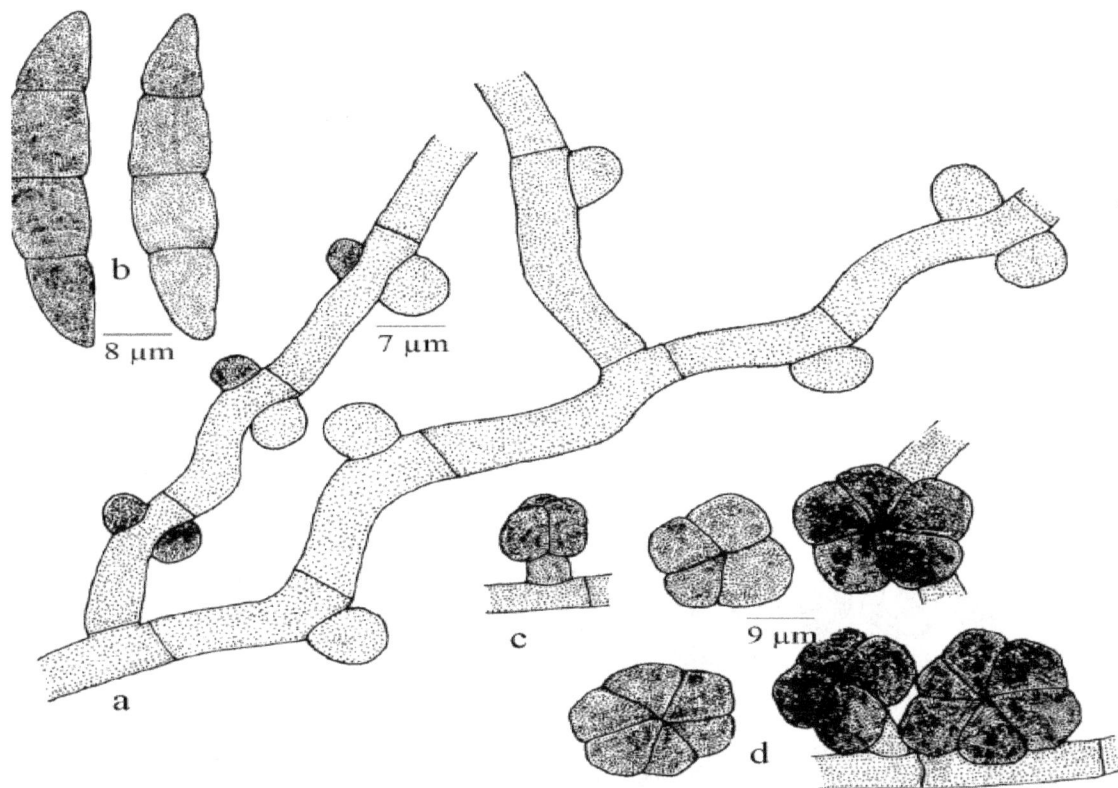

Figure 1. *Sarcinella loranthacearum* sp. nov. (a) Appressoriate mycelium, (b) Conidia of *Questieriella*, (c) *Sarcinella* conidium on conidiophore, (d) Conidia of *Sarcinella*.

leniter falcata, 3-septata, brunnea, 26-36 x 7-10 μm, parietus glabrus.

Colonies epiphyllous, dense, up to 2 mm in diameter, confluent. Hyphae brown, straight to sub-straight, branching alternate to opposite at acute to wide angles, closely reticulate, cells 16-29 × 4-7 μm. Appressoria alternate, unicellular, ovate to globose, entire, 9-12×7-10 μm. Conidiophores micronematous, mononematous, simple, straight, light brown, arise laterally from the hyphae, smooth, 9-15 × 7-10 μm; conidiogenous cells integrated, mostly terminal, monoblastic, determinate, cylindrical; conidia simple, solitary, dry, acrogenous, globose, smooth, brown to carbonaceous black, constricted at the septa, 5-8 celled, sarciniform, 24-29 μm in diameter. Conidia of *Questieriella* were present, straight to slightly falcate, 3-septate, pale brown, 26-36 × 7-10 μm, wall smooth.

Materials examined: On leaves of *Loranthus* sp. (Loranthaceae), Silent Valley National Park, Palakkad, July 12, 2008, HCIO 49041 (holotype), TBGT 3296 (isotype). The genus *Sarcinella* is reported here for the first time on the members of the family Loranthaceae (Hosagoudar, 2003; Hughes, 1987). Hence, it is accommodated in a new species.

Schiffnerula meliosmatis sp. nov. (Figure 2)

Coloniae epiphyllae, tenues, ad 2 mm diameter, confluentes. Hyphae brunneae, subrectae vel flexuosae, alternate vel irregulariter acuteque vel laxe ramosae, laxe vel arte reticulatae, cellulae 16-26 × 4-6 μm. Appressoria alternata vel unilateralis, unicellularis, globosae, ovatae, integrae, 7-10 × 7-10 μm. Conidiophorae macronematae, mononematae, simplices, rectae, leniter brunneae, glabrae, 7-10 × 6-8 μm; cellulae conidiogenae integratae, plerumque terminalis, monoblasticae, determinatae, cylindraceae; conidia simplices, solitarius, sicca, acrogena, globosa vel subglobosa, glabra, nigra, constrictus ad septatae, 5-8 cellulae, sarciniformes, 24-34 × 21-29 μm. Conidia Quesiteriella dispersa, raro hyphis affixus, leniter curvulae, 3-septata, leniter constrictus ad septata, germinans, leniter brunnea, 36-41 × 7-12 μm, parietus glabrus. Thyriothecia dispersa, orbicularis, radiatus initio, portionio ad centralis disintegratus ad maturitatae, ad 62 μm diam., cellulae marginalis radiatae; asci 1-4 per thyriotheciis, ovati vel subglobosi, octospori, 30-40 μm diam.; ascosporae oblongae, conglobatae, uniseptatae, fortiter constrictus ad septatae, brunneae, 15-25 × 7-12 μm.

Colonies epiphyllous, thin, up to 2 mm in diameter,

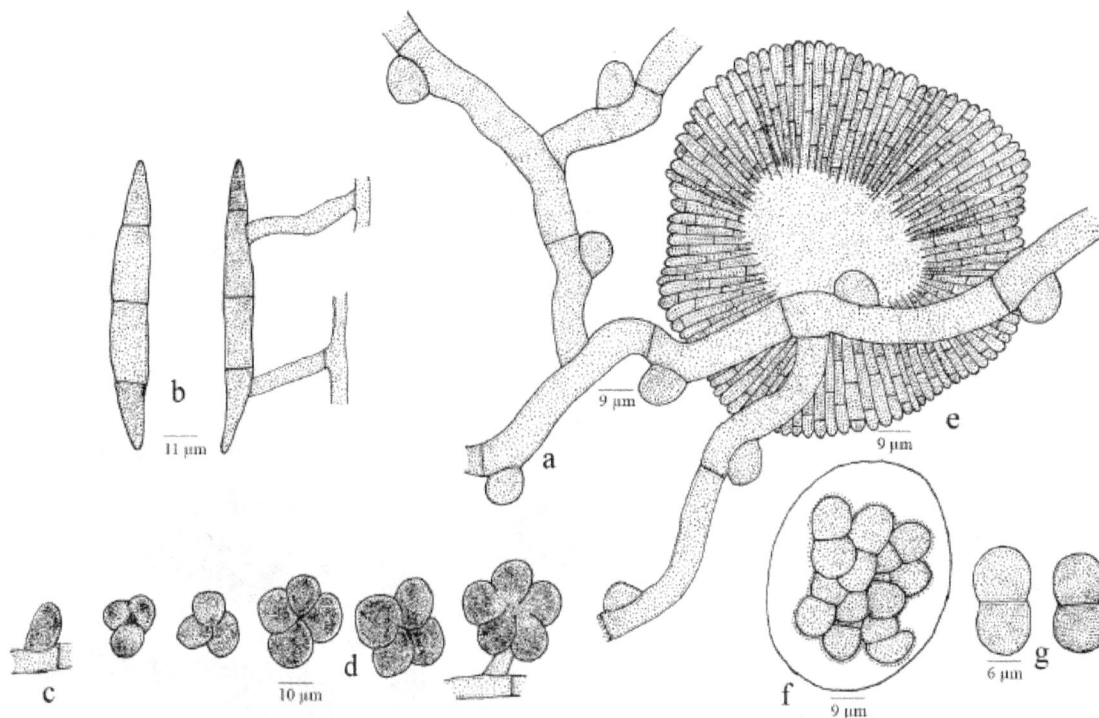

Figure 2. *Schiffnerula meliosmatis* sp. nov. (a) Appressoriate mycelium, (b) Conidia of *Questieriella*, c. *Sarcinella* conidium on conidiophore, d. Conidia of *Sarcinella*, e. Thyriothecium, f. Ascus, g.

confluent. Hyphae brown, substraight to flexuous, branching alternate to irregular at acute to wide angles, loosely to closely reticulate, cells 16-26 × 4-6 µm. Appressoria alternate to unilateral, unicellular, globose, ovate, entire, 7-10 × 7-10 µm. Conidiophores macronematous, mononematous, simple, straight, light brown, smooth, 7-10 x 6-8 µm; conidiogenous cells integrated, mostly terminal, monoblastic, determinate, cylindrical; conidia simple, solitary, dry, acrogenous, globose to subglobose, smooth, carbonaceous black, constricted at the septa, 5-8 celled, sarciniform, 24-34 × 21-29 µm. Conidia of Questieriella were scattered or attached to hyphae, slightly curved, 3-septate, slightly constricted at the septa, germinating, light brown, 36-41 × 7-12 µm, wall smooth. Thyriothecia scattered, orbicular, initially radiating, dehiscing by disintegrating the cells at the centre forming a wide opening and exposing the asci, up to 62 µm in diameter, marginal cells radiating; asci 1-4 per thyriothecia, ovate to subglobose, octosporous, 30-40 µm in diameter; ascospores oblong, conglobate, uniseptate, strongly constricted at the septum, brown, 15-25 × 7-12 µm.Material examined: On leaves of *Meliosma simplicifolia* (Roxb.) Walp. ssp. *pungens* (Wallich ex Wight and Arn.) Beus. (Sabiaceae), Near Peppara Dam, Peppara Wild life Sanctuary, Thiruvananthapuram, March 1, 2008, Jacob HCIO 48418 (holotype), TBGT 3139

(isotype). This is the first species of the genus *Schiffnerula* on the members of the family Sabiaceae (Hosagoudar, 2003; Hughes, 1987).

ACKNOWLEDGEMENTS

The authors thank Director, Tropical Botanic Garden and Research Institute, Palode, Thiruvananthapuram, Kerala for facilities and grateful to the Ministry of Environment and Forests, New Delhi for the financial support and to Forest Department, Govt. of Kerala for the forest permission.

REFERENCES

Hosagoudar VB (2003). The genus *Schiffnerula* and its synanamorphs. Zoos´ Print J., 18: 1071-1078.
Hughes SJ (1987). Pleomorphy in some hyphopodiate fungi, pp. 103-139. In: Sugiyama (ed.). Pleomorphic Fungi. The diversity and its Taxonomic Implications. Kodansha & Elswevier, Tokyo & Amsterdam.

16

Some sugar fungi in coastal sand dunes of Orissa, India

T. Panda

Department of Botany, S. N. College, Rajkanika 754 220, Kendrapara, Orissa, India.
E-mail: taranisenpanda@yahoo.co.in.

Occurrence and distribution of sugar fungi was studied from soil and leaf litter in coastal sand dunes of Orissa for a period of two years covering three distinct seasons. Fungal succession of litter was also studied. Microbial isolation and soil analysis was performed using standard procedures. Maximum population density was observed in the rainy season followed by winter and lastly summer. Higher microbial populations were encountered in plantation soil than the barren sand. They corresponded to the fluctuation of prevailing temperature, moisture and total organic carbon content of the said habitat. A total of 8 species of sugar fungi were isolated of which soil and the leaf litter had a share of 8 species each. Maximum population of sugar fungi was recorded from coastal sand dunes with *Casuarina* plantation which can be due to less competition with other fungi.

Key words: Sugar fungi, coastal sand dune, fungi, leaf litter.

INTRODUCTION

India has a rich diversity of fungi and forms an important geographical region for fungal distribution (Subramanian, 1962). The variety and galaxy of fungi not only occupy prime position in biodiversity but perform unique and indispensable activity in industry, agriculture, medicine, biogeochemical cycles (Cowan, 2001; Gates et al., 2005; Manoharchary et al., 2005) and many other ways on which other organisms including human depends. Sugar fungi, the members of Mucorales, are often the primary colonizers in a forest floor in tropical regions. They utilize the simplest carbohydrates and thereby play a pivotal role in initiation of cellulose decomposition in a soil eco-system. Though numerous species of fungi have been reported from forest soils (Behera et al., 1991; Behera and Mukherji, 1985; Mohanty and Panda, 1994) and the pattern of colonization and succession of fungi in leaf litter from different habitats have been studied by some workers (Chapela and Boddy, 1988; Mishra and Dickinson, 1984; Thomas and Ghattock, 1986). However, there appears to be no study in coastal sand dune which is considered as most unproductive and sterile habitat (Panda et al., 2007; Panda, 2009). It is especially true in case of Orissan coast with around 480 km long barren coast line filled with sand dunes only. Presently, uniculture plantations of *Casuarina equisetifolia* L. are created

along coast line to check wind blast and erosion of sand dunes. Although, it has solved the purpose to some extent, the effect of this plantation on occurrence and distribution of sugar fungi are yet to be studied. Hence a study was made with reference to sugar fungi in coastal sand dunes of Orissa.

MATERIALS AND METHODS

The study site was situated in Ganjam district of Orissa (19°15'N and 84°50'E) having 60 km of coastline along the Bay of Bengal at a height of 6 – 8 m above MSL. The unproductive uplands and coastal sand dunes are extensively covered by 30 - 40 rows of *C. equisetifolia* L. plants. Two sites of about one hectare each were selected for the present investigation. First site was on the sea shore without any vegetation and the second was along a coastal sandy bed with 6 - 8 years old uniculture plantation of *Casuarina* without any undergrowth. The study was conducted for a period of two years. Soil samples from surface and sub- surface (15 cm depth) were collected from two sites by random sampling method at monthly intervals in sterilized test tubes. Senescent leaf and three different types of litter that is, fresh litter, partially decomposed litter and highly decomposed litter were collected at monthly intervals by polythene bags. The samples were temporarily stored in an ice chest for isolation of microbes. The micro fungi were isolated by dilution plate (Waksman, 1927) and soil plate (Warcup, 1950) using PDA medium. Physico-chemical properties of soils were estimated as per Jackson (1967).

Table 1. Edaphic factors and fungal population of study site.

Sites		Temp (°c)	Moisture content (%)	pH	Total organic carbon (%)	Total nitrogen (%)	Total fungal population (10² g.dry wt.)	Total fungal species	Sugar fungi species
Site without vegetation	Surface soil	32.3	0.38	7.5	0.2	0.0108	36.47	91	6
	Sub-surface soil	30.8	0.96	7.5	0.17	0.0105	35.49	80	4
Site with Casuarina plantation	Surface soil	30.3	0.57	7.1	0.32	0.0143	41.84	78	7
	Sub-surface soil	28.9	1.21	7.4	0.24	0.0106	38.5	85	5

Average of 2 years data.

Table 2. Population of fungi and moisture content of leaf litter at site with Casuarina plantation.

Site with Casuarina plantation	Total fungal population (10³ g.dry wt.)	Moisture content (%)	Total fungal species	Sugar fungi species
Senescent leaf	340.8	50.7	81	6
Fresh litter	376	7.4	82	5
Partially decomposed litter	388.6	7.9	69	7
Highly decomposed litter.	400.8	12.3	60	7

Average of 2 years data.

RESULTS AND DISCUSSION

A comparative study on composition of soil status at two sites revealed that soil from site with Casuarina plantation had low temperature, high moisture and better nutrient status and therefore, harboured more fungi (Table 1). Micro fungi of both soils showed positive correlation with soil moisture and total organic carbon but were negatively correlated with soil temperature. The qualitative and quantitative differences of microbial population, genera and species at two sites indicated that surface vegetation as well as nutrient composition influenced micro fungal inhabitants of the soil (Mohanty et al., 1991; Panda et al., 2009). Similar results have been obtained from the soils of lower depth in all the sampling sites. Total population of fungi isolated from highly decomposed litter were more than that of the other three leaf litters (Table 2).The higher population associated with highly decomposed litter may be ascribed to the greater surface area available for microbial colonization.

The leaf surface mycoflora was richer in comparison to litter mycoflora even some species which were constantly recorded from senescent leaves never reported in highly decomposed litter (Table 3). The finding is akin to Mathur and Mukherji (1985). Moreover, the similarity in species composition between the highly decomposed litter and the soil with Casuarina plantation was found to be more akin than the soil without vegetation. The species composition in soil and leaf litter shows marked difference with change in habitat and surface vegetation (Table 4). A total of 141 species belonging to 69 genera from soil and 108 species belonging to 60 genera were isolated from leaf litter (Table 5). Species of Deutoromycotina contributed maximum followed by Zygomycotina and Ascomycotina. Their occurrence might be due to ability of the concerned group of fungi for survival in adverse condition and adjustment with the environment. Twenty four fungal species were common in all the samples. The occurrence of other species varied (Table 5). Total number of genera and species of sugar fungi isolated from soils and leaf litters (Table 6) during present study indicated that they never occur significantly at higher population levels in soils in comparison to leaf litters. It is noted that except a few genera, most of the Mucorales are never restricted to one

Table 3. Percentage contribution and ranks of some dominant fungi isolated from different samples at study sites different samples at study sites.

Fungi	Soil from site without vegetation		Soil from site with *Casuarina* plantation		Senescent leaf		Fresh litter		Partially decomposed litter		Highly decomposed litter	
	%	Rank	%	Rank	%	Rank	%	Rank	%	Rank	%	Rank
Absidia butleri	-	-	2.21	15	-	-		-	.64	19	1.76	16
A. glauca	-	-	-	-	-	-	-	-	5.3	6	10.04	03
A. spinosa	-	-	-	-	-	-	-	-	0.81	18	4.47	08
Alternaria alternata	-	-	-	-	5.05	06	3.18	07	-	-	-	-
Aspergillus awamori	8.99	2	3.21	10	9.65	03	7.96	03	12.72	02	9.22	04
A. candidus	-	-	2.1	17	-	-	-	-	-	-	-	-
A. flavus	2.04	16	6.86	03	-	-	1.06	18	.55	20	2.12	15
A. fumigates	6.95	03	3.32	09	6.99	14	2.86	08	4.83	09	3.8	10
A. nidulans	14.97	01	-	-	-	-	-	-	-	-	-	-
A. niger	5.38	5	6.09	04	4.88	07	6.68	04	9.03	05	8.86	05
A. terreus	2.76	13	3.77	7	0.78	15	2.02	12	-	-	-	-
Chaetomium homopilatum	3.36	11	-	-	-	-	-	-	-	-	-	-
Cladosporium cladosporoides	6.71	04	3.54	08	4.16	08	5.78	05	5.09	07	2.98	13
C. oxysporum	2.28	14	2.28	11	2.66	13	1.86	13	-	-	-	-
Cunninghamella verticilata	-	-	-	-	-	-	-	-	5.09	08	3.45	11
Curvularia eragrostidis	4.32	08	-	-	2.94	12	1.27	17	2.12	13	-	-
C. lunata	4.8	07	2.55	12	5.21	05	5.68	06	1.15	14	-	-
Cytosporina species	-	-	-	-	16.97	02	20.58	01	1.15	16	-	-
Cytosporella species	3.6	09	-	-	-	-	-	-	-	-	-	-
Drechslera australiensis	-	-	2.32	14	-	-	-	-	-	-	-	-
Fusarium species	-	-	5.2	06	4.16	09	1.8	14	4.37	10	6.9	06
Mucor species	-	-	-	-	-	-	-	-	-	-	4.27	09
Nigrospora sphaerica	-	-	-	-	3.05	11	0.64	20	-	-	-	-
Paecilomyces varioti	3.48	10	-	-	-	-	-	-	-	-	-	-
Penicillium citrinum	4.92	06	5.76	05	3.83	10	2.28	11	9.2	04	3.37	12
P. oxalicum	-	-	-	-	2.33	14	2.33	10	0.89	17	-	-
P. rubrum	-	-	2.43	13	-	-	-	-	-	-	-	-
P. verruculosum	2.16	15	6.98	02	-	-	1.59	15	10.64	03	11.57	02
Pestalotia species	-	-	-	-	17.14	01	15.0	02	1.31	15	-	-
Rhizopus nigricans	-	-	2.21	16	0.61	16	1.54	16	4.11	11	5.22	07
Syncephalastrum recemosum	-	-	-	-	-	-	0.95	19	2.42	12	1.57	17
Trichoderma viride	3.0	12	7.2	01	0.55	17	2.6	9	13.26	01	11.61	01

Table 4. Total count of fungi isolated during the study period.

Site	Total genera	Total species
Site without vegetation		
Surface soil	54	91
Sub surface soil	46	80
Site with *Casuarina* plantation		
Surface soil	36	78
Sub surface soil	38	85
Senescent leaf	43	81
Fresh litter	42	82
Partially decomposed litter	36	69
Highly decomposed litter	34	60

Table 5. List of fungi isolated from the study sites.

S/No.	Fungi	Soil from site without vegetation		Soil and leaf litter from site with Casuarina plantation					
		Surface	Sub surface	Surface	Sub surface	Senescent leaf	Fresh litter	Partially decomposed litter	Highly decomposed litter
1	Absidia butleri	+	+	+	+	+		+	+
2	A. glauca	+	+	+	+			+	+
3	A. spinosa	+		+	+			+	+
4	Acremonium furcatum	+	+	+		+	+	+	+
5	Alisidium resinae	+	+		+		+		+
6	Alternaria alternata	+	+			+	+	+	
7	A. padwickii	+				+	+	+	+
8	A. solani	+	+			+	+	+	+
9	Aphanocladium album		+	+	+	+	+		
10	Arachniotus terrestris	+	+	+	+	+	+	+	+
11	Arthrinium sacchari			+		+	+		
12	Aspergillus awamori	+	+	+	+	+	+	+	+
13	A. caepitosus	+	+		+				
14	A. candidus			+	+			+	+
15	A. carbonarius				+				
16	A. clavatus			+	+	+	+	+	+
17	A. fischeri			+	+				
18	A. flavus	+	+	+	+	+	+	+	+
19	Afoncecaceous				+				
20	A. fumigatus	+	+	+	+	+	+	+	+
21	A. funiculosus	+	+		+	+	+	+	
22	A. humicola			+	+				+
23	A. konongi			+	+				
24	A. luchuensis	+		+	+	+	+		
25	A. nidulans	+	+	+	+	+	+	+	
26	A. niger	+	+	+	+	+	+	+	+
27	A. quadrllineatus	+							
28	A. repens	+	+						
29	A. sparsus			+					
30	A. sulphureus			+		+	+	+	+
31	A. sydowi		+	+	+	+	+	+	+
32	A. tammari				+				
33	A. terreus	+	+	+	+	+	+	+	+

Table 5. Continues.

34	*A. terricola*	+	+	+	+	+	+	+	+
35	*A. ustus*	+	+		+	+			+
36	*A. variecolor*	+							+
37	*A. versicolor*	+	+	+	+	+	+	+	
38	*Asteromella species*	+	+			+			+
39	*Beltrania rhombica*	+	+	+	+			+	+
40	*Beltraniopsis esenbeckiae*	+	+						
41	*Bispora catenula*	+	+	+	+			+	
42	*Botryosphaeria species*	+					+		+
43	*Candida albicans*	+	+	+	+	+	+	+	+
44	*Catinula species*	+	+			+			+
45	*Cephalosporium roseogriseum*	+		+					+
46	*Cheatomium fimeti*	+							+
47	*C. funicola*	+	+		+	+			
48	*C. homopilatum*	+	+	+		+		+	
49	*C. magnum*	+	+		+				
50	*C. nigricolor*	+	+	+	+	+			+
51	*Choanephora cucurbitarum*	+	+	+			+		
52	*Chrysosporium tropicum*	+	+	+					
53	*Cladosporium cladosporoides*	+	+	+	+	+	+	+	+
54	*C. oxysporum*	+	+	+	+	+	+	+	+
55	*Cleistothecial form*	+					+		
56	*Cunninghamella verticilata*	+		+		+		+	+
57	*Curvularia clavata*	+	+		+		+	+	
58	*C. eragrostidis*	+	+	+	+	+	+	+	+
59	*C. lunata*	+	+	+	+	+	+	+	+
60	*C. lunata aeria*	+	+				+	+	
61	*C. ovoidea*	+	+	+		+	+	+	+
62	*C. pallescens*	+			+	+	+	+	
63	*C. tuberculata*	+			+	+		+	+
64	*Cytosporella species*	+	+	+	+	+	+		+
65	*Cytosporina species*	+	+	+		+	+	+	+
66	*Diplodina butleri*	+	+	+					
67	*Dichomera capparidis*	+	+	+					
68	*Drechslera australiensis*	+	+	+	+	+	+	+	+
69	*D. halodes*	+	+	+	+		+	+	
70	*D. hawaiensis*	+	+	+	+	+	+	+	+
71	*D. oryzae*	+						+	+
72	*D. iridis*	+						+	+
73	*Emericilopsis humicola*	+	+	+		+	+		
74	*Endocalyx indica*	+	+				+	+	
75	*Epicoccum nigrum*	+		+					
76	*Eurotium omstelodami*								+
77	*E. repens*	+	+		+			+	+
78	*Fusarium oxysporum*	+	+	+	+	+	+	+	+

Table 5. Continues.

		1	2	3	4	5	6	7
79	*Fusicoccum indicum*	+	+		+	+		
80	*Gilmaniella humicola*	+	+		+			
81	*Gliomastix species*			+			+	
82	*Haplosporangium accedens*	+	+		+	+	+	
83	*Humicola fuscoatra*	+		+		+	+	
84	*Isaria pulcherima*	+				+		+
85	*Lacellina graminicola*	+	+					
86	*Melanospora zamiae*	+			+	+		
87	*Monilia grisea*	+			+			+
88	*Monodictys antiqua*	+			+	+		
89	*M. fluctuata*	+						
90	*M. putredinis*		+					
91	*Mucor hiemalis*	+	+	+	+	+	+	+
92	*Myrothecium roridum*	+	+		+	+	+	+
93	*Neopeckia fulcita*		+		+	+		+
94	*Nigrospora oryzae*	+	+		+	+		
95	*N. sacchari*	+	+		+	+		
96	*N. sphaerica*	+	+	+	+	+	+	+
97	*Oidiodendron kalari*	+	+	+	+	+		
98	*Paecilomyces varioti*	+			+	+		+
99	*Penicillium adametezi*							
100	*P. citrinum*	+	+		+	+	+	+
101	*P. chermesinum*		+					
102	*P. chrysogenum*		+		+	+		
103	*P. cyaneum*		+					
104	*P. decumdens*	+	+	+	+	+	+	
105	*P. expansum*	+		+	+	+	+	
106	*P. fellutatum*				+	+		
107	*P. glabrum*		+					
108	*P. granulatum*							
109	*P. islandicum*	+	+	+	+	+	+	
110	*P. javanicum*	+	+	+	+	+	+	
111	*P. lanosum*	+	+	+	+	+	+	+
112	*P. minio-luteum*	+	+	+	+	+	+	+
113	*P. nigricans*	+	+	+	+	+	+	+
114	*P. oxalicum*	+			+		+	+
115	*P. purpurogenum*	+	+	+	+			
116	*P. resticulosum*	+	+	+				
117	*P. roseo-purpureum*	+	+	+	+	+	+	+
118	*P. rubrum*	+	+	+	+	+	+	+
129	*P. rugulosum*	+		+	+	+		
120	*P. variable*			+				
121	*P. verruculosum*	+	+	+	+	+	+	+
122	*Periconia byssoides*	+			+		+	+
123	*P. digitata*			+				

Table 5. Continues.

Table 5. Continues.

	Species						
124	*Pithomyces sacchari*	+					
125	*Pestalotia species*	+	+	+	+	+	+
126	*Phialophorophoma species*	+			+		
127	*Phoma species*	+	+	+	+	+	+
128	*Phylosticta acetosa*	+	+	+	+	+	
129	*Pyrenochaeta cajani*	+	+	+	+	+	+
130	*Pyronomella species*	+	+	+	+	+	
131	*Rhizopus nigricans*	+	+	+	+	+	+
132	*Rhynchophoma raduloid*			+	+		
133	*Scopulariopsis brumptii*	+	+	+	+	+	+
134	*Spegazzinia ornata*		+	+		+	
135	*Staphylotrichum coccosporium*	+	+	+	+	+	
136	*Syncephalastrum recemosum*	+	+	+	+	+	+
137	*Theilavia terricola*	+	+	+	+	+	+
138	*Theilaviopsis paradoxa*	+	+	+	+	+	
139	*Torula calligans*	+	+	+	+	+	+
140	*T. herbarum*	+	+	+	+		
141	*Trichocladium opacum*		+		+	+	+
142	*Trichoderma album*			+			
143	*T. koningi*	+	+	+	+		
144	*T. viride*	+	+	+	+	+	+
145	*Verticillium dahliae*	+	+	+	+		
146	*Black sterile*	+	+	+	+	+	+
147	*White sterile*	+	+	+	+	+	+

Table 6. Sugar fungi isolated from different sites.

Ascogenous fungi	Soil from site without vegetation		Soil and leaf litter from site with *Casuarina* plantation					
	Surface	Sub surface	Surface	Sub surface	Senescent leaf	Fresh litter	Partially decomposed litter	Highly decomposed litter
Absidia butleri	+	+	+	+	+	-	+	+
A. glauca	-	+	+	+	-	-	+	+
A. spinosa	-	-	+	-	+	+	+	-
Choanephora cucurbitarum	**+**	-	-	-	+	-	-	-
Cunninghamella verticilata	**+**	-	**+**	-	-	+	+	+
Mucor hiemalis	+	-	+	+	+	+	+	+
Rhizopus nigricans	+	+	+	+	+	+	+	+
Syncephalastrum recemosum	+	+	+	+	+	+	+	+

or neither the other samples nor they are common to all. This corroborates to the findings of Behera and Mukherji (1985). Their isolation is found to be dependant more on final growth and formation of mature colony in the culture plate than on the technique employed for the purpose. During the present study, it was observed that more varieties and higher population of Mucorales were recorded in partially decomposed litter and highly decomposed litter than the other samples. It may be due to low competition with other categories of fungi which are less abundant in this soil compared to unproductive virgin coastal sand dunes. Moreover, less number of sugar fungi was recorded in the present study looking their large varieties in tropical forest soils (Mohanty and Panda, 1998; Behera and Mukherji, 1985).

It can be concluded that species diversity of micro fungi was related to the particular habitats and ecosystem. Overall micro fungal diversity seems to be higher in unproductive coastal sand dunes without vegetation in spite of its low nutrient status which can be due to low competition with other categories of fungi. From the present study, it is clear that edaphic factors greatly influence the growth and development of micro fungi. It is suggested that a large number of permutations and combinations of media and technique should be employed to unravel the innumerable sugar fungi still unreported in coastal soils of Orissa.

REFERENCES

Behera N, Mukherji KG (1985). Seasonal variation and distribution of micro fungi in forest soils of Delhi. Folia Geo. Bot. Et. Phyto., 20: 291-312.

Behera N, Pati DP, Basu S (1991). Ecological study of soil micro fungi in a tropical forest soil of Orissa, India. Trop. Ecol., 32(1): 136-143.

Chapela IH, Boddy L (1988). The fate of early fungal colonizers in beech branches decomposing on the forest floor. Fems Microb. Ecol., 53: 273- 284.

Cowan A (2001). Fungi-life support for ecosystems. Essential ARB., 4: 1-5.

Gates GM, Ratkowsky DA, Grove SJ (2005). A comparison of macrofungi in young Silvicultural regeneration and mature forest at the Warra LTER siet in the southern forests of Tasmania. Tasforests. 16: 127-134.

Jackson ML (1967). Soil chemical analysis. Prentice Hall Pvt. Ltd. New Delhi., pp. 215-224.

Manoharchary C, Sridhar K, Singh R, Adholeya A, Rawat S, Johri BN (2005). Fungal biodiversity, distribution, conservation and prospecting of fungi from India. Cur. Sci., 89(1): 59-70.

Mathur, Mukherji KG (1985). Phylloplane fungi of *Crotolaria juncea* during growth and decomposing leaves. Ind. Phytopathol., 38:683-687.

Mishra RR, Dickinson CH (1984). Experimental studies on phylloplane and litter fungi on *Ilex equifolium* .Trans. Bri. Mycol. Soc., 82: 595-604.

Mohanty RB, Panda T, Pani PK (I991). Seasonal variation and distribution of microfungi in a tropical forest soil of south Orissa. J. Ind. Bot. Soc., 70: 267-271.

Mohanty RB, Panda T (1994). Ecological studies of the soil microfungi in a tropical forest of South Orissa in relation to deforestation and cultivation. J. Ind. Bot. Soc., 73: 213- 216.

Mohanty RB, Panda T (1998). Studies on the impact of deforestation and cultivation on the incidence of sugar fungi in a tropical forest soil of south Orissa, India. Trop. Ecol., 39(1): 149-150.

Panda T (2009). Diversity of sac fungi in coastal sand dunes of Orissa. J. Mycol. Pl. Pathol., 39(1): 94-98.

Panda T, Panda B, Mishra N (2007). A comparative study of Penicillia from soil, leaf, litter and air in a coastal sandy belt of Orissa. J. Phytol. Res., 20(2): 335-336. .

Panda T, Panda B, Prasad BK, Mishra N (2009). Influence of soil environment and surface vegetation on soil micro flora in a coastal sandy belt of Orissa, India. J. Hum. Ecol., 27(1): 69-73.

Subramanian CV (1962). The classification of hyphomycetes. Bull. Bot. Sur. Ind., 4: 249-259.

Thomas MR, Ghattock RC (1986). Filamentous fungal associations in the phylloplane of *Lolocum perenne*. Trans. Bri. Mycol. Soc., 87: 255-268.

Waksman SA (1927). Principles of soil microbiology. Williams and Willikins Co. Baltimore, p. 897.

Warcup JH (1950). The soil plate method for isolation of fungi from soil. Nature 166: 117-118.

17

Alternaria jacinthicola, a new fungal species causing blight leaf disease on water hyacinth [*Eichhornia crassipes* (Martius) Solms-Laubach]

Karim Dagno[1]*, Julien Crovadore[2], François Lefort[2], Rachid Lahlali[3], LudivineLassois[1] and M. Haïssam Jijakli[1]

[1]Unit of Plant Pathology, University of Liege, Gembloux Agro Bio Tech, Passage des Déportés 2, B-5030 Belgium.
[2]Plants and Pathogens Group, Institute Earth Nature and Landscape, University of Applied Sciences of Western Switzerland, 150 route de Presinge, Jussy, Geneva 1254, Switzerland.
[3]Agriculture and Agri-Food Canada, Saskatoon Research Centre, 107 Science Place, S7N0X2, Canada.

Water hyacinth (*Eichhornia crassipes*) causes environmental, agricultural and health problems in Mali. This is particularly severe in the District of Bamako and the irrigation systems of the "Office du Niger" area. During two years survey for fungal pathogens of water hyacinth infested areas, isolate Mlb684 was collected from diseased plant. This fungal isolate was identified as a potential mycoherbicide for sustainable management for water hyacinth. The aim of this study was to characterize isolate Mlb684. The characterization was based on a morphological description and a DNA sequence analysis. Various genes amplified from isolate Mlb684 were compared to those existing in Genbank. These genes were 18S ribosomal rDNA gene, ITS rDNA gene, elongation factor-1 alpha (EF1a) gene, calmodulin and actin genes. DNA sequence comparisons and morphological description provided enough evidences to show that isolate Mlb684 belonged to the *Alternaria* genus and was distinct from any other known *Alternaria* species. Based on these evidences, the new fungal isolate was called "*Alternaria jacinthicola* Dagno & M.H. Jijakli". A specimen culture has been deposited in the Gembloux Agro Bio Tech Plant Pathology unit fungal collection, with Mlb684 reference and in the Industrial Fungal and Yeast Collection (BCCM/MUCL, Belgium) under the accession number: MUCL 53159 and all DNA sequences were deposited in GenBank (NCBI).

Key words: ITS rDNA, 18S rDNA, actin, calmodulin, elongation factor, genetic characterization, *Alternaria jacinthicola, Eichhornia crassipes,* water hyacinth.

INTRODUCTION

Water hyacinth (*Eichhornia crassipes*) has spread throughout Africa causing widespread problems to millions of users of water bodies and water resources. The plant affects irrigation, water flow, water use, and navigation; it also poses a health risk by enabling the breeding of mosquitos, bilharzias, and other human parasites (Dagno et al., 2007). Biocontrol has been considered as the most adequate control strategy against water hyacinth (Charudattan, 2005). Among possibilities

offered by the biocontrol management, fungal pathogens could be an efficient control tool against this aquatic weed. Several research groups have identified promising microbial fungal agents notably *Alternaria eichhorniae* Nag Raj & Ponnappa and *Alternaria alternata* (Fr.) Keissler that might be developed and used as mycoherbicides (El-Morsy, 2006; Babu et al., 2002; Shabana, 1997). Research on fungal pathogens of water hyacinth began in 2006 in Mali and led to the isolation and identification of *Gibberella sacchari* Summerell & J.F. Leslie (isolate Mln799), *Cadophora malorum* (Kidd & Beaumont) W. Grams (isolate Mln715) and isolate Mlb684 (Dagno et al., 2011a). Among the 3 selected fungi, an unusual fungus, isolate Mlb684 applied in

*Corresponding author. E-mail: karimdagno@yahoo.fr.

Table 1. PCR primers for amplification of genes of *Alternaria* sp. isolate Mlb684.

Primers names	Sequences
[1]ITS1	5'TCCGTAGGTGAACCTGCGG3'
[1]ITS4	5'TCCTCCGCTTATTGATATGC3'
[1]ITS5	5'GGAAGTAAAAGTCGTAACAAG3'
[1]NS1	5'GTAGTCATATGCTTGTCTC3'
[1]NS2	5'GGCTGCTGGCACCAG.....TGC3'
[2]EF1-728F	5'CATCGAGAAGTTCGAGAAGG3'
[3]TEF1LLErev	5'AATTTGCAGGCAATGTGG3'
[2]CAL-228	5'GAGTTCAAGGAGGCCTTCTCCC3'
[2]CAL-737R	5'CATCTTTCTGGCCATCATGG3'
[2]ACT-783R	5'TACGAGTCCTTCTGGCCCAT3'
[2]ACT-512F	5'ATGTGCAAGGCCGGTTTCGC3'

[1]White et al. (1990), [2]Carbone and Kohn (1999), [3]Jaklitsch et al. (2005).

unrefined *Carapa procera* (L) oil and refined palm oils caused 87 to 90% of disease severity on water hyacinth 6 weeks after treatment respectively (Dagno et al., 2011b).

In 2007, the Industrial Fungal and Yeast Collection (BCCM/MUCL, Belgium) identified isolate Mlb684 as *Alternaria* sp. with reference BCCM/MUCL DIV/07-119C. However, morphological characters displayed by this isolate including the pattern and sporulation structure of this isolate hardly match with those currently used to describe known species in the genus *Alternaria* (E.G.S. characteristics, as described in the literature: Simmons, 2004; Simmons, 1999; Simmons and Roberts, 1993). It was therefore assumed that this isolate could belong to a novel species of *Alternaria*. Two *Alternaria* species were until now reported on water hyacinth. *A. eichhorniae* and *A. alternata* were recognized as virulent pathogens on *E. crassipes* species. They are best known as the causal agent of leaf blight disease on water hyacinth in Egypt and India (Shabana et al., 1995; Aneja and Sing, 1989).

The present study was designed to provide a taxonomic position of the isolate Mlb684 at the species level, using morphological characterization and DNA sequence comparisons.

MATERIALS AND METHODS

Fungal isolation and specimen collection

Infected parts of water hyacinth (petioles and leaves) were collected from the River Niger in the District of Bamako, the lake of Sebougou in Segou, and the central collector of Niono with GPS coordinates "12° 40' N, 7° 59' W" ; "13° 26' N, 6° 15' W" and "14° 15' N, 5° 59' W" respectively. Samples in clean plastic bags were brought to the laboratory and then stored at 4°C until examined. Stored plant parts were scrubbed under running water to remove surface debris, dissected into small segments (approximately 1 × 1 cm), and surface-sterilized by sequential immersion in 96% ethanol for 30 s and then in 14% hypochlorite for 30 s. The segments were rinsed in sterilised water for 1 min. Surface-sterilised segments (4

segments/plate) were plated on potato dextrose agar (PDA, Merck, Darmstadt, Germany) supplemented with 5 ppm streptomycin. Three plates were used for petiole and limb of water hyacinth plant. The plates were then incubated at 25°C for 5 to 7 days. Fungal mycelia that emerged from the plant fragments were isolated and pure cultures were obtained by the single-spore technique, cultures were then preserved at 4°C, for no more than 6 months, before use. Isolate Mlb684 was selected during a previous assay conducted to identify potential mycoherbicide for water hyacinth (Dagno et al., 2011a, b). Pure cultures were deposited in the Culture Collection of the Phytopathology Unit, Gembloux Agro Bio Tech (GxABT), University of Liege, Belgium. Duplicates of key isolates (specimens) were also deposited to the Industrial Fungal and Yeast Collection (BCCM[TM]/MUCL - Louvain-la-Neuve, Belgium). For mass production, isolate Mlb684 was incubated on V8 agar during 2 weeks at 25°C and 16 h photoperiod.

Molecular analysis

A liquid culture from the isolate Mlb684 was performed in 100 ml of potato dextrose broth (PDB) in a 250 ml Erlenmeyer for 5 days. Pure DNA was obtained from the resulting culture using a quick DNA extraction method (Lefort and Douglas, 1999). Universal oligonucleotide primers (Table 1) targeting 5 fungal genes were used for PCR amplification. Primers ITS1, ITS4 and ITS5 (White et al., 1990) for the 18S and 28S rDNA sequence; primers NS1 and NS2 (White et al., 1990) for a partial 18S rDNA sequence; primers EF1-728F (Carbone and Kohn, 1999) and TEF1LLEre (Jaklitsch et al., 2005) for the elongation factor-1 alpha (EF1a); primers CAL-228 and CAL-737R (Carbone and Kohn, 1999) for calmodulin gene; and primers ACT-783R and ACT-512F (Carbone and Kohn, 1999) for actin gene. PCR was carried out using the KAPA2G Robust PCR (KappaBiosystems, Japan) and each PCR amplification was performed in 20 µl reaction mixture consisting of 10 µl Maxima Sybr Green qPCR Master Mix 2X (Fermentas), 2 µl each of the forward and reverse primers (10 µM), 1 µl cDNA template (1 ng/µl), and 5 µl PCR-grade water. The cycling conditions were: pre-incubation for 10 min at 95°C, followed by 40 cycles, each consisting of 30 s denaturing at 95°C, 40s annealing at 52°C, and 45 s elongation at 72°C, the last cycle ending with a final 10-min extension at 72°C.

Total genomic DNA was extracted according to Lefort and Douglas (1999), and the concentration of the resulting DNA was determined with an ND-1000 UV/Vis spectrometer (NanoDrop Technologies, Wilmington, DE USA) version 3.1.0. Oligonucleotide primers (Table 1) were used to amplify and sequence the internal transcribed spacer (ITS) regions (including the partial 18 S and 28 S DNA genes) and regions corresponding to the genes encoding elongation factor-1 alpha (EF1a), calmodulin, and actin. DNA sequences were edited by Fasteris SA, Geneva, Switzerland (Lefort and Douglas, 1999). Resulting DNA sequences were deposited in GenBank (NCBI, Bethesda, MD, USA) and compared using the similarity search tool Blast.

DNA sequences recovered from GenBank for species close to the *Alternaria* isolate Mlb684 were aligned by ClustalW2 (Chenna et al., 2003) and used to generate molecular phylogenies with an optimal neighbour-joining method (Myers and Miller, 1988). Trees were then drawn with the software Jalview (Waterhouse et al., 2009), in addition, a Bayesian analysis was run to this one.

Morphology

Isolate Mlb684 were grown on V8 agar and potato-carrot agar mediums under strictly defined incubation conditions (Simmons, 1992) and examined for characteristics of the sporulation apparatus and conidium morphology to confirm species identity and compare morphological characters. Morphological description was performed

Table 2. Comparisons of DNA sequences between *Alternaria* sp. isolate Mlb684 and the closest fungal for which sequences were available in Genbank.

Fragment size	Species and sources	Genbank accession numbers	Homology percentage
ITS rDNA	*Alternaria* sp. (isolate MUCL 45333)	AY714488	100
	Alternaria sp. (isolate IA2448)	AY154699	100
	Alternaria sp. (isolate IA249)	AY154698	100
18S rDNA	*A. japonica*	U05199	100
	A. alternate	U05194	100
	A. brassicicola	U05197	100
EF1a	*Alternaria* sp. (isolate CBS 174.52)	DQ677911	98
	A. alternata (isolate AFTOL-ID 1610)	DQ677927	98
cmdA*repentis*	*Pyrenophora tritici- repentis* -(isolate Pt-1C-BFP)	XM00194109	97
act	*A. alternata*	GQ240307	97
	A. carotiincultae (isolates BMP0129)	EU141969	97
	A. carotiincultae (isolate BMP0095)	EU141972	97
	A. carotiincultae (isolate BMP0132)	EU141968	97
	A. radicina (isolate BMP0047)	EU141973	91
	A. radicina (isolate BMP0062)	EU141972	91
	A. radicina (isolate BMP0079)	EU141971	91

at 50x magnifications using 7 to 14 day-old cultures. It was based on observations concerning colony growth, color, type of mycelia, size, and form arrangement of conidia. Identification key E.G.S. 00.000 was used to record conidia of this fungal isolate as described by Simmons and Roberts (1993).

RESULTS

Molecular analysis

PCR resulted in the successful amplification of ITS rDNA (541 bp) gene using primers ITS1 and ITS4, 18S DNA (527 pb) gene with primers NS1 and NS2, elongation factor 1-alpha (1182 bp) gene with primers EF1-728F and TEF1LLErev, calmodulin (1182 bp) gene with primers CAL-228 and CAL-737R and actin (240 bp) gene with primers ACT-783R and ACT-512F for isolate Mlb684. All sequences determined in this study have been submitted to GenBank. Based on DNA sequences of the 5 studied genes, sequence comparisons were performed between *Alternaria* sp. isolate Mlb684 and the closest species or isolates for which sequences were available. Comparison of the ITS rDNA gene showed that *Alternaria* sp. isolate Mlb684 was 100% identical to 3 other *Alternaria* isolates (Table 2). *Alternaria* sp. isolate MUCL 45333 is pathogen on wheat crop, in opposite, *Alternaria* sp. isolates IA2448 and IA249 that are reported infect regularly *Hylocereus undatus* fruits in Iran.

Concerning the 18S rDNA gene, *Alternaria* sp. isolate

M1b684 showed 100% identity with 12 microorganisms among them 3 *Alternaria* species. *Alternaria japonica* and *A. alternata* isolated from disease plants of *Brassica rapa* ssp. *oleifera*, indeed, *Alternaria brassicicola* isolated from infected *Brassica oleracea* ssp. *capitata* in Canada. Similar to the ITS rDNA gene, the 18S rDNA gene sequence of *Alternaria* sp. isolate M1b684 showed than a 98 to 99% identity with those of 72 other *Alternaria* or unknown cultured fungi isolates making.

Comparing elongation factor 1-alpha gene sequences yielded 2 related *Alternaria* species (Table 2). *Alternaria* sp. isolate CBS 174.52 showed 98% identity along 64% of its sequence with *Alternaria* sp. isolate M1b684 while *A. alternata* isolate AFTOL-ID 1610 shared 98% identity along 61% of its sequence *Alternaria* sp. isolate M1b684.

Comparing the calmodulin gene sequence of *Alternaria* sp. isolate M1b684 to GenBank sequences yielded no *Alternaria* species sequences and the closest microorganism was the *Pyrenophora tritici-repentis* isolate Pt-1C-BFP. This fungal species shared 97% identity over 42% of its sequence. Finally when comparing the actin gene sequence of *Alternaria* sp. isolate M1b684 to GenBank sequences yielded 7 *Alternaria* species sequences ranging from 97% identity over 84% of its sequence to 91% identity along 83% of its sequence when compared. Table 2 illustrated these species. All isolates of *Alternaria carotiincultae* and *Alternaria radicina* were isolated from infected carrots. An ITS sequence (GenBank accession EU314716) for the

ITS rDNA gene of *A. eichhorniae* shared 97% of its sequence with the one of *Alternaria* sp. isolate M1b684 (GenBank accession HQ413695).

Figure 1A showed the genetic relationships between *Alternaria* sp. isolate M1b684 and the closest microorganisms for this elongation factor 1-alpha gene sequence, where it appears that it is quite distinct from the closest organisms. Genetic relationships are shown between closest calmodulin and actin gene sequences are shown respectively on Figure 1B and C. We have described phylogenetic relationships among *Alternaria* sp. isolate Mlb684 and *Alternaria* genera or isolate available in Genbank, based on sequences from five different genetic regions, ITS rDNA gene, 18S rDNA gene, EF1a gene, calmodulin gene and actin gene. ITS rDNA sequence analysis of *Alternaria* sp. isolate M1b684 presented 99% of identity to those of 93 other fungi isolates. Indeed, the 18S rDNA gene sequence of same isolate M1b684 showed also than a 98 to 99% identity with those of 72 other *Alternaria* or unknown cultured fungi isolates. In order to confirm the result obtained in phylogenies studies, a Bayesian analysis was run to this one.

Morphology

It produced an ash colony on V-8 agar and was well sporulated 2 weeks after incubation at 25°C with 16 photoperiods. Colony and conidia were easily recognizable as belonging to the *Alternaria* genus with relatively small and short conidia on branching chains. Similar descriptions were given for several species of *Alternaria* on V8 agar medium by Simmons and Roberts (1993) and Simmons (1999, 2004). Conidia of *Alternaria* sp. isolate Mlb684 are short ellipsoid to oval, tapering in the lower half into a narrow tail extension. The upper part which was materialized by a very short beak well rounded ending abruptly appears allowing the formation of new spores, thus furnishing evidence of catenulation. Primary conidiophores of *Alternaria* sp. isolate Mlb684 arise directly from hyphae at the V-8 agar surface; they can be simple or branched.

Mycelium is septate and the conidia (Figure 2A and B) are variable in size and shape, but most often short and ellipsoid to oval, tapering in the lower half. Sometimes a narrow tail extension is visible. The upper part bears a beak, but it is very short and rounded; catenulation is frequent. To our knowledge the sporulation pattern observed here, characterised by an unusually high percentage of relatively small conidia produced in non-disjunct series, and has not been observed previously in any *Alternaria* species (Figure 2C). The spores are often well formed, with septa. Most of them have a smooth wall like those of *Alternaria sesame*, *Alternaria sesamicola*, and *Alternaria simsimi* described by Simmons (2004). In cultures on potato-carrot agar, conidia E.G.S. 9-28 (-32) x

12-15 µm in size, with transverse septa and at least one longitudinal septum, were observed.

Taxonomy

Comparisons of DNA sequences, elongation factor 1-alpha gene, calmodulin gene and actin gene and morphology of isolate Mlb684 from Mali in the *Alternaria* genus revealed that the fungus from Mali represents a previously undescribed species in the genus. A new species are described as follows:

Alternaria jacinthicola Dagno & MH. Jijakli sp. nov. :Ex Colonies are described on V-8 agar. Conidiophores are abundant, heaped reaching a size range up to ca. 70 x 2-4 µm, geniculate apical growth extensions. Conidia are catenulate and produced in chain. Beak often very short rounded pouring out of the body from the conidium. Conidium bodies reach a size range of 9 to 28 (-32) x 12 to 15 µm, with 3 to 7 constricting transepta and 1 or 2 longisepta in 1 to 2 transverse sections of narrow conidia. Conidia have a smooth wall. Isolated from disease *E. crassipes* plant in Mali (Dagno et al., 2011a). Ex - Cult. Typ. BCCMTM/MUCL = MUCL 53159

Etymology- referring to the original host plant, water hyacinth. Description and observations on the ex-type culture are based on the isolate Mlb684, which was derived from infected water hyacinth plant. The material was collected in 2006 in Mali. Colony growth on V-8 and PCA is rapid, completely covering individual sectors of 90 mm diameter Petri dish within 2 weeks. Sporulation is dense on V-8, only slightly less so on PCA. Concentric rings of sporulation are evident. The conidia are obclavate (shaped like a bowling pin) and form single file chains. The spores have both 1 or 2 longitudinal and horizontal septae. Each conidium tapers into a narrow rounded protuberance. Conidium bodies reach a size range of 9 to 28 (-32) × 12 to 15 µm, with 3 to 7 constricting transepta and 1 or 2 longiseptum in 1 to 2 transverse sections of narrow conidia.

DISCUSSION

Two species of *Alternaria* genus (*A. eichhorniae* and *A. alternata*) were previously described as pathogenic fungi on water hyacinth (Aneja et al., 1989; Nag Raj and Ponnappa, 1970). This study records the discovery of an unknown species in the *Alternaria* genus. Comparisons of DNA sequences, elongation factor 1-alpha gene, calmodulin and actin of isolate Mlb684 and all *Alternaria* DNA sequences and other DNA sequences existing in GenBank, suggest that this fungus represents a new taxon, for which the name *A. jacinthicola* was provided. To date, no further examination of molecular relationships among *Alternaria* sp. isolate Mlb684 and other genera

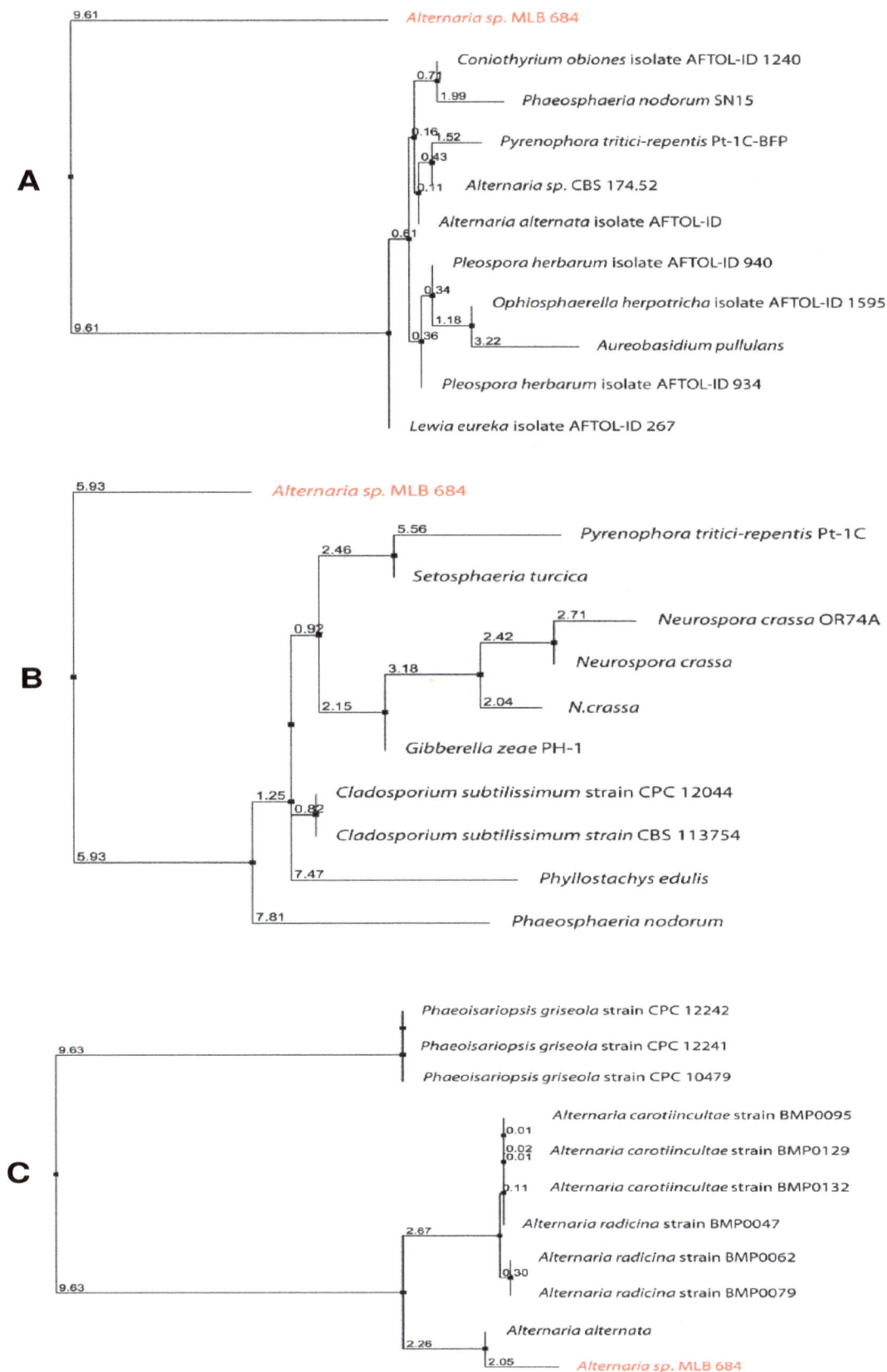

Figure 1. Molecular phylogenies of *Alternaria* sp. isolate Mlb684 based on 3 amplified and sequenced DNA regions. A) partial sequence of Elongation factor-1 alpha (EF1a) gene, B) partial sequence calmodulin gene, C) partial sequence of actin gene.

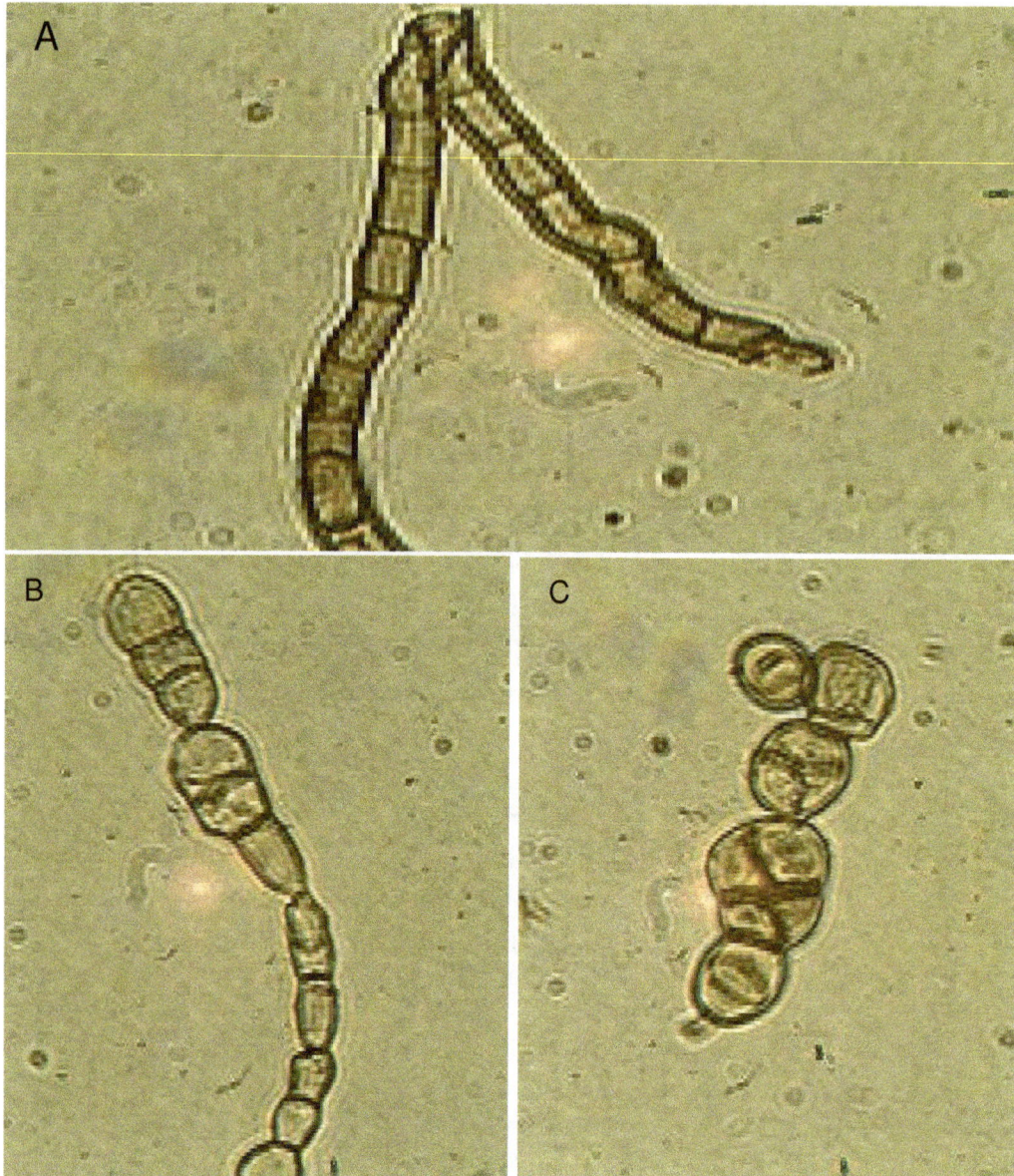

Figure 2. *Alternaria jacinthicola.* Septa mycelia (A) Conidiophore, (B) and Arrangement of conidia (C) and ex representatives of isolate Mlb684; from development of foliar disease and colony developed on V8 agar (100x magnification).

has been explored. Moreover, no analysis of this isolate Mlb684 phylogeny has ever been conducted. This work provides the first systematic examination of isolate Mlb684 as they relate to the hypothesized related taxa of *Alternaria* (Simmons, 1992). Moreover, it should be noted that some *Alternaria* species were already shown to infect water hyacinth and have been assessed as potential biocontrol agents against this weed. These are *A. eichhorniae* in Egypt (Shabana, 2005) and *A. alternata* in India (El-Morsy et al., 2006; Babu et al., 2003) which caused severe disease on the plant in greenhouse test conditions. Genetic comparisons of *Alternaria* sp. isolate Mlb684 with DNA sequences from these *Alternaria*

isolates had not been possible in absence of genetic characterization of these organisms, at the exception of one sequence of *A. eichhorniae*. Additionally, isolate of *A. eichhorniae* in GenBank is not the one originally described by Shabana (1995) but by Nag Raj and Ponnappa on water hyacinth in India (Nag Raj and Ponnappa, 1970).

The objective of this phylogenetic study is to examine the relationships among *Alternaria* sp. isolate Mlb684 and the *Alternaria* genera and the closest species or isolates based on mitochondrial rDNA sequences for which sequences were available in Genbank. Results of Bayesian analysis were confirmed by the phylogenetic data obtained in this study. Because ITS sequence of *Alternaria* sp. isolate Mlb684 was then 99% identical to those of 93 other isolates, making it not pertinent to build a molecular phylogeny. In addition, the 18S rDNA gene sequence of *Alternaria* sp. isolate Mlb684 showed also than a 98 to 99% identity with those of 72 other *Alternaria* or unknown cultured fungi isolates making so it not pertinent to build his molecular phylogeny. For morphological description, the taxon of isolate Mlb684 does not appear to be one that is identifiable with those currently recognized by E.G.S. as described in the literature (Simmons, 1999, 2004; Simmons and Roberts, 1993). Genetic results confirmed that the fungal isolate Mlb684 belonged to the genus *Alternaria* and was distinct from any *Alternaria* species and isolates which had been previously characterised.

ACKNOWLEDGMENTS

This study is part of the PhD thesis presented to Gembloux Agro-Bio-Tech, University of Liege, Belgium. We are grateful to Dr. E.G. Simmons (Crawfordsville, U.S.A) for his opinion about isolate Mlb684 and Dr. C. Decock (BCCM™/MUCL, Belgium) for the first identification of the isolate Mlb684 as *Alternaria* sp.

REFERENCES

Aneja KR, Singh K (1989). *Alternaria alternata* (Fr.) Keissler a pathogen of water hyacinth with biocontrol potential. Trop. Pest Manage., 35: 354-356.

Babu MR, Sajeena A, Seetharaman K, Vidhyasekaran P, Rangasamy P, Som PM, Senthil RA, Biji KR (2002). Host range of *Alternaria alternate,* a potential fungal biocontrol agents for water hyacinth in India. Crop Prot., 22: 1005-1013.

Babu MR, Sajeena A, Seetharaman K, Vidhyasekaran P, Rangasamy P, Som PM, Senthil RA, Ebenezer G (2003). *Alternaria alternata* toxin detection by fluorescence derivatization and separation by high performance liquid chromatography. Phytoparasitica, 31: 61-68.

Carbone I, Kohn LM (1999). A Method for Designing Primer Sets for Speciation Studies in Filamentous Ascomycetes. Mycology, 91: 553-556.

Charudattan R (2005). Ecological, practical and political inputs into selection of weed targets: what makes a good biological control target? Biol. Cont., 35: 183-196.

Chenna R, Sugawara H, Koike T, Lopez R, Gibson TJ, Higgins DG, Thompson JD (2003). Multiple sequence alignment with the Clustal series of programs. Nucleic Acids Res., 31:3497-3500.

Dagno K, Diourté M, Lahlali R, Jijakli MH (2011a). Fungi occurring on water hyacinth [Eichhornia crassipes (Martius) Solms-Laubach] in Niger River in Mali and their evaluation as mycoherbicides has been accepted by the J. Aquat. Plant Manag for the January 2011 issue.

Dagno K, Diourté M, Lahlali R, Jijakli MH (2011b). Production and oil-emulsion formulation of *Cadophora malorum* and *Alternaria jacinthicola,* two biocontrol agents against water hyacinth (*Eichhornia crassipes*). Afr. J. Microbiol. Res., 5: 924-929.

Dagno K, Lahlali R, Friel D, Bajji M, Jijakli MH (2007). Synthèse bibliographique : problématique de la Jacinthe d'eau, *Eichhornia crassipes* dans les régions tropicales et subtropicales du monde, notamment son éradication par la lutte biologique au moyen des phytopathogènes. Biotechnol. Agron. Soc. Environ., 11: 299-311.

El-Morsy EM, Dohlob SM, Hyde KD (2006). Diversity of *Alternaria alternata* a common destructive pathogen of *Eichhornia crassipes* in Egypt and its potential use in biological control. Fungal Divers., 23: 139-158.

Jaklitsch WM, Komon M, Kubicek CP, Druzhinina IS (2005). *Hypocrea voglmayrii* sp. nov. from the Austrian Alps represents a new phylogenetic clade. Hypocrea/ Trichoderma. Mycology, 97:1391-1404.

Lefort F, Douglas GC (1999). An efficient micro-method of DNA isolation from mature leaves of four hardwood tree species Acer, Fraxinus, Prunus and Quercus. Ann. For. Sci., 56: 259-263.

Myers EW, Miller W (1988) Optimal alignments in linear space. Comput. Applic. Biosci., 4: 11-17.

Nag Raj TR, Ponnappa KM (1970). Blight of water hyacinth caused by *Alternaria eichhorniae* sp.nov. Trans. Br Mycol. Soc., 55: 123-130.

Shabana YM (2005). The use of oil emulsions for improvising the efficacy of *Alternaria eichhorniae* as a mycoherbicide for water hyacinth (*Eichhornia crassipes*). Biol. Cont., 32: 78-89.

Shabana YM (1997). Formulation of *Alternaria eichhorniae*, a mycoherbidide for water hyacinth, in invert emulsions averts dew dependence. J. Plant Dis. Prot., 104: 231-238.

Shabana YM., Charudattan R, Elwakil MA (1995). First record of *Alternaria eichhorniae* and *Alternaria alternata* on water hyacinth in Egypt. Plant Dis., pp. 279-319.

Simmons EG (1992). *Alternaria* taxonomy: current status, viewpoint, challenge. In: Chelkowski J, Visconti A., eds., eds. *Alternaria*: biology, plant disease and metabolites. Amsterdam: Elsevier, pp. 1-35.

Simmons EG (1999). *Alternaria* themes and variations (236-243). Host-specific toxin producers. Mycotaxon, 70: 325-369.

Simmons EG (2004). Novel dematiaceous hyphomycetes. Stud. Mycol., 50: 09-118.

Simmons EG, Roberts RG (1993). *Alternaria* themes and variations (73). [Sporulation patterns]. Mycotaxon, 48: 109-140.

Waterhouse AM, Procter JB., Martin DM, Clamp M, Barton GJ (2009). Bioinformatics. Jalview Version 2-a multiple sequence alignment editor and analysis workbench, 25: 1189-1191.

White TJ, Bruns T, Lee S, Taylor J (1990). Amplification and direct sequencing of fungal ribosomal RNA genes from phylogenetics. In: M.A. Innis, D.H. Gelfand, J.S. Sninsky and T.J. White, Editors, PCR Protocols, Academic Press, London, U.K. pp. 315-322.

18

First qualitative survey of filamentous fungi in Dal Lake, Kashmir

Suhaib A. Bandh[1]*, Azra N. Kamili[1], Bashir A. Ganai[2], Samira Saleem[1], Bashir A. Lone[1] and Humera Nissa[1]

[1]Centre of Research for Development, University of Kashmir Srinagar-190006, India.
[2]Department of Biochemistry, University of Kashmir Srinagar-190006, India.

Filamentous fungi comprehend a heterogeneous group of heterotrophic microorganisms that act as saprobes or parasites or, less frequently as symbionts living in association with other organisms. Water samples obtained seasonally from April 2010 to March 2011 at sixteen different sites of Dal Lake, Kashmir were serially diluted five folds followed by spread plate technique for the isolation of filamentous fungi, spreading 0.1 ml inoculum from the serial dilution tubes on the Petri dishes containing Rose-Bengal Streptomycin Agar medium. Twenty three (23) species of fungi namely *Penicillium caseicolum, P. commune, P. chrysogenum, P. funiculosum, P. lilacinum, P. olivicolor, P. dimorphosporum, Penicillium* sp. I, *Penicillium* sp. II, *Penicillium* sp. III, *Penicillium* sp. IV, *Aspergillus flavus, A. fumigatus, A. japonicus, A. niger, A. terreus, A. versicolor, A. wentii, Aspergillus* sp. *Fusarium* sp. *Rhizopus* sp. *Acremonium* sp. and *Mucor* sp. belonging to five genera were recovered from the Lake water samples. *Penicillium* and *Aspergillus* were the most dominant genera with a total of 11 and 8 species respectively. The most prevalent species was *P. chrysogenum* with its occurrence at all sixteen (16) sampling stations and a highest total of seventeen species was recorded at site 16 (Pokhribal Nallah II).

Key words: Filamentous fungi, Dal Lake, serial dilution, qualitative.

INTRODUCTION

Fungi are a diverse group of organisms belonging to the kingdom Eumycota. This kingdom comprises five phyla namely Ascomycota, Basidiomycota, Zygomycota, Chytridiomycota, and Glomeromycota (Kirk et al., 2001; Schußler et al., 2001). As a practical approach to classification, fungi have been divided into groups, such as the filamentous fungi, also called moulds, the yeasts, and the mushrooms. Some fungi are primarily adapted to aquatic environments, and will therefore, naturally be found in water. These fungi are zoosporic and may belong in phyla Chytridiomycota. Fungi belonging to the other phyla in Eumycota are primarily adapted to terrestrial environments. They are present in soil, organic material, and air, and anything in contact with air (Kirk et

al., 2001). These fungi can also enter water bodies from various locations, although this is considered an 'unnatural' habitat for them. The knowledge of the occurrence of fungi in water was limited, but has increased due to the various studies performed. The filamentous fungi are ubiquitous group of organisms which explore almost all ecological niches on earth. They are estimated to be responsible for the spoilage of up to 25% of all plant-derived foods produced annually (Geisen, 1998).

Filamentous fungi or moulds are vital for the maintenance of ecosystems. By breaking down dead organic material, they continue the cycle of nutrients through ecosystems. Some of them act as plant pathogens causing severe crop losses and post-harvest food spoilage. In the reagent industry and medicine areas, filamentous fungi are the source of commercial enzymes, organic acids, and numerous drugs, such as

*Corresponding author. Email: suhaibbandh@gmail.com.

antibiotics (e.g. penicillin, cefalosporin). *Penicillium* species have been frequently recovered from water in the various studies performed. Several of the species in genus *Penicillium* and *Aspergillus* are known to produce mycotoxins in other substrates, such as food and beverages (Moreau 1979; Pitt and Hocking, 1999). Interestingly, detection of aflatoxins produced by *A. flavus* in water from a cold water storage tank was demonstrated by Paterson et al., (1997). *Aspergillus* species is one of the more commonly isolated genus in water. *A. niger* and *A. flavus* are common allergens and may cause opportunistic invasive infections (De Hoog et al., 2000; Denning, 1998).

Predominant fungal genera and species in treated and untreated water are: *Aspergillus, Cladosporium, Epicoccum, Penicillium, Trichoderma, Arthrinium phaeospermum, A. flavus, C. cladosporioides, Fusarium culmorum, Mucor hiemalis* and *Trichoderma harzianum* (Kinsey et al., 1999). Many other fungal genera isolated from Danube river water in Europe include: *Mortierella, Absidia, Rhizopus, Acremonium, Beauveria, Doratomyces, Monilia, Rhizopus arrhizus, Acremonium strictum, Fusarium oxysporum* and *Stemphyllium botryosum* (Tothova, 1999)

The ecology of aquatic fungi affects their distribution both locally and globally, and the factors influencing the fungi are complex and vary depending on the aquatic environment. What governs the distribution of freshwater fungi is difficult to determine, although some species appear to be more common either in temperate or tropical regions (Shearer et al., 2007; Raja et al., 2009).

MATERIAL AND METHODS

Location and site description: Dal Lake, located at 34° 07′ N, 74° 52′ E, 1584 m a.s.l in Srinagar, Jammu and Kashmir, India- a multi-basined lake with Hazratbal, Bod Dal, Gagribal and Nageen as its four basins, having two main inlets as Boathall Nallah and Tailbal Nallah and two main outlets as Dal Lock Gate and Pokhribal Nallah, was taken up for the current study. Sixteen (16) sites viz., Hazratbal Open, Hazratbal littoral, Nageen Open, Nageen near Houseboats, Gagribal Open, Gagribal near Houseboats, Nishat Open, Near Centeur, Boathall Nallah-I, Boathall Nallah-II, Tailbal Nallah-I, Tailbal Nallah-II, Dal Lock Gate-I, Dal Lock Gate-II, Pokhribal Nallah-I and Pokhribal Nallah-II were selected with eight (8) sites from the four basins, Four (4) sites from two inlets and Four (4) others from the two outlets.

Collection of water samples

The water samples were collected on seasonal basis for a period of 12 months between April 2010 to March 2011, from Dal Lake in white plastic containers, which were previously sterilized with 70% alcohol and rinsed with distilled water. At the lake, the containers were rinsed thrice with the lake water before being used to collect the samples.

Isolation of fungi

Water samples obtained from different sites were serially diluted five folds followed by spread plate technique for isolation of filamentous fungi, spreading 0.1 ml inoculum from the serial dilution tubes on the Petri dishes containing Rose-Bengal Streptomycin Agar medium. Growing colonies were transferred to Petri dishes containing different culture media like Potato Dextrose Agar (PDA) (MERCK, Germany), Malt Extract Agar (MEA) (Acumedia, USA), Czapek's dox Agar (CZ) (MERCK, Germany) and Czapek's Yeast Agar (CYA) (MERCK, Germany), 25% Glycerol nitrate Agar (G25A) for identification, and then transferred everything to PDA for stock cultures. Plates were incubated at 25 to 37 °C for one week in dark.

Identification of fungi

Identification of fungi was performed mainly on the basis of the micro- and macromorphological features, reverse and surface coloration of colonies grown on CZ, MEA, CYA and PDA media. Fungi were identified to genus level using Barnett and Hunter (1999). Cultures were identified to species level using various mycological texts: *Penicillium* LINK, species were identified using colony diameters, macro- and micromorphology according to the standardized conditions of PITT's monograph (2000). These species were grown on various differential media all prepared according to the recipes of Pitt (2000). Each *Penicillium* culture was inoculated in triplicate on each medium and incubated at three different temperatures (5, 25 and 37 °C) for a period of 7 days in the dark. The monograph by Raper and Fennell (1965) was used for identification of *Aspergillus* species. In addition to these the morphological characteristics of these and various other species were studied by making slide cultures obtained by inoculating microfungi directly on a small square of agar medium.

RESULTS

Although there are many methods such as filtration, direct plating, baiting etc. for sampling fungi from water (Kinsey et al., 1999) we used direct plating method for its isolation in our study. Twenty three (23) species (Table 1) of filamentous fungi; *P. caseicolum, P. commune, P. chrysogenum, P. funiculosum, P. lilacinum, P. olivicolor, P. dimorphosporum, Penicillium* sp. I, *Penicillium* sp. II, *Penicillium* sp. III, *Penicillium* sp. IV, *A. flavus, A. fumigatus, A. japonicus, A. niger, A. terreus, A. versicolor, A. wentii, Aspergillus* sp. *Fusarium* sp. *Rhizopus* sp. *Acremonium* sp. and *Mucor* sp. belonging to five genera were recovered from sixteen sampling stations of the Lake. The prevailing genera were *Aspergillus, Penicillium, Fusarium, Rhizopus, Acremonium* and *Mucor*. The most frequent genera obtained were *Penicillium* (11 species) and *Aspergillus* (8 species). However, *Fusarium* (1 species), *Rhizopus* (1 species), *Acremonium* (1 species) and *Mucor* (1 species) were also reported during the study. The most prevalent species was *P. chrysogenum* with its occurrence at all sixteen (16) sampling stations followed by *P. funiculosum, A. niger and A. flavus* from 14 stations each, *A. fumigatus* and *Fusarium* sp. from thirteen (13) stations each, *P. caseicolum, P. lilacinum, P. olivicolor, A. terreus, A. wentii,* and *Mucor* sp. from eight (8) stations each, *P. commune, P. dimorphosporum, A. japonicus,* and *A. versicolor* from seven (7) stations each, *Penicillium* sp. II,

Table 1. Occurrence of filamentous fungi in different seasons at different sites.

S/N	Name of Fungi	Seasons			
		Spring 2010	Summer 2010	Autumn 2010	Winter 2010
1.	*Penicillium caseicolum* Bain.	5, 9, 10, 14	3, 11, 12	-	2
2.	*Penicillium commune* Thom.	1, 2, 11	14	6, 13, 16	-
3.	*Penicillium chrysogenum* Thom.	4, 6, 7, 9, 12, 13, 16	1, 2, 4, 6, 9, 11, 12, 14, 15, 16	3, 5, 8, 12, 14, 16	3, 4, 9, 10, 11, 13, 14, 16
4.	*Penicillium funiculosum* Thom.	1, 2, 5, 6, 7, 8, 9, 12, 15, 16	2, 3, 4, 6, 12, 14, 16	11, 15, 14	3, 4, 6
5.	*Penicillium lilacinum* Thom.	3, 14, 15	5, 6, 11, 12, 14	7, 6, 11, 12	-
6.	*Penicillium olivicolor* Pitt	1, 2, 3, 4, 9, 11, 12	3, 9, 10	5	-
7.	*Penicillium dimorphosporum* Swart	1, 2, 3, 4, 5	5, 6, 8	-	-
8.	*Penicillium* sp. I	-	-	-	9, 10
9.	*Penicillium* sp. II	16, 14, 15	2, 3, 4	-	-
10.	*Penicillium* sp. III	-	14, 15, 16	-	-
11.	*Penicillium* sp. IV	9	11, 12	-	-
12.	*Aspergillus flavus* Link: Fr	4, 5, 8, 16	1, 2, 3, 5, 6, 9, 10, 11, 15, 16	9, 13, 16	3, 4, 5, 14, 15
13.	*Aspergillus fumigatus* Fresenius	3, 4, 6, 9, 10, 11, 14, 16	1, 6, 8, 11, 14, 15, 16	3, 6, 14	8, 12, 13, 15
14.	*Aspergillus japonicus* Saito	-	3, 4, 15	-	5, 12, 13, 16
15.	*Aspergillus niger* Van Tieghem	6, 12, 13, 14, 16	2, 3, 4, 10, 11, 14, 16	1, 5, 9, 11, 15, 16	5
16.	*Aspergillus terreus* Thom.	-	4, 6, 9, 10, 12, 14	16	2
17.	*Aspergillus versicolor* gr.	2, 3, 13, 15, 16	1, 10, 16	-	-
18.	*Aspergillus wentii* gr.	4, 6, 13, 14, 16	5, 8, 13, 16	7	8
19.	*Aspergillus* sp.	-	13, 16	-	8, 9
20.	*Fusarium* sp.	1, 3, 5, 6, 10, 14, 16	2, 3, 5, 6, 7, 8, 16	1, 9, 15	1, 4, 9, 10 14, 15
21.	*Rhizopus* sp.	1, 9, 14, 16	1, 4, 9, 14	-	-
22.	*Acremonium* sp.	2, 6, 7	7, 16	-	-
23.	*Mucor* sp.	4, 6, 13, 14, 15	13, 14	5, 6, 7, 16	15, 16

1= Hazratbal open, 2= Hazratbal littoral, 3= Nageen open, 4= Nageen near houseboats, 5= Gagribal open, 6= Gagribal near houseboats, 7= Nishat open, 8= near Centeur, 9= Boathall Nallah-I, 10= Boathall Nallah-II, 11= Tailbal Nallah-I, 12= Tailbal Nallah-II, 13= Dal Lock Gate-I, 14= Dal Lock Gate-II, 15= Pokhribal Nallah-I, 16= Pokhribal Nallah-II.

from six (6) stations, *Rhizopus* sp. from five (5) stations, *Aspergillus* sp. and *Acremonium* sp. from four (4) stations each, *Penicillium* sp. III and *Penicillium* sp. IV from three (3) stations each and *Penicillium* sp. I from two (2) stations. The highest total of seventeen fungal species was recorded from site 16 (Pokhribal Nallah II), followed by site 14 (Dal Lock Gate II) with fifteen species, site 4 (Nageen Near Houseboat) with fourteen species, site 3 (Nageen Open) site 6 (Gagribal near Houseboats) site 9 (Boathall Nallah) with thirteen species each, site 2 (Hazratbal littoral) with twelve species, site 5 (Gagribal open) site 15 (Pokhribal Nallah II) with eleven species each, site 10 (Boathall Nallah) site 11 (Tailbal Nallah I) site 12 (Tailbal Nallah II) site 13 (Dal Lock Gate I) with ten species each, site 8 (Near Centeur) with eight species and site 7 (Nishat Open) with six species.

DISCUSSION

The overwhelming presence of these terrestrial moulds in water supports the paradigm that their deposition is attributable to contamination of the water body due to the entry of sewage from the catchment areas, as they survive conventional treatment strategies and enter the distribution through the sewage coming out from the sewage treatment plants (Neimi et al., 1982). It can be attributed to the entry of sewage from the drains into the lake, as these genera have been reported frequently from the drain waters with maximum densities during higher pollution (Khulbe and Drugapal, 1994) and can therefore be inferred that these species are good indicators of pollution. The genera and species isolated in the present study were previously isolated, but with various numbers and frequencies, from different substrata in Saudi Arabia such as rainfall water and mud (El-Nagdy et al., 1992), soil (Abdel-Hafez, 1982a) and ferns (Abdel-Hafez, 1984). Almost all of the filamentous fungal genera recovered in this study had been found in various habitats in India (Saju 2011; Shafi et al., 2011, Bandh et al., 2011a) Egypt (Abdel-Hafez and Bagy, 1985; El-Hissy et al., 1990; Moharrum et al., 1990; El-Nagdy and AbdelHafez, 1990) Brazil (Gomes et al., 2008) and other countries (Barlocher and Kendrick, 1974; Bettucci et al., 1993; Bettucci and Roquebert, 1995). Terrestrial fungi in aquatic habitats are likely to originate from air (Sparrow, 1968), as well as from living or dead animal and plant, soil and litter being in contact with water (Park, 1972). These species have also been isolated from soils, water and other substrata in Saudi Arabia, Egypt and other countries (Abdel-Hafez et al., 1978; Abdel-Hafez, 1982b; Abdel-Kader et al., 1983; Abdel-Hafez and Bagy, 1985; Bandh et al., 2011b).

Aspergillus and *Penicillium* spores are the most widespread aeroallergens in the world. According to qualitative and quantitative reports, the former is the dominant species in tropical regions whilst the latter is dominant all over the world (Rosas et al., 1992).

A. fumigatus, found in our study, is one of the most ubiquitous airborne saprophytic fungi. Water fungi can play a vital role in the decomposition of some organic materials such as dead leaf and stem litter. The decomposition of fallen leaves and other detritus in streams is dominated by fungi (Garnett et al., 2000).

Penicillium was the most frequent and predominant genus detected in our study, followed by *Aspergillus*. According to the Kinsey et al. (1999), certain fungi such as *Aspergillus*, *Cladosporium*, *Epicoccum*, *Penicillium* and *Trichoderma* species appear more frequently than others in water. Our results concur with theirs except that we did not find *Epicoccum*, *Cladosporium* and *Trichoderma*. *A. fumigatus*, *A. niger*, *P. chrysogenum* and many other species belonging to the two genera observed in the current study with a high occurrence at different sampling stations have also been found to be widespread in Turkey and have been reported in many studies (Asan, 2000). *Aspergillus* spp and *Penicillium* spp are major contaminants of the environments and occur as ubiquitous saprophytes, with their spores able to survive

and reproduce in water as well. The present results are confirmed by a Brazilian study on filamentous fungi of sand and water from "Bairro Novo" and "Casa Caiada" beaches in which *Aspergillus* and *Penicillium* were the most frequent genera in both sand and water, with a total of 11 and 19 species, respectively (Gomes et al., 2008)

Conclusion

The mycoflora of Dal Lake with reference to filamentous fungi investigated in the present work showed that the genus *Penicillium* was found to be widespread in the water samples indicating that the spores of this genus are most widespread in nature.

ACKNOWLEDGEMENTS

This work was supported by Centre of research for Development, University of Kashmir. We would like to thank Department of Microbiology, Sheri-Kashmir Institute of Medical Sciences (SKIMS) Soura and Agharkar Research Institute, Pune for their valuable and insightful guidance in the identification of various strains.

REFERENCES

Abdel-Hafez SII, Bagy MMK (1985). Survey on the terrestrial fungi of Ibrahimia canal water in Egypt. Proc. Egyptian Bot. Soc., 4: 106-123.

Abdel-Hafez SII, Moubasher AH, Abdel-Fattah HM (1978). Cellulose-decomposing fungi of salt marshes in Egypt. Folia Microbiol., 23: 37-44.

Abdel-Hafez SLI (1982a). Survey of the mycoflora of desert soils in Saudi Arabia. Mycopathologia, 80: 3-8.

Abdel-Hafez SLI (1982b). Cellulose-decomposing fungi of desert soils in Saudi Arabia. Mycopathologia, 78: 73-78.

Abdel-Hafez SLI (1984). Rhizosphere and phyllosphere fungi of four ferns plants growing in Saudi Arabia. *Mycopathologia* 85: 45-52.

Abdel-Kader MLA, Abdel-Hafez ALI, Abdel-Hafez SII (1983). Composition of the fungal flora of Syrian soils. II-Cellulose-decomposing fungi. Mycopathologia, 81: 167-171.

Asan A (2000). Check list of *Aspergillus* and *Penicillium* species reported from Turkey. Turk. J. Bot., 24: 151-167.

Bandh SA, Kamili AN, Ganai BA, Saleem S (2011a). Isolation, Identification and Seasonal Distribution of *Penicillium* and *Aspergillus* Species in Dal Lake, Kashmir. Int. J. Curr. Res., 3(10): 038-042.

Bandh SA, Kamili AN, Ganai BA (2011b). Identification of some *Penicillium* species by traditional approach of morphological observation and culture. Afr. J. Microbiol. Res., 5(21): 3493-3496.

Barlocher F, Kenderick B (1974). Dynamics of the fungal population on leaves in a stream. J. Ecol., 62: 761-790.

Barnett HL, Hunter BB (1999). Illustrated Genera of Imperfect Fungi (fourth ed.). APS Press, St. Paul, Minnesota, USA, 218 pp.

Bettucci L, Roquebert M (1995) Studies on microfungi from tropical rain forest litter and soil. A preliminary study. Nova Hedwigia, 61: 111-118.

Bettucci L, Rodriguez C, Indarte R (1993). Studies on fungal communities of two grazing-land soils in Uruguay. Pedobiologia, 37: 72-82.

De Hoog GS, Guarru J, Gene J. Figueras MJ (2000). *Atlas of Clinical fungi*. Centraalbureau voor Schimmel cultures. Mycopathologia, pp. 159-160.

Denning DW (1998) .Invasive aspergillosis. Clin. Infect. Dis., 26: 781-805.

El-Hissy ET, Moharrum AM, El-Zayat SA (1990). Studies on the mycoflora of Aswan High Dam Lake, Egypt; monthly variation. *J. Basic Microbiol.*, 30: 231-236.

El-Nagdy MA, Abdel-Hafez SI (1990). Occurrence of zoosporic and terrestrial fungi in some ponds of Kharga Oases, Egypt. J. Basic Microbiol., 30: 233-240.

Garnett H, Barloche F, Giberson D (2000). Aquatic hyphomycetes in Catamaran Brook: Colonization dynamics, sasonal patterns, and logging effects. Mycologia, 92: 29-41.

Geisen R (1998). PCR methods for the detection of mycotoxin-producing fungi. In: Bridge, P.D.; Arora, D.K.; Reddy, C.A.; Elander, R.P. (Ed.). Applications of PCR in mycology. Oxon, London: CAB International, pp. 243-266.

Gomes DNF, Cavalcanti MAQ, Fernandes MJS, Lima DMM, Passavante JZO (2008). Filamentous fungi isolated from sand and water of "Bairro Novo" and "Casa Caiada" beaches, Olinda, Pernambuco, Brazil. Braz. J. Biol., 68(3): 577-582

Khulbe RD, Drugapal A (1994). Sewage mycoflora in relation to pollutants in Nainital, Kumaun Himalaya. Poll. Res., 13(1): 53-58.

Kinsey GC, Paterson RR, Kelley J (1999) Methods for the determination of filamentous fungi in treated and untreated waters. J. Appl. Microbiol., 85: 214S-224S.

Kirk PM, Cannon PF, David JC, Stalpers JA (2001). Ainsworth & Bisby's Dictionary of the fungi, 9th edn. CAB International, Wallingford.

Moharrum AM, El-Hissy FT, El-Zayat SA (1990). Studies on the mycoflora of Aswan High Dam Lake, Egypt: Vertical fluctuations. J. Basic Microbiol., 30: 197-208.

Moreau C (1979). Moulds, Toxins and Food, 2nd edn. John Wiley & Sons, New York.

Neimi R, Knuth S, Lundstrom K (1982). Actinomycetes and fungi in surface waters and in potable water. Appl. Environ. Microbiol., 43: 378-388.

Park DE (1972). On the ecology of heterotrophic microorganisms in freshwater. Trans. Br. Mycol. Soc., 58: 291-299.

Paterson RRM, Kelley J, Gallagher M, (1997) Natural occurrence of aflatoxins and *Aspergillus flavus* (Link) in water. Lett. Appl. Microbiol., 25: 435-436.

Pitt JI, Hocking AD (1999). Fungi and Food Spoilage, 2nd edn. Aspen Publishers, Gaithersburg, MD.

Pitt JI (2000). A laboratory guide to common *Penicillium* species. Food Science.

Raja HA, Schmit JP, Shearer CA (2009). Latitudinal, habitat and substrate distribution patterns of freshwater ascomycetes in the Florida Peninsula. Biodivers. Conserv., 18: 419-455.

Raper KB, Fennell DI (1965). The Genus *Aspergillus*. The Williams & Wilkins Comp. Baltimore, USA, 686 pp.

Rosas I, Calderon C, Escamilla B, Ulloa M (1992) Seasonal distribution of *Aspergillus* in the air of an urban area: Mexico City. Grana, 31: 315-319.

Saju DS (2011). Occurrence of Fungi in Pond Water (Dumaratarai Talab) of Raipur City, C.G., India J. Phytol., 3(4): 30-34.

Schußler A, Schwarzott D, Walker C, (2001) A new fungal phylum, the Glomeromycota: phylogeny and evolution. Mycol. Res., 105: 1413-1421.

Shearer CA, Descals E, Kohlmeyer B, Kohlmeyer J, Marvanova L, Padgett D, Porter D, Raja HA, Schmidt JP, Thornton HA, Voglymayr H (2007). Fungal biodiversity in aquatic habitats. Biodivers. Conserv., 16: 49-67.

Shafi S, Bandh SA, Kamili AN, Shah MA, Ganai BA, Shameem N (2011). A Preliminary Microbiological Study of Sindh, a Glacier fed River of Sonamarg Kashmir. New York Sci. J., 4(10):58-62.

Sparrow FK (1968). Ecology of freshwater fungi. In: The Fungi, an Advanced Treatise. Val. 3rd G.C. Ainsworth and A.S. Sussman). Academic Press, London, U.K, pp. 41-93.

Tothova L (1999) Occurrence of microscopic fungi in the Slovak section of the Danube River. Biologia, 54: 379-385.

19

Halophilic fungi in a polyhaline estuarine habitat

Valerie Gonsalves, Shweta Nayak and Sarita Nazareth*

Department of Microbiology, Goa University, Taleigao Plateau, Goa-403206, India.

Halophilic mycobiota was isolated from Mandovi estuary and it was dominated by *Aspergillus* and *Penicillium* species. *Cladosporium* and *Eurotium* were found in lesser numbers while obligate halophiles were found only amongst the aspergilla and they were all identified as *Aspergillus penicillioides*. Some aspergilli and all the isolates of *Penicillium*, *Cladosporium* and *Eurotium* were facultative halophiles. There were significant differences in growth of each isolate at different salt concentrations. Most of the isolates were euryhaline, having a wide range of salt tolerance; a few were stenohaline, with a narrow range of halotolerance. The isolates were mainly moderate halophiles, with a very few slight halophiles. Isolation of obligate halophilic fungi from polyhaline environment of an estuary is hereby reported for the first time.

Key words: *Aspergillus penicillioides*, obligate halophiles, facultative halophiles.

INTRODUCTION

An estuary is under the influence of marine conditions such as tides, waves, influx of saline water, as well as the flow of fresh water and sediment of the river (Manoharachary et al., 2005). Studies on the mycobiota in estuarine ecosystems have focused on the diversity or metabolic activities of fungi from estuarine waters, sediments, wood and litter, marshes and mangroves (Borut and Johnson, 1962; Shearer, 1972; Cooke and Lacourse, 1975; Rai and Chowdhery, 1978; D'souza et al., 1979; Maria and Sridhar, 2002; da Silva et al., 2003; Tsui and Hyde, 2004; Anita et al., 2009; Karamchand et al., 2009; Nambiar and Raveendran, 2009; Rani and Panneerselvam, 2009; Pearman et al., 2010; Mohamed and Martiny, 2011). The halophilic nature of these fungi has not been described.

Hypersaline environments have been a focus of study for halophilic organisms that are able to survive in these environments. *Gymnascella marismortui* was the first halophilic fungi to be isolated from the Dead Sea (Buchalo et al., 1998). The black yeasts and closely related dematiaceous *Cladosporium* were among the first halophilic fungi to have been isolated from salterns in Secovlje, Slovenia-Adriatic (Gunde-Cimerman et al.,

2000; Butinar et al., 2005a) and subsequently from Cabo Rojo, Puerto Rico (Diaz-Munoz and Montalvo-Rodriguez, 2005). The isolation of other filamentous fungi from salterns (Cantrell et al., 2006; Nayak et al., 2012), as well as from the Dead Sea (Kis-Papo et al., 2003a, b; Wasser et al., 2003; Nazareth et al., 2012), Mono Lake, California (Steiman et al., 2004), coastal environments of Arctics (Gunde-Cimerman et al., 2005) and from saline soils of Soos, Czech Republic (Hujslova et al., 2010) has followed.

Microbes that inhabit high-salt environments may be of halotolerant or halophilic nature, being adapted to high levels of ions, as well as to low a_w (Grant, 2004). Halophiles have been further classified as facultative or obligate (Kushner, 1978; Nazareth et al., 2012).

The occurrence of halophilic filamentous fungi, particularly that of obligate halophilic aspergilli, in a polyhaline estuarine environment, is recorded herein for the first time.

METHODOLOGY

Samples

Sampling was done along the estuary of the Mandovi (EM) which flows into the Arabian Sea, on the West Coast of the Indian Peninsula. Samples of top and bottom water (wt) and (wb), and of sediment(s) were obtained using the Niskin and Grab samplers,

---

*Corresponding author. E-mail: saritanazareth@yahoo.com.

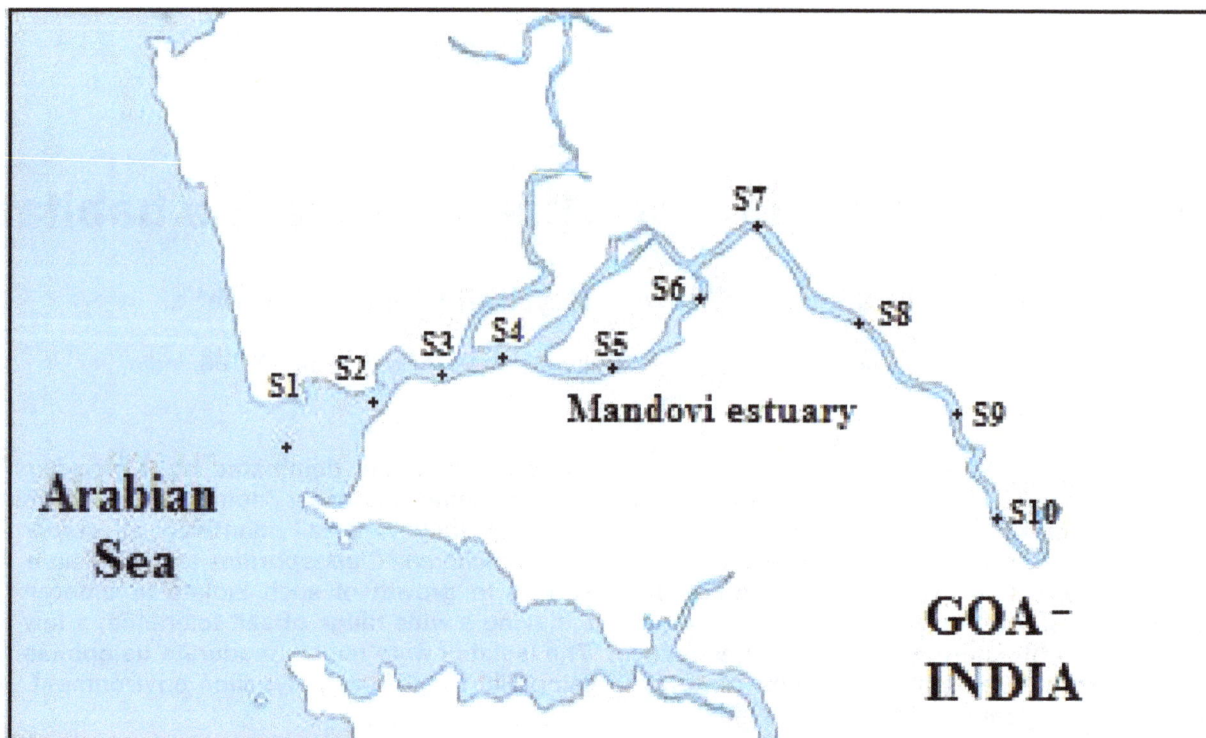

Figure 1. Map indicating the stations of sampling along the Mandovi estuary.

respectively, from 10 Stations, S1 to S10 (Figure 1), beginning at the mouth and moving hinterland between 73°46.65' to 74°2.5' longitude (courtesy G. N. Nayak, Goa University), as shown in Figure 1.

The salinity and pH were measured as described by Nazareth et al. (2012).

Isolation and identification of fungi

The samples were processed under sterile conditions. The water samples and suspensions of sediment were each plated on to Czapek-Dox Agar + 20% solar salt (20% S-CzA) for isolation of halophilic fungi, which were then incubated at room temperature of about 30°C, up to 30 days.

The isolates were picked based on dissimilarity in the colony characteristics, purified and numbered according to the station and the sample of top or bottom water and of sediment.   Purified isolates were maintained on 10% S-CzA. Fungal identification was done on the basis of colony and micro morphology characteristics with reference to identification keys (Raper and Fennell, 1965; Ellis, 1971; Domsch et al., 1980).

The identification of an obligate halophilic isolate was confirmed by ITS rDNA sequence analysis (Merck-GeNei Services), using consensus primers for 18S rRNA, ITS1, 5.8S rRNA, ITS2 and 28S rRNA gene fragment. A GenBank accession number was obtained. Sequence similarities were obtained using NCBI BLAST. Alignment and phylogenetic tree were constructed in Clustal X version 2 and the NJ distance method.

Halophilic nature of the isolates

Salt tolerance of the isolates was checked by inoculating the cultures in triplicate on CzA amended with salt up to concentrations

of 30% w/v (Nazareth et al., 2012). Growth was recorded after 7 days incubation in terms of colony diameter; plates that did not show growth in 7 days were further incubated till 15 days to check for delayed growth and then sub-cultured onto corresponding media to confirm the salt tolerance level. Salt tolerance curves were obtained by plotting the arithmetic mean of colony diameter with the standard error.

Statistically significant difference (P<0.05) in the effect of different salt concentrations on growth of all isolates, as well as in growth within the species, was analyzed by two-way ANOVA.

RESULTS

Sample Salinity and pH

The salinity of the surface water was comparable to that of the bottom water at all stations. The salinity at S1 at the mouth of the estuary was 37‰, close to that of sea water, with a gradual decrease to approximately 30 to 31‰ at S5; it decreased drastically to approximately 24 to 25‰ at S6 and S7 and further there was a sequential decrease to approximately 10 to 16‰ till the last station. The sediment had a salinity of 5‰ at S6 and 10‰ at the rest of the stations, which was comparatively much lower than that of the water column at all stations.

The pH of the samples was around neutral.

Fungal isolates

The isolates obtained belonged to the genera of

Table 1. Species identification of isolates from different stations.

Species	Isolate number
A. flavus	EM2s111, EM7w$_b$140
A. fumigatus	EM2s112, EM4w$_t$119, EM6w$_b$135
A. nidulans	EM5w$_t$129
A. penicillioides	EM2w$_b$107, EM4w$_t$118, EM4w$_t$120, EM4w$_b$121, EM4w$_b$123, EM4w$_b$125, EM5w$_t$130, EM6w$_t$133, EM6s137, EM7w$_t$138, EM7w$_t$139, EM7w$_b$141, EM7w$_b$142, EM7w$_b$143, EM7s144, EM7s145, EM8w$_t$146, EM8w$_t$147, EM8w$_t$148, EM8w$_b$149, EM9w$_b$155, EM8s153, EM9s156
A. sydowii	EM8s 152
A. versicolor	EM2w$_t$104, EM2w$_t$105, EM2w$_t$106, EM2w$_b$108, EM2s110, EM6w$_b$136
P. asymetrica sub-section fasciculata	EM5w$_b$131, EM8s151
P. canescens	EM1w$_b$101
P. chrysogenum	EM1s103, EM6wt134
P. corylophilum	EM2s109, EM3w$_b$114, EM5s132, EM8s150
P. steckii	EM1w$_t$102, EM3w$_t$113, EM3s115, EM4w$_t$117
E. amstelodami	EM4w$_b$124
E. repens	EM3s116, EM4s126, EM4s127
C. carpophilum.	EM9w$_b$154
C. cladosporioides	EM4w$_b$122, EM4s128

Aspergillus, Penicillium, Cladosporium and Eurotium (Table 1). Culturally dissimilar isolates from each sample were picked and identified as Aspergillus: A. flavus, A. fumigatus, A. nidulans, A. penicillioides, A. sydowii, A. versicolor; Penicillium: P. asymetrica subsection fasciculata, P. canescens, P. chrysogenum, P. corylophilum, P. steckii; Cladosporium: C. cladosporioides and C. carpophilum and Eurotium: E. amstelodami and E. repens.

Halophilic nature of the isolates

The isolates that did not grow in absence of salt were characterized as obligate halophiles and those that grew in absence of added salt, but showed enhanced growth at increased levels of salt were termed as facultative halophiles. The halotolerance curves of the isolates are presented according to their obligate or facultative halophilic nature as shown in Figures 2a and b, respectively.

Obligate halophiles

Twenty-three obligate halophiles were obtained, all identified as A. penicillioides. These required a minimum of 2, 5 or 10% salt for growth (Figure 2a). Most were moderate halophiles, with optimal growth at 10% salt, and two at 5% salt. Most showed a euryhaline nature, having a capacity to grow on a wide range of salt concentrations up to 20% or even 30% salt; three showed a stenohaline characteristic, with a narrow range of salt tolerance of 5 to 15%. Significant difference (P<0.05) was obtained in the growth at different salt concentrations, as well as in the halotolerance curves of the isolates.

The isolates were obtained mainly from samples of top or bottom water, or sediment of S4 – S9. A. penicillioides had a very different appearance in colony characteristics from the other aspergilli, forming comparatively small colonies with a compact, tough felt growth and furrowed appearance; this also ensured that such isolates were picked from each of the samples in which they were found. Identification of a representative isolate, EM6s137, was confirmed by gene sequence analysis as A. penicillioides, the GenBank accession number given as JQ240645. The phylogenetic relationship of this isolate is shown in Figure 3.

Facultative halophiles

Thirteen isolates of aspergilli and all isolates of Penicillium, Cladosporium and Eurotium were facultative halophiles, randomly distributed throughout the stations; Cladosporium was isolated only from bottom water and sediment samples.

The isolates of A. sydowii and A. penicillioides showed maximal growth at 10% salt and could grow at salt concentrations up to 25 and 30%, respectively; A. fumigatus and A. nidulans isolates grew maximally at salt concentration of 5% and A. versicolor isolates showed

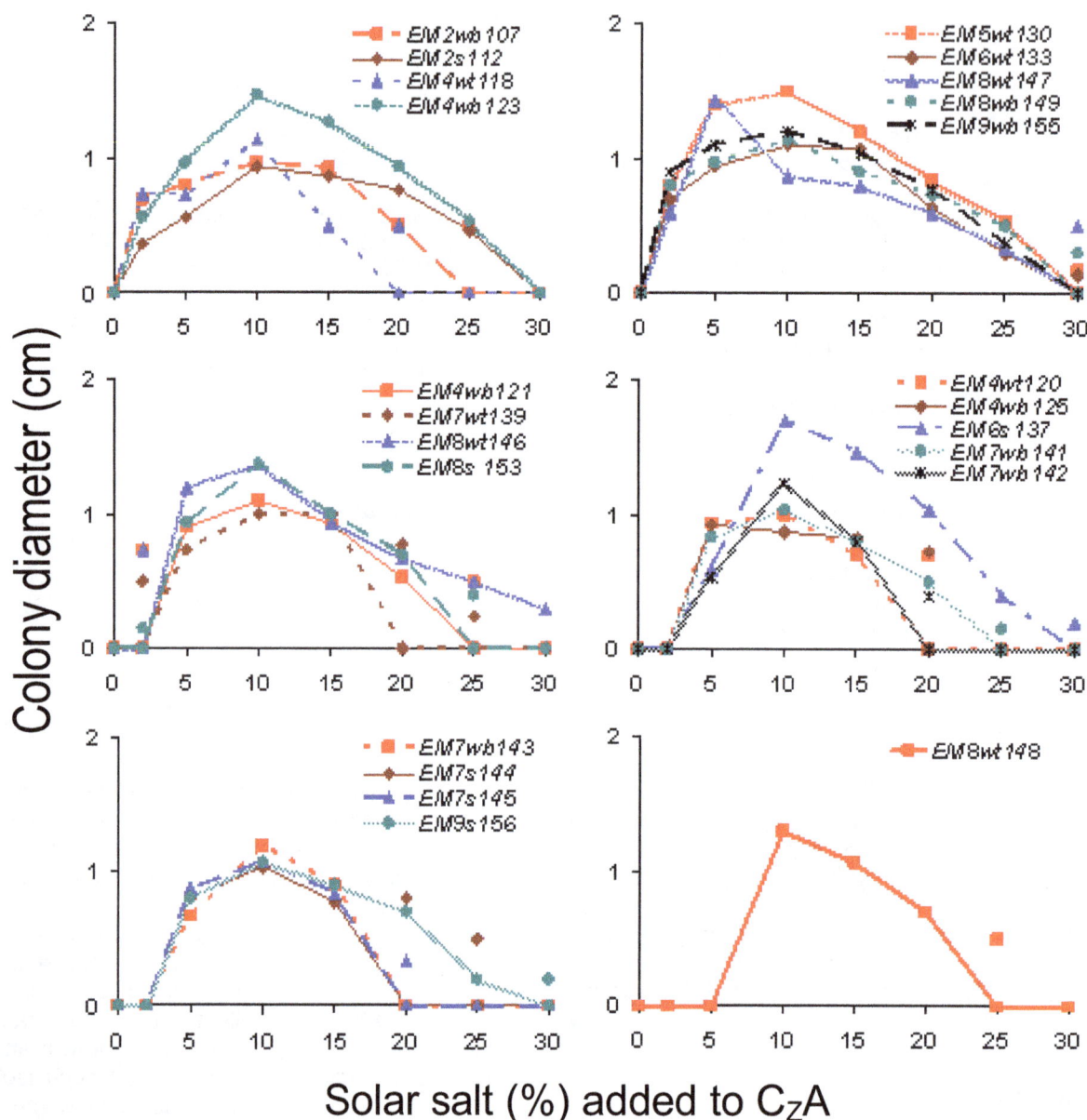

Figure 2a. Salt tolerance curves of obligate halophilic *A. penicillioides* as recorded after 7d incubation; unconnected symbols indicate delayed growth at respective salt concentrations, after 15d incubation.

optimum growth at 2 to 10% salt, with a tolerance level of 20 to 25% salt; *A. flavus* isolates had optimal growth in presence of 2 and 10% salt and tolerated a concentration of 15 and 25% salt, respectively.

The isolates of *P. asymetrica subsection fasciculata, P. canescens* and *P. chrysogenum* grew optimally at 5% salt and exhibited a tolerance level of 20 to 30% salt; *P. corylophilum* showed optimal growth with 2 or 5% salt, and *P. steckii* with 5 or 10% salt, all able to grow in presence of 20 to 25% salt.

Salt concentrations of 2 and 5% supported optimal growth of *C. cladosporioides* and *C. carpophilum*, respectively; these species tolerated a maximum of 20%

salt. Isolates of *Eurotium*: *E. amstelodami* and *E. repens* showed maximum growth in presence of 2 or 5% salt, and tolerance to 20 to 25%.

The isolates were euryhaline in nature, most of which were moderate halophiles; some grew optimally with 2% salt and were termed as slight halophiles. Significant difference (P<0.05) was obtained in the growth of a given species at different salt concentrations, as well as between the strains of species of *A. versicolor, P. corylophilum* and *P. steckii*; however, similarity was found in the growth of strains of *P. asymetrica subsection fasciculata*, of *P. chrysogenum*, of *E. repens* and of *C. cladosporioides*.

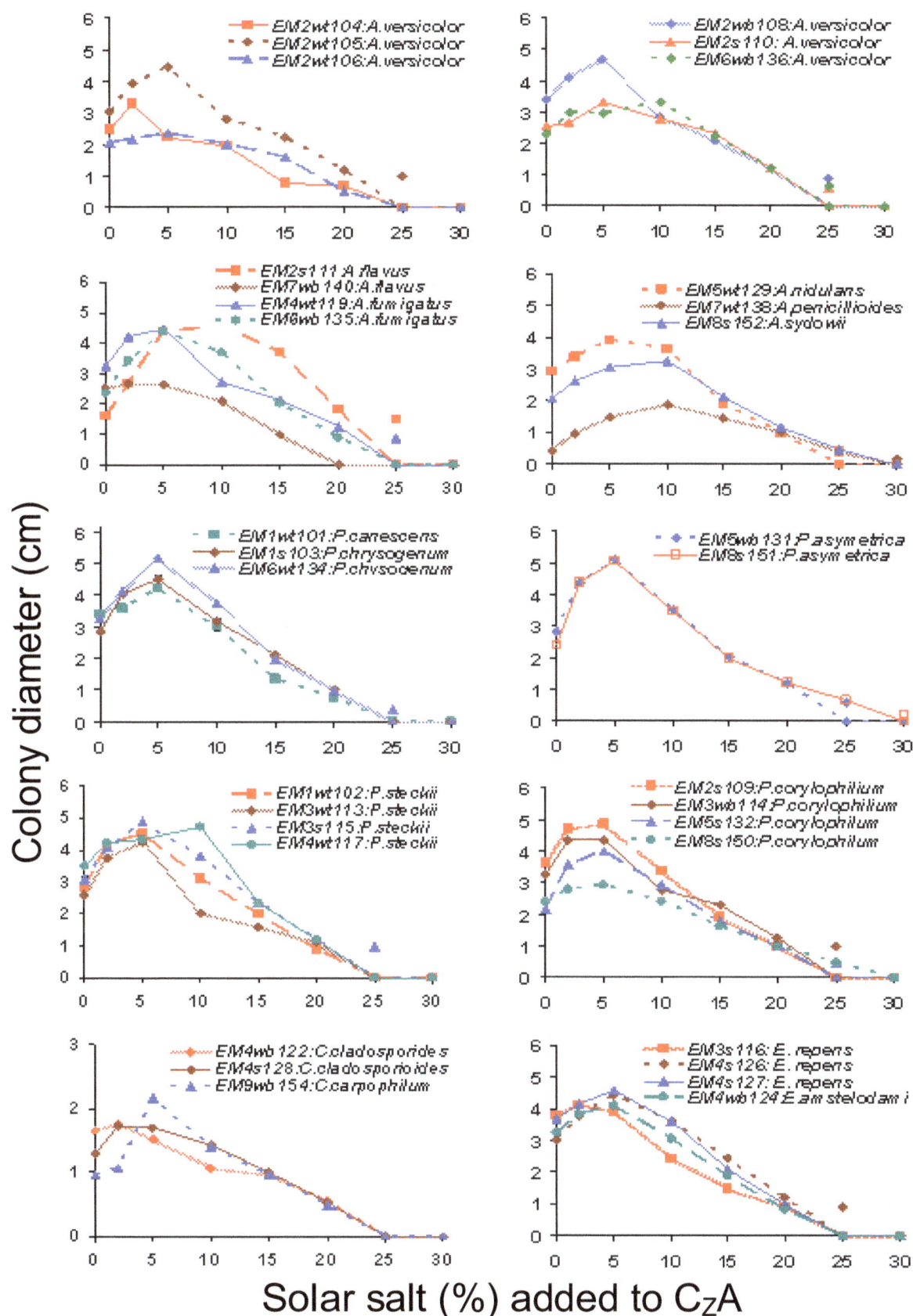

Figure 2b. Salt tolerance curves of facultative halophililc *Aspergillus*, *Penicillium*, *Cladosporium* and *Eurotium* as recorded after 7d incubation; unconnected symbols indicate delayed growth at respective salt concentrations, after 15d incubation.

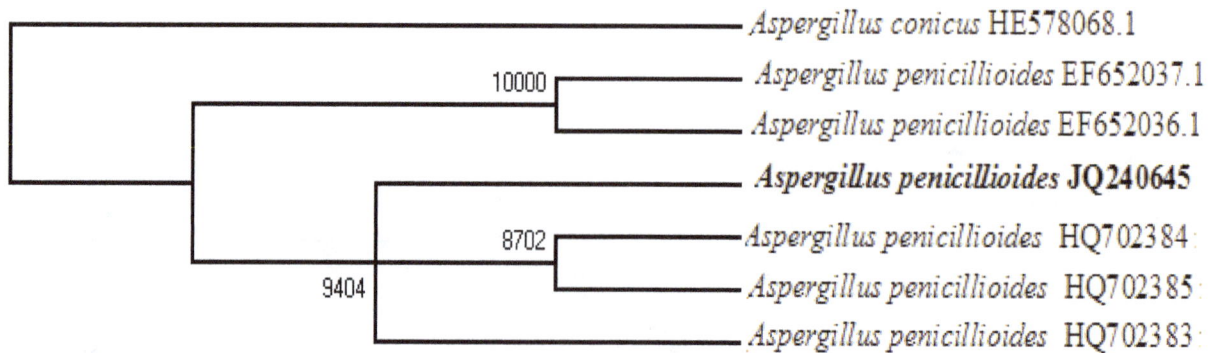

Figure 3. Phylogenetic tree obtained from the alignment of ITS region of rDNA of *A. penicillioides* species; the isolate *A. penicillioides* JQ240645 was obtained during the present study.

DISCUSSION

The work demonstrates the presence of moderate halophiles, in particular, the obligate halophile *A. penicillioides,* as well as other species belonging to the genera *Aspergillus, Penicillium, Eurotium* and *Cladosporium,* in the polyhaline environment of the estuary. The high concentration of salt used in the media for isolation, helped in the selective isolation of halophilic fungi over that of the non-halophiles.

Aspergillus and *Penicillium* were the dominant halophilic genera; *Eurotium* and *Cladosporium* were isolated in lesser numbers. While the genus *Aspergillus* had both obligate and facultative halophiles, *Penicillium, Cladosporium* and *Eurotium* species were exclusively facultative halophiles. These results corroborate earlier findings (Nazareth et al., 2012; Nayak et al., 2012). The obligate halophiles, by means of their absolute requirement for salt, are truly of marine origin. The facultative halophiles could be from terrestrial or fresh water environment, which have adapted so as to grow and sporulate in marine environment.

It was observed that the obligate halophilic *A. penicillioides* were found in greater numbers at the stations hinterland, although the salinity was lower, as compared to that at the mouth, where the salinity was nearly equal to that of sea water. Similar results were reported by Borut and Johnson (1962) wherein marine fungi were isolated from fresh and brackish sediments of an estuary, the irregular pattern of water currents giving rise to changes in environmental factors as well as determining the direction and extent of spore transport.

The presence of obligate halophiles upstream could be due to their ability to synthesize or take up compatible solutes from the estuarine environment, which is abundantly lined with mangroves. Tolerance of salinity has been viewed as a partial function of nutrient level (Borut and Johnson, 1962). It has been reported that fungi associated with standing litter of macrophytes could adapt to daily fluctuations of water availability through adjustment of their intracellular solute concentrations by

degradation of organic matter (Kuehn et al., 1998). It is known that in estuarine ecosystems, the detritus and marsh vegetation constitute a major part of the organic content (Manoharachary et al., 2005). The mangroves bordering the estuary, are more abundant upstream, where there is also a decrease in the cross-sectional area, and would serve as a nutrient source for the obligate halophiles. Earlier results have in deed shown the presence of the obligate halophile *A. penicillioides* in mangroves bordering the Mandovi estuary (Nayak et al., 2012). Tidal movement which extends to a distance of about 50 km (Sundar and Shetye, 2005), a point just beyond station 10, would aid the availability of the nutrient, the receding tide bringing particulate material from upstream of the estuary, and the rising tides carrying it back, thereby forming a nutrient reservoir (Verma and Agarwal, 2007). Interestingly, no halophile was obtained from the samples from station 10 hinterland where there were no mangroves.

Earlier reports have focused on isolation of halophiles only from hypersaline econiches (Buchalo et al., 1998; Gunde-Cimerman et al., 2000; Kis Papo et al., 2003a, b; Wasser et al., 2003; Steiman et al., 2004; Butinar et al., 2005a,b; Diaz-Munoz and Montalvo-Rodriguez, 2005; Gunde-Cimerman et al., 2005; Cantrell et al., 2006; Gunde-Cimerman et al., 2009; Hujslova et al., 2010; Nazareth et al., 2012; Nayak et al., 2012), which are expected to support the growth of such organisms. However, recent work records the finding of obligate halophiles from brackish waters of mangroves, with a salinity of 32‰ (Nayak et al., 2012). In this work, obligate halophiles were found in the polyhaline waters of the estuary, at a salinity range of 12 to 33‰.

This work demonstrates the occurrence of obligate moderately halophilic aspergilli in estuarine environment having salinity lower than that of the sea, thus indicating that hypersaline environments are not the sole econiches for true halophiles.

This is a first report on the isolation and description of an obligate halophilic *A. penicillioides* species from a polyhaline environment of an estuary.

ACKNOWLEDGEMENT

Authors are grateful to Mofeeda Gazem for help in the identification of the fungi.

REFERENCES

Anita DD, Sridhar KR, Bhat R (2009). Diversity of fungi associated with mangrove legume Sesbania bispinosa (Jacq.) W. Wight (Fabaceae). Livest. Res. Rur. Dev., 25(1): 67.

Borut SY, Johnson Jr. TW (1962). Some biological observations on fungi in estuarine sediments. Mycologia, 54(2): 181-193.

Buchalo AS, Nevo E, Wasser SP, Oren A, Molitoris HP (1998). Fungal life in the extremely hypersaline water of the Dead Sea: first records. Proc. R. Soc. Lon., 265: 1461-1465.

Butinar L, Sonjak S, Zalar P, Plemenitas A, Gunde-Cimerman N (2005a). Melanized halophilic fungi are eukaryotic members of microbial communities in hypersaline waters of solar salterns. Bot. Mar., 48: 73–79.

Butinar L, Zalar P, Frisvad JC, Gunde-Cimerman N (2005b). The genus Eurotium – members of indigenous fungal community in hypersaline waters of salterns. FEMS Microbiol. Ecol., 51: 155–166.

Cantrell SA, Casillas-Martinez L, Molina M (2006). Characterization of fungi from hypersaline environments of solar salterns using morphological and molecular techniques. Mycol. Res., 110: 962-970.

Cooke JC, LaCourse JR (1975). Preliminary Study of Microfungi from the Connecticut River Estuary. B. Torrey Bot. Club, 102(1): 1-6.

da Silva M, Umbuzeiro GA, Pfenning LH, Canhos VP, Esposito E (2003). Filamentous fungi isolated from estuarine sediments contaminated with industrial discharges. Soil Sediment Contam., 12(3): 345-356.

Diaz-Munoz G, Montalvo-Rodriguez R (2005). Halophilic black yeasts Hortaea werneckii in the Cabo Rojo Solar Salterns: its first record for this extreme environment in Puerto Rico. Caribb. J. Sci. 41: 360-365.

Domsch KH, Gams W, Anderson TH (1980). Compendium of Soil Fungi IHW-Verlag, Eching, p. 1.

D'Souza J, Araujo A, Karande A, Freitas YM (1979). Studies on fungi from coastal waters of Bombay and Goa. Ind. J. Mar. Sci., 8(2): 98-102.

Ellis MB (1971). Dermatiaceous hyphomycetes. Commonwealth mycological institute, Kew Surrey England.

Grant WD (2004). Life at low water activity. Phil. Trans. R. Soc. Lond. B., 359: 1249–1267.

Gunde-Cimerman N, Zalar P, De Hoog GS, Plemenitas A (2000). Hypersaline waters in salterns: natural ecological niches for halophilic black yeasts. FEMS Microbiol. Ecol., 32: 235-240.

Gunde-Cimerman N, Butinar L, Sonjak S, Turk M, Uršic V, Zalar P, Plemenitaš A (2005). Halotolerant and halophilic fungi from coastal environments in the Arctics. In: Gunde-Cimerman N, Oren A, Plemenitaš A (eds) Adaptation to life at high salt concentrations in Archaea, Bacteria and Eukarya. Springer, Netherlands, pp. 397-423.

Gunde-Cimerman N, Ramos J, Plemenitas A (2009). Halotolerant and halophilic fungi. Mycol. Res. 113: 1231-1241.

Hujslova M, Kubatova A, Chudickova M, Kolarik M (2010). Diversity of fungal communities in saline and acidic soils in the Soos National Natural Reserve, Czech Republic. Mycol. Progr., 9: 1–15.

Karamchand KS, Sridhar KR, Bhat R (2009). Diversity of fungi associated with estuarine sedge Cyperus malaccensis Lam. J. Agr. Tech., 5(1): 111-127.

Kis-Papo T, Grishkan I, Gunde-Cimerman N, Oren A, Wasser SP, Nevo E (2003a). Spatiotemporal patterns of filamentous fungi in the hypersaline Dead Sea. In: Nevo E, Oren A, Wasser SP (eds) Fungal Life in the Dead Sea. Gantner Verlag, Ruggel, pp. 271-292.

Kis-Papo T, Oren A, Wasser SP, Nevo E (2003b). Survival of filamentous fungi in hypersaline Dead Sea water. Microbial. Ecol., 45: 183-190.

Kuehn KA, Churchill PF, Suberkropp K (1998). Osmoregulatory Responses of Fungi Inhabiting Standing Litter of the Freshwater Emergent Macrophyte Juncus effuses. Appl. Environ. Microb., 64(2): 607–612.

Kushner DJ (1978). Life in high salt and solute concentrations. In: Kushner DJ (ed) Microbial Life in Extreme Environments. Academic Press, London, pp. 317–368.

Manoharachary C, Sridhar K, Singh R, Adholeya A, Suryanarayanan TS, Rawat S, Johri BN (2005). Fungal biodiversity: Distribution, conservation and prospecting of fungi from India. Curr. Sci. Ind., 89(1): 58-71.

Maria GL, Sridhar KR (2002). Richness and diversity of filamentous fungi on woody litter of mangroves along the west coast of India. Curr. Sci. Ind., 83(12): 1573-1580.

Mohamed DJ, Martiny JBH (2011). Patterns of fungal diversity and composition along a salinity gradient. Int. Soc. Microb. Ecol., 5: 379–388.

Nambiar GR, Raveendran K (2009). Frequency and abundance of marine mycoflora in mangrove ecosystem of North Malabar, Kerala (India). Acad. Plant Sci., 2(2): 65-68.

Nayak S, Gonsalves V, Nazareth S (2012). Isolation and salt tolerance of halophilic fungi from mangroves and solar salterns in Goa – India. Indian J. Mar. Sci. (In Press).

Nazareth S, Gonsalves V, Nayak S (2012). A first record of obligate halophilic aspergilli from the Dead Sea. Indian J. Microbiol., 52: 22-27.

Pearman JK, Taylor JE, Kinghorn JR (2010). Fungi in aquatic habitats near St Andrews in Scotland. Mycosphere, 1: 11-21.

Rai JN, Chowdhery HJ (1978). Microfungi from mangrove swamps of West Bengal. India Geophytology, 8(1): 103-110.

Rani C, Panneerselvam A (2009). Diversity of lignicolous marine fungi recorded from Muthupet environs, east coast of India. J. Agr. Biol. Sci., 4(5): 1-6.

Raper KB, Fennell DI (1965). The Genus Aspergillus. Williams and Wilkins Co, Baltimore.

Shearer CA (1972). Fungi of the Chesapeake Bay and Its Tributaries. III. The Distribution of Wood-InhabitingAscomycetes and Fungi Imperfecti of the Patuxent River. Am. J. Bot., 59(9): 961-969.

Steiman R, Ford L, Ducros V, Lafond JL, Guiraud P (2004). First survey of fungi in hypersaline soil and water of Mono Lake area (California). Antonie van Leeuwenhoek, 85: 69–83.

Sundar D, Shetye SR (2005). Tides in the Mandovi and Zuari estuaries, Goa, west coast of India. J. Earth Syst. Sci., 114(5): 493-503.

Tsui CKM, Hyde KD (2004). Biodiversity of fungi on submerged wood in a stream and estuary in the Tai Ho Bay, Hong Kong. Fungal Divers., 15: 205-220.

Verma PS, Agarwal VK (2007). Cell biology, genetics, molecular biology, evolution and ecology. S Chand and Company Ltd, New Delhi.

Wasser SP, Grishkan I, Kis-Papo T , Buchalo AS, Paul AV, Gunde-Cimerman N, Zalar P, Nevo E (2003). Species diversity of the Dead Sea fungi. In: Nevo E, Oren A, Wasser SP (eds) Fungal Life in the Dead Sea. Gantner Verlag, Ruggel, pp. 203-270.

20

Imaging flow cytometric analysis of *Schizosaccharomyces pombe* morphology

Radha Pyati[1]*, Lindsay L. Elvir[1], Erica C. Charles[1], Umawattee Seenath[1] and Tom D. Wolkow[2]

[1]Department of Chemistry, University of North Florida, Jacksonville, FL 32224, USA.
[2]Department of Biology, University of Colorado at Colorado Springs, Colorado Springs, CO 80933, USA.

The morphology of *Schizosaccharomyces pombe* can be rapidly monitored in asynchronous, G2-rich populations using imaging flow cytometry (IFC). Cell morphology was analyzed in terms of length and aspect ratio before and after exposure to several toxins. The toxins target the DNA (hydroxyurea and phleomycin) or cytoskeletal elements (thiabendazole, carbendazim and latrunculin A) and exert well-characterized effects on the morphology. Using IFC and yeast mutants, predictable morphological changes were detected accompanied with loss of gene products required during cellular responses to these toxins. IFC is a sensitive tool for accurate detection of subtle morphological changes in large, asynchronous *S. pombe* populations.

Key words: Imaging flow cytometry, fission yeast, *Schizosaccharomyces pombe*, cell cycle, morphology, genotoxin, cytoskeletal toxin.

INTRODUCTION

Changes in cell morphology accompany the development of prokaryotic and eukaryotic cells (Singh and Montgomery, 2011; Watanabe and Takahashi 2010). Changes in cell morphology also accompany genomic instability, as morphology is linked to mechanisms that influence chromosome stability of both cell types (Gisselsson, 2002; Moseley and Nurse, 2010). Therefore, identifying the molecular processes that influence morphology is important for understanding normal and abnormal developmental processes (Lleonart et al., 2000). *Schizosaccharomyces pombe*, or fission yeast, is a very good model system for studying cell morphology (Moseley and Nurse, 2010; La Carbona et al., 2006). *S. pombe* are cylindrically shaped cells that grow from both ends as they proceed through the cell cycle (La Carbona et al., 2006). During the cell cycle, cell length increases from roughly 7 µm to 12 -15 µm while width remains

constant at 3.5 µm (Nasim et al., 1989a). The cell cycle takes ~2.5 h to complete during exponential growth in liquid culture at 29.5°C. The majority of cells in an exponentially growing culture are in the G2 phase of the cell cycle, because it requires 70% of the cell cycle (1.75 h) to complete. Mitosis follows G2 and lasts about 5% of the cell cycle (7.5 min). Next, cytokinesis and septation divide the cytoplasm and result in the construction of new cell wall material called a septum. The G1 phase of both cells is completed in the next 15% of the cell cycle (22.5 min), before the septum is dissolved to produce two separate cells. The S-phase of both cells requires ~10% of the cell cycle (15 min) for completion and like G1, is initiated before septum dissolution (Nasim et al., 1989b).

Microtubules and actin are cytoskeletal components that play two critical roles during the cell cycle (La Carbona et al., 2006). First, they establish and maintain the linear growth axis of the yeast. Microtubules help localize actin to both ends of the cell, and actin incorporates new cell wall material at each end of the cell. Second, they are required to segregate DNA to daughter cells. Microtubules are major components of the mitotic spindle that pulls replicated sister chromatids to opposite ends of the cell. Actin is a major component of the contractile ring that then divides the cytoplasm at a

---
*Corresponding author. E-mail: radha.pyati@unf.edu.

Abbreviations: AR, Aspect ratio; **IFC,** imaging flow cytometry; **HU,** hydroxyurea; **MBC,** carbendazim; **Phleo,** phleomycin; **TBZ,** thiabendazole.

location between the segregated sister chromatids. Together, these cytoskeletal proteins have great influence over the growth and division patterns of *S. pombe*. Successful completion of a typical cell cycle results in the formation of two cells that inherit healthy genomes. Toxins that challenge the structure of DNA or the cytoskeleton, threaten genomic stability. Genotoxins such as phleomycin and hydroxyurea compromise DNA structure and are highly mutagenic. Toxins that compromise the structures of microtubules and actin affect linear growth, as well as the division machinery that segregates sister chromatids. Microtubules are damaged by toxins such as methylbenzimidazol-2-ylcarbamate (MBC) and thiabendazole (TBZ), and actin is damaged by Latrunculin A (LatA).

To protect against the threats of such substances, fungi and other eukaryotic cells are equipped with cell cycle checkpoints (Hartwell and Weinert, 1989; Ciccia and Elledge, 2002; Pietenpol and Stewart, 2002; Carr, 1995). These checkpoints are signal transduction pathways that delay cell cycle progression following insult to the genome or the cytoskeleton. Cell cycle delay allows extra time to address damage, so that it is fixed before final separation of the two daughter cells. Checkpoints operate during the entire cell cycle stage and can delay cell cycle progression at many points, including G1/S, intra S, G2/M, intra M and cytokinesis. Checkpoint signaling and the resulting delay to cell cycle progression do not, however, inhibit cell growth, which continues during the cell cycle delay and results in abnormally long cells. Mutations in checkpoint *rad* genes like *rad26⁺* and *rad24⁺*, compromise these cell cycle delays and permit cell cycle progression in the presence of DNA damage or cytoskeletal damage (al-Khodairy et al., 1994; Ford et al., 1994; Carr, 1995). Toxins that target the cytoskeleton activate cell cycle checkpoints as well (Gorbsky, 1997; Lew, 2000; Gachet et al., 2006). Clearly, responses to toxins that target DNA and cytoskeletal proteins affect cell morphology. Studying cell morphology has traditionally been accomplished using optical microscopic methods. But observing the characteristics of large cell populations using optical microscopy has presented challenges in imaging thousands of particles, such as experimenter fatigue. Imaging flow cytometry (IFC) combines the high particle throughput capacity of flow cytometry with a two-dimensional imaging system to acquire images of very large cell populations in a short period of time. Both traditional and imaging flow cytometers are extremely versatile and can perform multispectral analyses (Calvert et al., 2008). However, by virtue of its ability to capture images, IFC offers a straightforward way to conduct morphological characterization of cells, even in the absence of multispectral modes.

The aim of this work is to observe and quantify morphological changes caused by cell growth toxins of both wild-type and mutant *S. pombe*, and to do so in large, asynchronous cell populations that are difficult to

study using optical microscopy. This study reports the successful detection by IFC of predictable changes in cell length and shape that occur when asynchronous cultures of wild type and checkpoint-defective *rad26Δ* and *rad24Δ* strains were treated with a variety of genotoxins and cytoskeletal toxins. These results, and the promise that IFC holds for identifying compounds and defining the genes that influence cell shape in eukaryotes, are discussed. To our knowledge, this is the first report of an IFC study of fission yeast morphology.

MATERIALS AND METHODS

Reagents, strains and growth conditions

YE5S media was used in this study and made from yeast extract (Becton Dickinson Bacto), dextrose (anhydrous, BDH), adenine hemisulfate dehydrate, (MP Biomedicals, Solon, Ohio), L-histidine free base (MP Biomedicals), L-leucine (MP Biomedicals), and uracil (MP Biomedicals) (Moreno et al., 1991). The toxins and stock solutions used in this study were 200 mM hydroxyurea in water (HU; MP Biomedicals), 8 mg/ml carbendazim (MBC; Aldrich, Belgium), 20 mg/ml thiabendazole (TBZ; MP Biomedicals), 5 mg/ml phleomycin (phleo, Invivogen, San Diego, CA) and 100 µM latrunculin A (LatA, Biomol International, Plymouth Meeting, PA). All stock solutions were in DMSO, unless otherwise noted. Five strains were used in this study. Two *rad26⁺* wild-type strains were used; these were designated 236, with genotype *leu1-32, ura4-d18,h⁻* and 696, with genotype *leu1-32, ura4-d18,h⁺*. Two *rad26Δ* strains were used, designated 257, with genotype *rad26::ura⁺, ura4-d18, h⁺*; and 1001, with genotype *rad26::ura⁺, ura4-D18, leu1-32, ade6-704, h⁻*. One *rad24Δ* strain was used, designated 466 with genotype *rad24::ura4+, ade⁻, leu⁻, ura⁻, h⁻*.

Cells stored on agar at 4°C were initially cultured overnight in liquid YE5S media at 30°C and 120 rpm for 24 h. After incubation, optical density (OD) at 595 nm was measured using liquid YE5S media as a blank. Following starter culture, transfer volumes and growth times for final samples were carefully controlled in order to obtain OD_{595} values between 0.2 and 0.9, so that log-phase growth and sufficient cell populations could emerge, but plateau-stage growth and the accompanying senescence and exhaustion of media nutrients would not take place. Cell populations were asynchronous. The transfer volume ($V_{transfer}$) was calculated using the OD_{595} obtained from the starter culture and the formula as follows:

$$V_{transfer} = \frac{0.3}{actualOD}(\frac{1}{2^{N-1}}) * V_{final} \quad where \ N = \frac{t_{incub}}{t_{gen}}$$

The transfer volume was then halved or cut into thirds as needed. This $V_{transfer}$ was added into 5 ml of YE5S media and incubated at 30°C and 120 rpm for a minimum of 12 h. Following this uniform growth period, OD_{595} was measured again. If OD_{595} exceeded 0.3, the sample was diluted with media in order to achieve an OD_{595} of 0.3, to ensure that plateau growth and senescence did not occur during the 3-h period of toxin exposure. After any necessary dilution, samples were left untreated or treated with 10 mM HU, 8 µg/ml MBC, 20 µg/ml TBZ, 0.2 µM LatA, or 7.5 µg/ml phleo for 3 h. The untreated samples served as control samples that did not have any toxin added but underwent the same OD measurement, dilution, and 3-h growth period. OD_{595} was measured once more to ensure that all final OD_{595} values lay between 0.2 and 0.9. The samples were centrifuged at 5000 rpm for 2 min, washed in

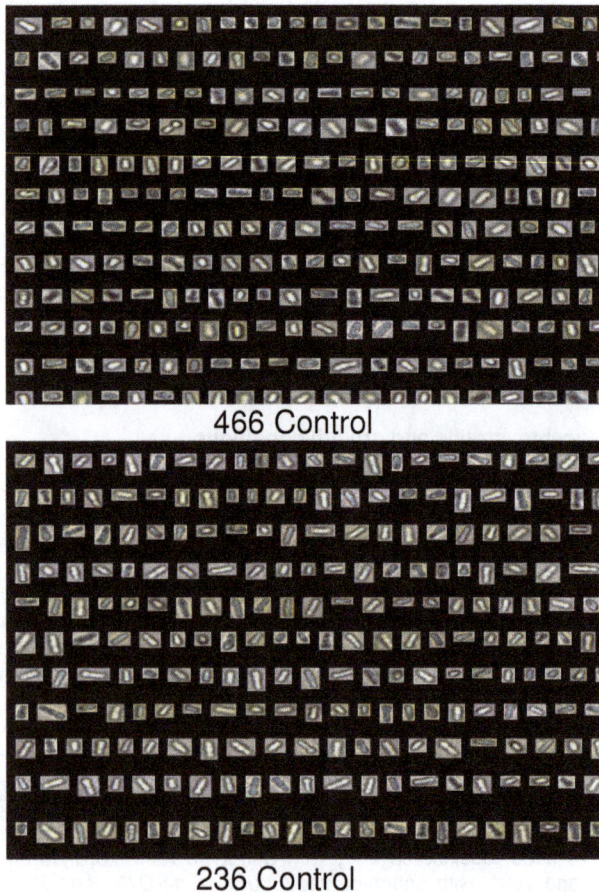

Figure 1. Representative cell image data for a minimum of 10,000 cell images per run captured by FlowCAM imaging flow cytometer (IFC). The IFC directs the cell suspension through a 50-μm flow cell positioned in front of a 20X objective and a camera that captures real-time images of the particles in the fluid as they pass through the flow cell. Each image is stored separately along with up to 26 individual particle measurements, including length and aspect ratio, for each image. Each run was set to stop upon reaching 40,000 particle images. These were then filtered to remove undesirable images like partial cells, blurry images, and multiple cells to leave a minimum of 10,000 images for each experiment. A table containing 26 parameters for each cell was exported to Microsoft Excel, within which median values of length and aspect ratio were calculated.

phosphate buffered saline (0.2 M Phosphate, 1.5 M NaCl), aspirated, centrifuged again, and resuspended in 70% cold ethanol to preserve samples. The samples were stored at 4°C until ready for flow cytometric analysis.

Imaging flow cytometry

Samples in ethanol were resuspended in PBS for analysis using the image flow cytometer. Samples underwent two cycles of the following: centrifugation at 5,000 rpm for 2 min, removal of supernatant, dilution in PBS, aspiration, and vortexing to break-up cell clumps. Final samples in PBS were incubated for 30 min prior to flow cytometric analysis. Five to seven drops of the sample

solution was placed into the opening of the FlowCAM imaging flow cytometer (Fluid Imaging Technologies, Yarmouth, ME). Figure 2 depicts the FlowCAM imaging flow cytometer (IFC). The fluidics system uses a peristaltic pump to pull the fluid sample through a flow cell perpendicular to the optical path. In this study, the pump was set to a value of 8. Each run was set to stop upon reaching 40,000 particle images. Fluorescence detection was not used in this study.

Data analysis

In each experiment, 40,000 images were rapidly collected and then filtered to remove undesirable images like partial cells, blurry images, and multiple cells to leave a minimum of 10,000 images for each experiment. One more filtering step was done by creating a library of cell images that possessed the general shape of fission yeast cells but included a wide range of lengths and aspect ratios. A table containing 26 parameters for each cell was exported to Microsoft Excel. The two parameters of interest in this work were length and aspect ratio; median values of these were calculated in Excel. Triplicate trials of each experiment yielded average values of these, along with standard deviations for each.

RESULTS AND DISCUSSION

Cell length and aspect ratio

Figure 1 shows two representative examples of image data collected in this study. The two collages of cell images show a random sample of cell images for strains 236 (wild-type) and 466 (*rad24Δ*). Table 1 shows results for cell length and aspect ratio for all strains, including untreated-control and toxin-exposed. Comparison between control and toxin-exposed was done by calculating percent change in a given parameter relative to control. For example, percent change in median cell length (PC_L) relative to the control was calculated by the formula:

$$PercentChange = \left(\frac{L_{toxin} - L_{control}}{L_{control}} \right) * 100$$

Standard deviations for PC_L were obtained using the standard deviations for both L_{toxin} and $L_{control}$ and a partial derivative method that incorporates both errors (Mortimer, 1981). Aspect ratio was handled in the same manner. Percent change values are also displayed in Table 1. Figures depicting these results are shown with error bars consisting of that calculated standard deviation above and below the median; these appear large because the plot shows percent change, not percent of $L_{control}$.

Cell length

IFC was used to measure the cell length of five different strains: two *rad⁺* strains, two *rad26Δ* strains and one

Table 1. Cell lengths and aspect ratios, and percent change in median cell length and median aspect ratio of toxin study.

Sample	Median cell length (um)	Percent change in median cell length relative to control (%)	Median aspect ratio	Percent change in median aspect ratio relative to control (%)
Rad⁺ (wild-type)				
236Control	7.20 ± 0.37	--	0.47 ± 0.03	--
236Phleo	11.23 ± 1.26	56.0 ± 19.2	0.29 ± 0.07	-38.7 ± 15.3
236HU	7.87 ± 0.20	9.3 ± 6.3	0.45 ± 0.01	-5.6 ± 5.6
236TBZ	7.59 ± 0.05	5.4 ± 5.5	0.45 ± 0.02	-4.2 ± 6.7
236MBC	7.20 ± 0.16	0.0 ± 5.6	0.48 ± 0.03	0.7 ± 7.5
236LatA	8.31 ± 1.20	15.4 ± 17.7	0.45 ± 0.06	-5.6 ± 13.0
696Control	8.01 ± 0.06	--	0.43 ± 0.01	--
696Phleo	11.84 ± 0.43	47.8 ± 5.5	0.22 ± 0.02	-48.5 ± 3.6
696HU	8.95 ± 0.23	11.7 ± 3.0	0.39 ± 0.02	-10.0 ± 4.8
696TBZ	8.29 ± 0.09	3.5 ± 1.4	0.38 ± 0.00	-12.3 ± 1.2
696MBC	8.15 ± 0.08	1.7 ± 1.3	0.40 ± 0.01	-6.9 ± 1.8
696LatA	10.44 ± 0.07	30.3 ± 1.3	0.38 ± 0.01	-12.3 ± 2.6
Rad26Δ				
257Control	7.47 ± 0.08	--	0.46 ± 0.00	--
257Phleo	7.12 ± 0.03	-4.7 ± 1.1	0.47 ± 0.00	2.2 ± 0.0
257HU	7.04 ± 0.07	-5.8 ± 1.4	0.49 ± 0.01	6.5 ± 2.2
257TBZ	7.57 ± 0.10	1.3 ± 1.7	0.46 ± 0.01	-0.7 ± 1.3
257MBC	7.32 ± 0.09	-2.0 ± 1.6	0.46 ± 0.01	0.0 ± 2.2
257LatA	9.06 ± 0.12	21.3 ± 2.1	0.43 ± 0.01	-7.2 ± 1.3
1001Control	7.03 ± 0.59	--	0.51 ± 0.05	--
1001Phleo	6.94 ± 0.16	-1.3 ± 8.6	0.53 ± 0.02	4.6 ± 11.1
1001HU	7.12 ± 0.10	1.3 ± 8.6	0.58 ± 0.01	13.1 ± 11.7
1001TBZ	7.19 ± 0.21	2.3 ± 9.1	0.52 ± 0.02	2.0 ± 10.9
1001MBC	7.09 ± 0.45	0.9 ± 10.6	0.55 ± 0.03	8.5 ± 12.7
1001LatA	9.72 ± 0.32	38.3 ± 12.5	0.46 ± 0.01	-9.2 ± 9.3
Rad24Δ				
466Control	6.68 ± 0.09	--	0.58 ± 0.01	--
466Phleo	6.64 ± 0.07	-0.6 ± 1.7	0.60 ± 0.01	2.9 ± 2.7
466HU	6.03 ± 0.13	-9.7 ± 2.3	0.63 ± 0.01	8.6 ± 2.5
466TBZ	6.37 ± 0.04	-4.6 ± 1.4	0.66 ± 0.02	13.2 ± 4.1
466MBC	6.24 ± 0.55	-6.6 ± 8.3	0.61 ± 0.00	5.2 ± 1.8
466LatA	8.86 ± 0.03	32.6 ± 1.8	0.50 ± 0.01	-14.4 ± 1.8

Values given are means of three replicates ± S.D. (each trial contains 10,000 cells minimum).

rad24Δ strain. Asynchronous cultures of these strains were grown in complete liquid medium to an OD between 0.2 and 0.9 before processing them for IFC. Overall, median cell length fell between 6.85 and 8.23 μm, although slight differences between strains were observed. For example, the rad⁺ strains differ by 0.81 μm and the two rad26Δ strains differ by 0.44 μm. Strain backgrounds therefore affect cell length. Next, asynchronous cultures of each strain were treated for 3 h with five different toxins, each of which activates a cell cycle checkpoint mechanism that delays cell cycle progression but not cell growth. These toxins may be categorized by similar mechanisms of action, but within each category, differences exist between toxins. Their effects are shown in Figure 2A. Phleo and HU are both genotoxins that delay progression from G_2 into mitosis by rad⁺-dependent mechanisms (Enoch and Nurse, 1990; Baschal et al., 2006) however, they operate in different ways. Phleo causes DNA breaks (Sleigh and Grigg, 1977) and damages DNA throughout the 2.5 h cell cycle, including G_2, where ~70% of cells in an asynchronous culture reside. Following treatment with phleo, those cells closest to the G_2/M boundary will delay entry into mitosis sooner than those cells that are positioned earlier in the cell cycle. After 3 h of treatment, the cells closest to the G_2/M boundary when the drug was added should therefore be longer than those that were in early G_2 or S-phase when the drug was added. The ~5% of cells in mitosis, when phleo was added, will display very little elongation, since these require another ~2 h 22 min (95% of 2.5 h) of growth before they reach the G_2/M boundary. HU, on the other hand, alters deoxyribonucleoside triphosphate pools and blocks DNA synthesis (Bianchi et al., 1986). In an asynchronous culture, HU only affects ~10% of cells that are in S-phase and causes them to delay at the G_2/M boundary. Those cells in G_2 at the time of HU addition will divide and enter S-phase of the next cell cycle, at which point HU will affect their DNA synthesis. Therefore, HU should have a less pronounced effect on cell length when compared to phleo.

As anticipated, phleo caused wild type cells to elongate 45 to 55% of control length while HU caused them to elongate ~10% of control length. Also as anticipated, the three $rad\Delta$ strains lacking a functional G_2/M checkpoint failed to display significant elongation. LatA is a toxin that inhibits polymerization of actin microfilaments (Coue et al., 1987) and at low concentration, blocks actomyosin-based contractile ring function in S. pombe without significantly affecting elongation (Liu et al., 2000). Cells treated with LatA complete mitosis, but due to an inability to complete cytokinesis, are delayed for entry into mitosis of the following cell cycle. In liquid culture, this delay eventually results in a large population of elongated cells with two G_2 nuclei. Since this delay is dependent upon Rad24, $rad24\Delta$ cells undergo multiple nuclear divisions during LatA treatment but continue to elongate due to an inability to perform cytokinesis. Therefore, elongated and multinucleated cells are produced when a culture of $rad24\Delta$ cells is treated with LatA. As expected, in this study, LatA treatment caused elongation of all strains to, 15 to 38% of the control. The fact that LatA's effect on length was less than phleomycin and greater than hydroxyurea is consistent with expectation, because LatA blocks cytokinesis, a cell cycle event that occurs between G_2 and S-phase.

TBZ and MBC are both microtubule drugs that disrupt microtubule polymerization in dose-dependent manners. Snaith and Sawin (2003) observed that MBC is a more potent inhibitor that almost completely disrupts microtubule architecture at doses above 15 to 20 μg/ml, while TBZ fails to completely disrupt this architecture at the extremely high doses of 100 μg/ml. MBC is also more specific than TBZ, which disrupts actin polarity determinants that remain unperturbed during MBC treatment (Snaith and Sawin, 2003). Using elutriation to produce a synchronous population of wild-type cells, Snaith and Sawin observed that TBZ inhibited cell elongation and MBC did not. In the present study, little change in the length of wild-type or $rad\Delta$ strains was

observed after both 3 h treatments. These inconsistencies most likely reflect experimental procedures; much lower concentrations of MBC (8 μg/ml) and TBZ (20 μg/ml) were used in this work. Overall, phleo produced the greatest effect on wild-type cell length because it can affect cells in G_2. LatA produced the next greatest effect on cell length, because it targets the actin machinery required for cytokinesis, followed by HU that targets S-phase cells. TBZ and MBC produced little change in cell length.

Cell aspect ratio

Cell aspect ratio (AR) is also a useful measure of cellular morphology. AR is defined in this work as a unitless quantity equal to the width of the cell divided by its length. The pattern recognition software in the FlowCAM instrument identifies length as the longest dimension of a cell, and width as the shortest dimension. Generally cells like S. pombe that are shaped like a cylinder have an aspect ratio of less than one. An AR of one represents a circular or square object. The higher the AR, or the closer it approaches one, the more rounded the cell morphology. The lower the AR, or the further from one, the more elongated the morphology. Rad24 is a 14-3-3 protein required during responses to DNA damage and both microtubule and actin cytoskeleton damage (Ford et al., 1994; Mishra et al., 2005; Baschal et al., 2006; van Heusden and Steensma, 2006). It also functions to help regulate a pheromone signaling pathway that regulates a switch from mitotic to meiotic division (Ozoe et al., 2002). Due to one or a combination of these roles, loss of $rad24\Delta$ results in a striking cone-shaped spherical morphology (Ford et al., 1994). Rad26 is a coiled-coil protein that responds to structural abnormalities of DNA and the microtubule cytoskeleton (al-Khodairy et al., 1994; De Souza et al., 1999; Herring et al., 2010). Loss of its microtubule-related role by the $rad26.4a$ allele results in the production of shorter cells that retain their cylindrical shape but have slightly altered length/width ratios (L/W = 2.06, AR = W/L = 0.485) when compared to wild-type cells (L/W = 2.53, AR = W/L = 0.395) (Herring et al., 2010).

Therefore, both $rad24\Delta$ and $rad26\Delta$ cells have altered morphologies although $rad24\Delta$ cells have a more rounded appearance. AR measurements confirmed that $rad24\Delta$ cells were the most spherical (AR = 0.58), while strain differences prevent a comparison of the ARs of the two wild-type and $rad26\Delta$ strains. As expected, both wild-type strains retain their cylindrical appearances following all five toxin treatments, although strain differences were apparent (Figure 2B). The $rad26\Delta$ and $rad24\Delta$ strains become more spherical following phleo and HU, likely because a population of the asynchronous cells fail to arrest division and instead divides to produce short, dead cells in the presence of these drugs. It is not clear why

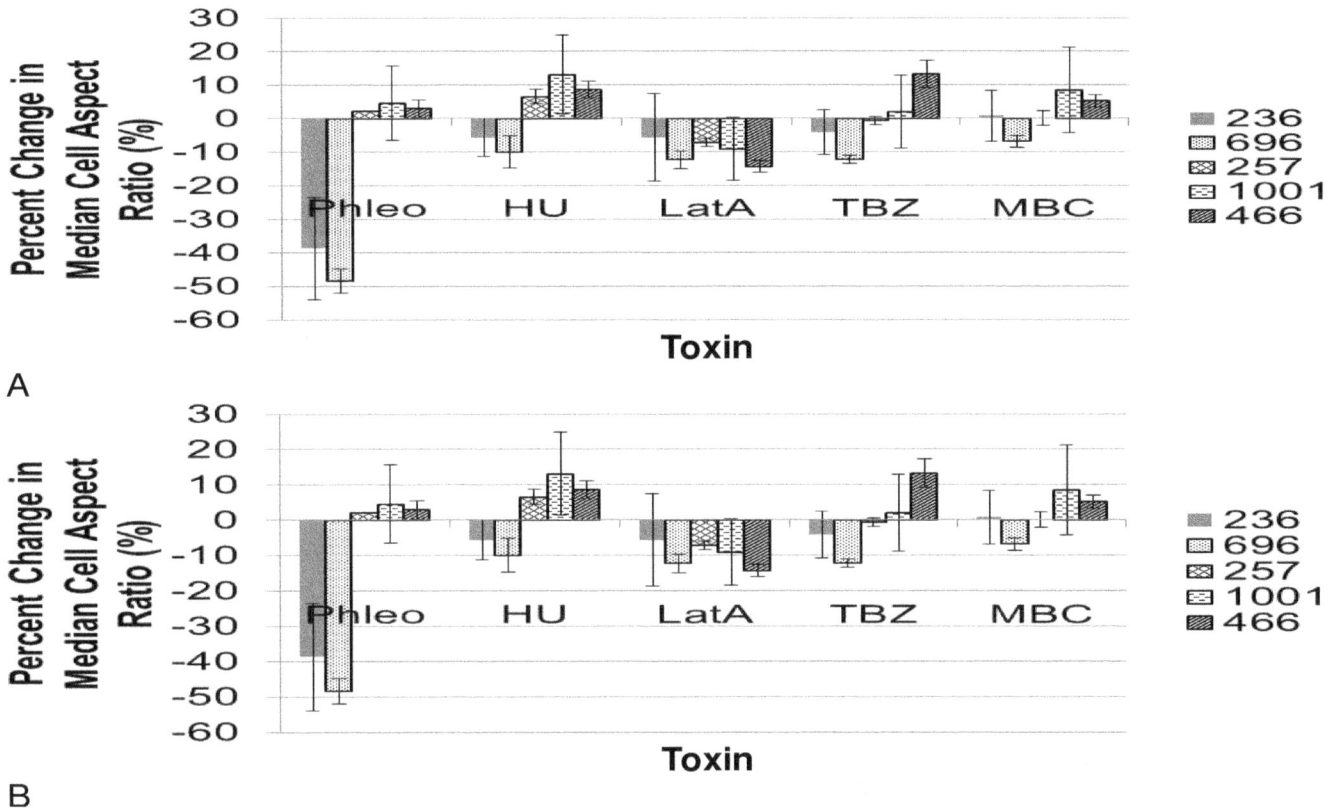

Figure 2. A) Percent change in median cell length versus toxin. Error bars represent one calculated standard deviation above and below the median. B) Percent change in median aspect ratio versus toxin. Error bars represent one calculated standard deviation above and below the median.

they become more spherical following HU treatment compared to phleo treatment. These genes may be required to preserve cylindrical morphology after replication forks are stalled, considering that budding yeast checkpoint proteins have been shown to control cell morphology during HU treatment (Enserink et al., 2006). Changes in the ARs of the *radΔ* strains differed following MBC and TBZ treatments, consistent with the observation that TBZ targets processes in addition to those targeted by MBC. For example, TBZ increases the AR of the *rad24Δ* by 13%, while MBC increases it by 6%. On the other hand, TBZ only increases the AR of the 1001 *rad26Δ* by 2%, while MBC increases the AR of this strain by almost 9%. The AR of the 257 *rad26Δ* strain does not change following either TBZ or MBC treatment. Not surprisingly, none of the strains becomes more spherical during LatA treatment, which at this low dose targets the cytokinetic machinery only.

In summary, the purpose of this study was to determine if IFC can be used to monitor two morphological characteristics, length and AR, of *S. pombe* cells grown in asynchronous cultures. To address both characteristics, a variety of different strains were treated with a variety of different toxins in an effort to produce a range of predictable cell lengths and ARs. With respect to monitoring length, IFC was able to capture differential increases in cell length that occur when wild-type and *radΔ* cells are treated with drugs that target different cell structures at different cell cycle positions (Figure 2A). Likewise, IFC also captured the spherical appearances of untreated *rad24Δ* cells and toxin-treated *radΔ* cells (Figure 2B). We conclude that IFC is a reliable technique that can monitor these two morphological characteristics of *S. pombe*.

Applications of IFC using *S. pombe*

IFC may prove useful for linking molecular characteristics of cells to their morphology in two different ways. First, it may be used to rapidly screen synthetic or natural product chemical libraries for small molecules that affect cell length or AR. Those that affect length will likely target essential cellular structures, like DNA or the contractile ring, and result in cell cycle delays and concomitant increases in cell length. The extent of length increase should also help define the point in the cell cycle where drugs act. Routine genetic screens can then be performed to identify the target of this small molecule. Such small molecules may have use in basic and translational biomedical research. IFC may also be used to rapidly screen *S. pombe* mutant collections in search of those with elevated ARs.

This would lead to the identification of non-essential genes, like *rad24⁺*, that influence cell shape and would otherwise be very difficult to identify using laborious genetic methods (Verde et al., 1995). While IFC cannot replace sophisticated high-throughput image capture platforms used to study more complex organisms (Shamir et al., 2010). It can be used to study the relatively simple morphology of single celled model organisms like yeast in search of evolutionarily conserved mechanisms that regulate eukaryotic morphology.

ACKNOWLEDGEMENT

The authors thank the University of North Florida for supporting this work.

REFERENCES

al-Khodairy F, Fotou E, Sheldrick KS, Griffiths DJ, Lehmann AR, Carr AM (1994). Identification and characterization of new elements involved in checkpoint and feedback controls in fission yeast. Mol. Biol. Cell., 5(2): 147-60.

Baschal EE, Chen KJ, Elliott LG, Herring MJ, Verde SC, Wolkow TD (2006). The fission yeast DNA structure checkpoint protein RAD26$^{ATRIP/LCDI/UVSD}$ accumulates in the cytoplasm following microtubule destabilization, BMC Cell Biol., 7:32.

Bianchi V, Pontis E, Reichard P (1986). Changes of deoxyribonucleoside triphosphate pools induced by hydroxyurea and their relation to DNA synthesis. J. Biol. Chem., 261: 16037-16042.

Calvert MEK, Lannigan JA, Pemberton LF (2008). Optimization of yeast cell cycle analysis and morphological characterization by multispectral imaging flow cytometry. Cytometry Part A, 73A: 825-833.

Carr AM (1995). DNA structure checkpoints in fission yeast. Semin. Cell Biol., 6: 65-72.

Ciccia A, Elledge SJ (2002). The DNA Damage Response: Making It Safe to Play with Knives. Mol. Cell., 40: 179-204.

Coue M, Brenner SL, Spector I, Korn ED (1987). Inhibition of actin polymerization by latrunculin A. FEBS Lett., 213: 316-318.

De Souza CP, Ye XS, Osmani SA (1999). Checkpoint defects leading to premature mitosis also cause endoreplication of DNA in Aspergillus nidulans. Mol. Biol. Cell., 10: 3661-3674.

Enoch T, Nurse P (1990). Mutation of fission yeast cell cycle control genes abolishes dependence of mitosis on DNA replication. Cell, 60: 665-73.

Enserink JM, Smolka MF, Zhou H, Kolodner RD (2006). Checkpoint proteins control morphogenetic events during DNA replication stress in *Saccharomyces cerevisiae*. J. Cell Biol., 175: 729-741.

Ford JC, al-Khodairy F, Fotou E, Sheldrick KS, Griffiths DJ, Carr AM (1994). 14-3-3 protein homologs required for the DNA damage checkpoint in fission yeast. Science, 265: 533-535.

Gachet Y, Reyes C, Goldstone S, Tournier S (2006). The fission yeast spindle orientation checkpoint: A model that generates tension? Yeast, 23(13): 1015-1029.

Gisselsson D (2002). Tumour morphology – interplay between chromosome aberrations and founder cell differentiation. Histol. Histopathol., 17: 1207-1212.

Gorbsky GJ (1997). Cell cycle checkpoints: arresting progress in mitosis. Bioessays, 19(3): 193-197.

Hartwell LH, Weinert TA (1989). Checkpoints: controls that ensure the order of cell cycle events. Science, 246: 629-634.

Herring M, Davenport N, Stephan K, Campbell S, White R, Kark J, Wolkow TD (2010). Fission yeast RAD26ATRIP delays spindle-pole-body separation following interphase microtubule damage. J. Cell Sci., 123(Pt 9): 1537-1545.

La Carbona S, Le Goff C, LeGoff X (2006). Fission yeast cytoskeletons and cell polarity factors: connecting at the cortex. Biol. Cell, 98: 619-631.

Lew DJ (2000). Cell-cycle checkpoints that ensure coordination between nuclear and cytoplasmic events in Saccharomyces cerevisiae. Curr. Opin. Genet. Dev., 10(1): 47-53.

Liu J, Wang H, Balasubramanian, MK (2000). A checkpoint that monitors cytokinesis in Schizosaccharomyces pombe. J. Cell Sci., 113: 1223-1230.

Lleonart ME, Martin-Duque P, Sanchez-Prieto R, Moreno A, Ramon y, Cajal S (2000). Tumor heterogeneity: morphological, molecular and clinical implications. Histol. Histopathol., 15: 881-898.

Mishra M, Karagiannis J, Sevugan M, Singh P, Balasubramanian MK (2005). The 14-3-3 protein rad24p modulates function of the cdc14p family phosphatase clplp/flplp in fission yeast. Curr. Biol., 15: 1376-1383.

Moreno S, Klar A, Nurse P (1991). Molecular genetic analysis of fission yeast Schizosaccharomyces pombe. Methods Enzymol., 194: 795-823.

Mortimer RG (1981). Mathematics for physical chemistry. Macmillan Publishing Co,. Inc., New York, p. 280.

Moseley JB, Nurse, P (2010).Cell division intersects with cell geometry. Cell, 142: 184-188.

Nasim A, Young PG, Johnson BF (1989a). Editors, Molecular Biology of the Fission Yeast (Cell Biology Series). New York: Academic Press, New York, p. 332.

Nasim A, Young PG, Johnson BF (1989b). Editors, Molecular Biology of the Fission Yeast (Cell Biology Series). New York: Academic Press; New York, 1989, p 217.8a.

Ozoe F, Kurokawa R, Kobayashi Y, Jeong HT, Tanaka K, Sen K, Nakagawa T, Matsuda H, Kawamukai M (2002). The 14-3-3 proteins Rad24 and Rad25 negatively regulate Byr2 by affecting its localization in Schizosaccharomyces pombe. Mol. Cell. Biol., 22: 7105-7119.

Pietenpol JA, Stewart ZA (2002). Cell cycle checkpoint signaling: cell cycle arrest versus apoptosis. Toxicology, 181-182: 475-481.

Shamir L, Delaney JD, Orlov N, Eckley DM, Goldberg IG (2010). Pattern recognition software and techniques for biological image analysis. PLoS Comput. Biol., 6(11): e1000974.

Singh SP, Montgomery BL (2011). Determining cell shape: regulation of cyanobacterial cellular differentiation and morphology. Trends Microbiol, 19: 278-285. DOI: 10.1016/j.tim.2011.03.001.

Sleigh MJ, Grigg GW (1977). Sulphydryl-mediated DNA breakage by phleomycin in *Escherichia coli*. Mutat. Res., 42: 181-190.

Snaith HA, Sawin KE (2003). Role of microtubules and tea1p in establishment and maintenance of fission yeast cell polarity. J. Cell Sci., 117(5): 689-700.

van Heusden GP, Steensma HY (2006). Yeast 14-3-3 proteins. Yeast, 23: 159-171.

Verde F, Mata J, Nurse P (1995). Fission yeast cell morphogenesis: Identification of new genes and analysis of their role during the cell cycle. J. Cell Biol., 131(6 Pt 1): 1529-1538.

Watanabe T, Takahashi Y (2010). Tissue morphogenesis coupled with cell shape changes. Curr. Opin. Genet. Dev., 20: 443-447. DOI: 10.1016/j.gde.2010.05.004.

21

Isolation and physiological characterization of indigenous yeasts from some Algerian agricultural and dairy products

Meriem Amina Rezki[1,2], Laurent Benbadis[2], Gustavo DeBillerbeck[2,3], Zoubida Benbayer[1] and Jean Marie François[2]

[1]Laboratoire de Biotechnologie des Rhizobia et Amélioration des Plantes (LBRAP), Département de Biotechnologie, Faculté des Sciences, Université d'Es-Senia, BP 1524 El Mnaouar 31 000 Oran, Algérie.
[2]Université de Toulouse; INSA, UPS, Laboratoire d'Ingénierie des Systèmes Biologiques et des Procédés, INRA-UMR792 and CNRS-UMR5504; 135 Avenue de Rangueil; F-31400 Toulouse, France.
[3]INP-ENSAT, Avenue de l'Agrobiopole, F-31326 Castanet-Tolosan Cedex, France.

The purpose of this study was to isolate yeasts that may express original characteristics notably in terms of flavour from traditional Algerian agricultural and dairy products. A total of eighteen yeast isolates were recovered from dates, melons and gherkins, and from milk of camel, goat, sheep and cow. Molecular taxonomy based on the sequences of the D1/D2 domain of the 26S ribosomal RNA gene, grouped these isolates into nine yeast species. The identification of *Clavispora lusitaniae*, *Hanseniaspora uvarum*, *Kodamaea ohmeri*, *Pichia kudriavezii*, *Zygosaccharomyces baili*, *Zygosaccharomyces rouxii* and *Yarrowialipolytica* was in accordance with the physico-chemical characteristics of agricultural and dairy products analyzed. The *P. kudriavezii* species showed resistance to ethanol and osmotic stress, as well as production of alcohols that was remarkably higher than the laboratory *S. cerevisiae* strain CEN.PK. In conclusion, this study draws attention to the use of this yeast species for biotechnological application in the field of flavour production.

Key words: Yeast taxonomy, microbial physiology, stress, fermentation, flavours.

INTRODUCTION

For millenniums, traditional industrial biotechnology has used yeasts and filamentous fungi in the production of large variety of consumer goods like beer, wine and fermented foods. Nowadays, there is a shift towards modern industrial biotechnology, also termed white biotechnology, to harness energy and produce bio-based chemicals from renewable carbon sources using microorganisms (Octave and Thomas, 2009). There is also a growing interest in the biodiversity and ecology of yeasts associated with different foods either as biocontrol agents of food spoilage (Fleet, 2007) or as new vectors for producing metabolites of high added value such as flavours and aromas (Berger, 2009). Potential relevance

of these studies could be to exploit these original traits expressed by these 'wild' yeasts in biotechnological applications, such as microbial production of flavour compounds (Lomascolo et al., 1999), isoprenoids derivatives for use in biofuel (Kirby and Keasling, 2008) or to transfer genes bearing specific traits into genetically tractable hosts to enhance their physiological performance (Fortman et al., 2008).

The remarkable development in molecular techniques associated with the ease of DNA sequencing allows a rapid investigation of the biodiversity and ecology of microorganisms isolated from traditional food, beverages or other natural biotopes. Current methods for yeasts identification are based on restriction-fragment length polymorphism of 5.8S-ITS rDNA region or sequencing of the D1 and D2 domains in the 26S rDNA gene (Kurtzman and Robnett, 1998). The purpose of this work was to explore the biodiversity of yeasts in traditional Algerian food and dairy products. We reported on the isolation of 9 different yeast species isolated from these traditional Algerian agricultural products, and investigated five out of these nine isolates for their growth performance on various sugars, resistance to high ethanol and osmotic pressure as well as for their capacity to produce specific aroma and flavours in comparison to the well-studied laboratory yeast *Saccharomyces cerevisiae* CEN.PK122-2N (van Dijken et al., 2000).

MATERIALS AND METHODS

Biological sampling and isolation of yeasts

The following products were collected from local markets in different Algerian cities: three varieties of dates (Deglet-Nour from Ghardaïa; Boufakous-gharess from Bechar and Hamira from Ourgla), milk of cow, goat, sheep and camel were from Ghardaïa; melon and gherkins from a local market at Oran and honey from Mostaganem. Yeasts were isolated from dates, gherkins and melons after grounding them in a sterile mortar, collecting one gram of the paste and mixing it with 9 ml of 2% (w:v) sodium citrate. The resulting solution was serially diluted in sterile water and 0.1 ml was spread on a solid OGA containing 2% (w/v) glucose, 0.5% (w/v) yeast extract, 2% (w/v) agar supplemented with 0.5 mg/ml chloramphenicol, 0.1 mg/ml oxytetracycline and 3 mg/ml thiabendazole. The plates were incubated at 30°C for three days. Isolation of yeast species from milk of camel, cow, goat and sheep was obtained after its coagulation by lactic acid fermentation. Following this step, the samples were checked for the apparition of yeasts under the microscope and using gentian violet crystal. Yeast colonies showing different size, shape and/or color were picked up, examined under the microscope, and finally isolated by an enrichment technique in OGA plates as above. Isolated colonies were streaked and stored at 4°C on Sabouraud agar slants (2% (w/v) glucose, 1% (w/v) bactopeptone, 1.5% (w/v) agar and 0.5 mg/ml chloramphenicol. After molecular identification, the yeast species were stored at -80°C in YEPD medium containing 25% (v/v) glycerol.

Identification of yeast isolates

The molecular identification of yeast isolates was carried out by the "Centre International de Ressources Microbienne" (CIRM; http://www7.inra.fr/cirmlevures/page.php?lang=fr&page= home) in Thiverval Grignon, Paris. The molecular taxonomy employed the procedure developed by Kurtzmann and Robnett (1997, 1998). The method is based on PCR amplification of the D1 and D2 domain of the large subunit (26S) rRNA gene followed by sequencing and comparison of the nucleic sequences using a BLAST search (Altschul et al., 1997) with the GenBank database of NCBI.

Culture media and growth conditions

Unless otherwise stated, yeast cultures were carried in 1 L Erlenmeyer flasks containing 0.30 L of either glucose rich medium

(YEPD) (20 g glucose, 10 g yeast extract, 10 g Bactopeptone per liter), glucose-synthetic medium (SD) (20 g glucose, 1.7 g Yeast nitrogen base without amino acids and ammonium; 5 g ammonium sulphate per liter) set at pH 5.0 with succinic acid (1.35% w/v) and NaOH (0.65% w/v); or glucose-synthetic mineral medium (SMM) made according to Verduyn et al. (1992). The flasks were agitated at 30°C on a rotary shaker set at 200 rpm. For solid media, agar was added to YEPD or SD medium to reach 2% (w/v). Anaerobic growth was carried out in 50 ml-capped vials containing 25 ml of SD medium complemented with 0.05% (v/v) Tween 80; 0.01% (w/v) ergosterol (from a 1000x stock solution made in isopropanol and in the presence of 1 g/l Resazurin sodium salt as oxygen indicator).

Phenotypic analysis

The growth capacity on various carbon sources was carried out using the commercial available API:ID 32C gallery (purchased from BioMérieux SA, France). This micro-method allowed a rapid survey of the carbon assimilation and hence on the metabolic ability of the yeast isolates. Phenotypic assays for various stresses were performed on YEPD or SD agar plates. Prior to the assays, yeast cells were cultivated in YEPD at 30°C for 24 h. After reading the optical density (OD) at 600 nm, 1 ml of the culture was collected and concentrated to 8.0 OD_{600}. Then, 5 µl from a serial dilution made in sterile water was spotted on the agar plates containing NaCl, sorbitol, ethanol or 2-phenylethanol at concentration indicated in the figures. Unless otherwise stated, the plates were incubated at 30°C, and the growth was scored after two days.

Analytical methods

Glucose, glycerol, ethanol and acetate were determined in the cell-free supernatant with commercial biochemical kits or by high performance liquid chromatography. In the latter case, the supernatant was filtered through 0.22 µm- pore-size nylon filters prior to loading on a HPX-87H Aminex ion exclusion column. The column was eluted at 48°C with 5 mM H_2SO_4 at a flow rate of 0.5 ml/min and the concentration of the compounds was determined using a Waters model 410 refractive index detector. For quantification of higher alcohols, these volatile compounds were first extracted by mixing 1 ml of the culture medium with 500 µl of diethylether in the presence of octanol at 0.1% (v/v) as internal standard. After vigorous vortexing and centrifugation at 1000 g for 5 min, 2 µl of the organic phase was injected on an Agilent 6890N gas chromatograph system equipped with a flame ionisation detector (FID). A HP5 capillary GC column (Agilent Technologies) with dimensions of 30 m length x 0.32 mm internal diameter with a 0.25 µm film thickness was used for separation. The injector temperature was set a 250°C with a split ratio at 1:10. carrier gas was helium. The oven temperature program was as follows: 50°C for 1 min; rise to 250°C at 30°C per min and set at 250°C for 5 min. Detection with FID was performed at 270°C with 30 ml/min hydrogen and 300 ml/min air flow rates. Data acquisition and processing were performed using the Chromeleon Chromatography Management System (Thermo Scientific Dionex). For each compounds analysed, a calibration curve was constructed using known amount of the pure standards (purchased from Aldrich-Sigma).

RESULTS AND DISCUSSION

Diversity of yeasts in Algerian food and dairy products

About 18 yeast-like colonies were recovered from food

Table 1. Identification of yeast isolates from Algerian fruits and dairy products.

Isolates code	Food or dairy origin	Species attributed	Identity match with NCBI
AH1 (S1)	Dates (Ghardaia)	*Clavispora lusitaniae*	500/500 (100%)
AJ3 (S2)	Dates (Bechar)	*Hanseniaspora uvarum*	525/527 (99%)
BFK (S3)	Dates (Ourgla)	*Kodamaea ohmeri*	487/487 (100%)
MJ2 (S5)	Melon (Oran)	*Candida parapsilosis*	535/535 (100%)
LVG/PC (S6)	Cow melk (Ghardaïa)	*Yarrowia lipolytica*	525/525 (100%)
LVB/PR (S7)	Cow melk (Oran)	*Yarrowia lipolytica*	525/525 (100%)
LBN/CR (S10)	sheep milk (Ghardaia)	*Yarrowia lipolytica*	525/525 (100%)
LCT/CV (S11)	Goat milk (Ghardaia)	*Yarrowia lipolytica*	525/525 (100%)
LCS/ CS (S14)	Camel milk (Ghardaïa)	*Pichia kudriavezii (formely Issatchenkia orientalis)*	526/526 (100%)
LCW/GC (S15)	Camel milk (Bechar)	*Trichosporon asahii*	563/563 (100%)
C/PC2 (S17)	Gherkins (oran)	*Zygosaccharomyces bailii*	545/545 (100%)
MM2 (S18)	honey (Mostaganem)	*Zygosaccharomyces rouxii*	543/543 (100%)

The molecular taxonomy of the yeast isolates was based on the sequencing of the domain D1/D2 of the large subunit rRNA gene 26S and comparison of the sequences with NCBI database, as described in Materials and Methods.

and dairy products collected in local markets of different Algerian cities, grouped into nine yeast species according to genomic analysis of the PCR- amplicon sequence from domain D1/D2 of the 26S rDNA gene (Kurtzman and Robnett, 1998). *Pichia kudriavezvii* (synonymus of *Issatchenkia orientalis)* and *Trichosporon asahii*were isolated from camel milk, *Clavispora lusitaniae,* *Hanseniaspora uvarum* and *Kodamaera ohmeri*from dates, *Zygosaccharomyces bailii* from gherkins, *Candida parapsilosis* from melons, *Zygosaccharomyces rouxii* from honey, and *Yarrowia lipotytica* species was the prominent yeast in cow, goat and sheep milk (Table 1). Since *T. asahii* is considered as an opportunistic pathogen whose main habitat is the skin of humans and animals (Ebright et al., 2001), probably this yeast arose during the isolation procedure. In contrast, the isolation of *Y. lipolytica* in milk of camel, cows and sheep was not unexpected due to the high content in these dairy products of lipids, a preferred carbon source for this yeast species. Also, the presence of the sugar tolerant *Z. rouxii* in honey was expected as it is very rich in sugars and so highly osmotic.

The presence of the other yeast species, notably *P. kudriavezii* in camel milk and *C. lusitaniae* in dates were less expected. *P. kudriavezii* species is one of the indigenous yeast present in wine (Clemente-Jimenez et al., 2004) and is characterized by high acidic and ethanol tolerance (Okuma et al., 1986). Thus, the presence of this species might be explained by the fact that the isolation procedure came after lactic fermentation of the camel milk. *C. lusitaniae* is a yeast species that can be found in a broad array of plants and animals substrates, as well as in industrial wastes (Lachance et al., 2003), but has been also isolated in clinical specimens (Gargeya et al., 1990). The isolation of *H. uvarum* from dates maybe in accordance with the fact that this yeast species

is commonly found on grapes skin that are very sweet fruit, very rich in glucose and fructose. On the other hand, the presence of *K. ohmeri* in dates in less obvious since this yeast-like fungus is known to be vectored by insects and bees during their visit on ephemeral flowers (Lachance et al., 2001).

There is also recent report on the emergence of this yeast species as an important etiologic agent of fungemia in immuno-compromised patients (Al Sweih et al., 2011). This leaves the finding of this yeast species from dates questionable as to whether it is a truly indigenous yeast species present in dates or it has been isolated by accident during the isolation procedure. This question remains to be answered.

Growth behaviour of representative yeast isolates from Algerian agricultural and dairy products

As a result of good genetic, genomic and physiological knowledge on *Y. lipolytica* (Barth and Gaillardin, 1997), *Z. bailii* and *Z.rouxii* (Merico et al., 2003; Martorell et al., 2007), we addressed some characterization of the 5 remaining yeast isolates, namely *C. lusitaniae, H. uvarum, K. ohmeri, P. krudiavezii* and *T. asahii*. Growth studies were carried out on API: ID 32C galleries (Table 2). As expected, all yeast species grew very actively on glucose and galactose as carbon source. *C. lusitaniae* and *T. asahii* were also able to grow on C5-sugars such as arabinose, xylose and ribose, while the *H. uvarum* had the shortest sugars spectrum for growth, as it did not grow on mannose, sucrose, maltose, trehalose or raffinose (Table 1). This qualitative analysis showed that the carbon sources spectrum of the *P. kudriavezii* species for growth was similar to that of the laboratory *S. cerevisiae* CEN.PK122-2N. Also, the *P. kudriavezii* yeast

Table 2. Growth tests of yeast species on several carbon source using API-galleries.

ID name	CEN.PK122-2N	Clavispora lusitaniae	Hanseniaspor auvarum	Kodamaea ohmeri	Pichia kudriavezii	Trichosporo nasahii	Substrate (from ID name)
GAL	++	+++	++	+++	++	+++	D-Galactose
ACT	+	-	++	-	-	+++	Actidione
SAC	++	+++	-	+++	+++	+++	D-Saccharose
NAG	+	+++	-	+++	+++	+++	N-Acetyl-Glucosamine
LAT	++	+++	-	+	+++	+++	Lactique acid
ARA	+	++	-	-	-	++	L-Arabinose
CEL	+	+++	++	+++	-	+++	D-Cellobiose
RAF	++	++	+	+++	-	-	D-Raffinose
MAL	++	+++	-	+++	++	+++	D-Maltose
TRE	++	+++	-	+++	-	+++	D-Trehalose
2KG	+	++	++	+++	-	+++	Potassium 2-ketoglutamate
MDG	+++	+++	-	+++	-	+++	Methyl-αD-Glucopyranoside
MAN	+	+++	-	+++	+	+	D-Mannitol
LAC	+	-	-	-	-	+++	D-Lactose
INO	-	-	-	-	-	-	Inositol
O	0	0	0	0	0	0	No substrat
SOR	-	++	-	+++	-	-	D-Sorbitol
XYL	-	+++	-	-	-	+++	D-Xylose
RIB	-	++	++	+	-	+++	D-Ribose
GLY	-	+++	-	+++	++	-	Glycerol
RHA	-	+++	-	+	-	+	L-RhamnosE
PLE	+++	+++	-	+++	++	+++	Palatinose
ERY	-	-	-	-	-	++	Erythritol
MEL	-	-	-	-	-	-	D-Melibiose
GRT	-	++	-	-	-	+++	sodium GlucURONATE
MLZ	+++	+++	-	+	++	+	D-MELEZITOSe
GNT	-	++	++	-	++	+++	Potassium Gluconate
LVT	-	++	-	+	-	-	Levulinicacid
GLU	+++	+++	++	+++	+++	+++	D-glucose
SBE	-	+++	-	++	-	+	L-SorBOSE
GLN	-	++	-	++	-	++	Glucosamine
ESC	-	++	++	++	-	-	Esculine (fecitrate)

The growth was scored after 24 h at 32°C. The scoring legend is: -, no growth; +, weak biomass; ++, good biomass; +++ , high biomass.

isolated in this work was likely genetically different from the one isolated from cane syrup by Gallardo et al. (2011) which was unable to grow on sucrose and from the DMKU-3ET15 strain isolated from pork sausage fermentation which could not grow on galactose, maltose, sucrose or melizitose (Yuangsaard et al., 2013).

It was also found that all yeast species except *T. asahii* could grow under anaerobiosis (data not shown). Based on these qualitative growth assays, we have determined the maximal growth rate of these five yeast species culti-

vated on a glucose-rich (YEPD) medium (Table 1). With the exception of *T. asahii*, they grew all at the rate of about 10 to 25% faster than *S. cerevisiae* CEN.PK122-2N. In addition, they fermented glucose into ethanol, with *P. kudriavezii* and *C. lusitaniae* showing products fermentation yields very close to that of *S. cerevisiae* CEN.PK122-2N, whereas *T. asahii* showed an apparent respiratory metabolism of glucose (Table 3).

Similar results were found using a glucose synthetic mineral medium made according to Verduyn et al. (1992),

Table 3. Maximal growth rate (μ), and main byproducts formation of the various yeast species cultivated in a glucose-rich medium (YEPD).

Strain	μ (h⁻¹)	Ethanol (g/l)	Yield (ethanol/glucose) (g/g)	Glycerol (g/l)	Acetate (g/l)
S. cerevisiae CEN.PK122-2N	0.45	8.5	0.42	0.55	0.22
C.lusitaniae	0.58	6.3	0.32	0.4	0.11
H. uvarum	0.50	5.3	0.27	0.65	0.24
K.ohmeri	0.56	5.9	0.30	0.4	0.07
P.kudriavezii	0.68	6.8	0.40	2.2	0.05
T. asahii	0.30	0.55	0.03	0.54	0.01

Growth was carried at 30°C. Samples were taken at the time of complete exhaustion of glucose (initial amount was 20 g/l) for analysis of the byproducts by HPLC as described in Materials and Methods. The results are the mean of 2 independent experiments.

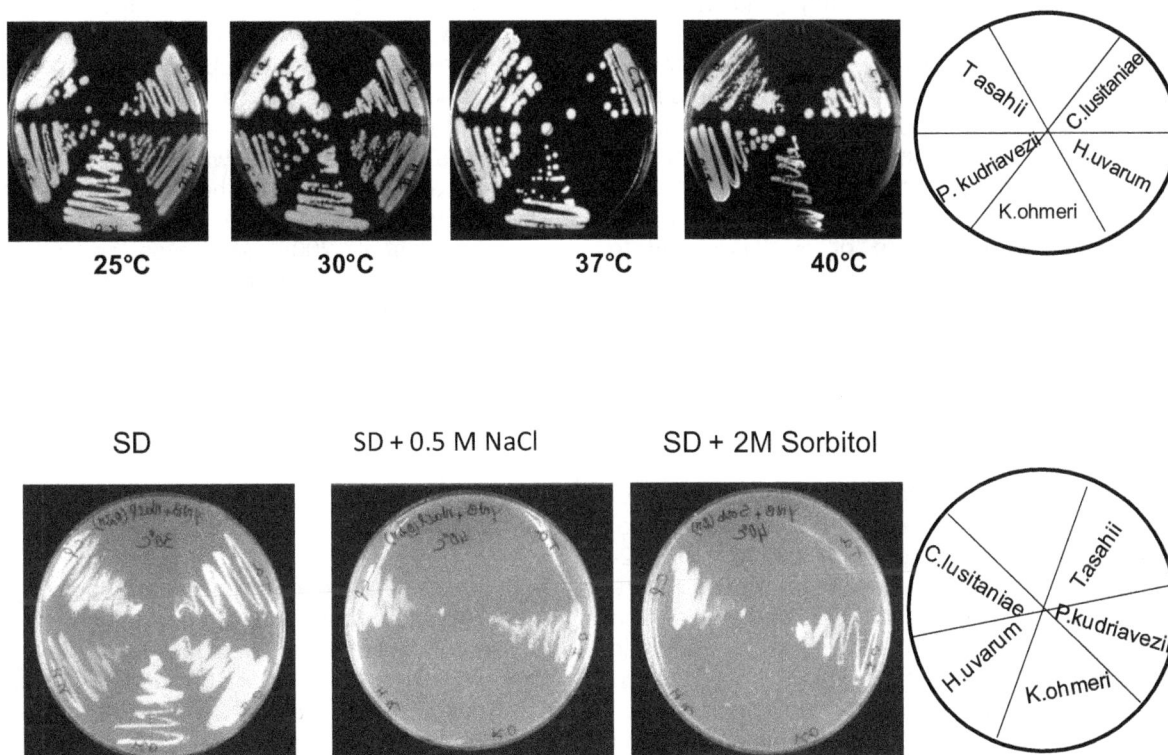

Figure 1. (A) Growth of yeast isolates on YEPD agar plates at different temperature; (B) Effect of NaCl 0.5 M or sorbitol 2 M on the growth of yeast isolates on SD agar plates at 40°C. Growth was scored after two days

except that the growth of *H. uvarum* on this medium required the supplementation of vitamins B2 (riboflavin), B3 (niacin) and B9 (folic acid) (data not shown).

Analysis of tolerance to stressing conditions

Growth dependencies on temperature and pH were determined for the 5 yeast species retained in this study. Except *H. uvarum* that was unable to sustain growth above 30°C, the other four yeast species grew pretty well up to 40°C (Figure 1A). We noticed that the growth of *P. kudriavezii* was the least affected at high temperature, in agreement with the thermotolerant properties of this yeast species reported in previous studies (Gallardo et al., 2011; Kwon et al., 2011; Yuangsaard et al., 2013). In addition, this yeast species displayed the widest range of pH for growth (pH 2.5 to 7), whereas *T. asahii* and *K ohmeri* were unable to grow at pH equal or lower than 2.5

(data not shown). These yeast species were also tested for salt and osmotolerance. It was found that only *C. lusitaniae* and *P. kudriavezii* were capable of growing on glucose medium supplemented with 0.5 M NaCl or 2 M sorbitol even at 40°C (Figure 1B). *P. kudriavezii* was also found to be the most resistant yeast species to ethanol toxicity since it could still grow in the presence of 18% ethanol, whereas the growth of *S cerevisiae* CEN.PK122-2N strain started to be impaired at 12% ethanol, and that of the other yeast species was inhibited at concentration equal or above 8% ethanol (Figure 2). In line with the high resistance of *P. kudriavezii* to ethanol, we found that this yeast species was about two times more resistant than *S. cerevisiae* to 2-phenylethanol (2-PE), a relevant rose-like flavour produced by the Erlich pathway (Hazelwood et al., 2008) (Figure 3). Like ethanol, 2-PE can diffuse freely in and out of the cell and can disturb cell membrane integrity (Silver and Wendt, 1967; Lloyd et al., 1993; Hua and Xu, 2011). Although, the molecular mechanism underlying ethanol resistance is not yet fully understood (Stanley et al., 2010), a role of membrane composition and fluidity has been proposed to be crucial for this resistance (Ding et al., 2009). Thus, investigation of the membrane fluidity and determination of the phospholipids membrane composition in this yeast species might provide some clues to a better understanding of the resistance of *P. kudriavezii* to ethanol and 2-PE and to eventually exploit this knowledge to increase the tolerance of the yeast *S. cerevisiae* to these alcoholic molecules.

Analysis of the production of higher alcohols

We were also interested in determining the capacities of the yeast isolates for the production of biotechnological relevant products, such higher alcohols and 2-phenylethanol (Berger, 2009). The yeast species were cultivated on three different media, namely a rich (YEPD), a synthetic mineral (SMM) and a synthetic mineral enriched (SMM-AA) supplemented with amino acids leucine, isoleucine, valine and phenylalanine, each added at 1% (w/v), and higher alcohols and corresponding esters was determined at the time glucose was exhausted from the medium. As shown in Table 4, 2-methyl-2-butanol (amyl-alcohol, 2-MB), 3-methylbutanol (isoamyl alcohol, 3-MB) and 2-phenylethanol (2-PE) were identified in the fermentation broth of these yeasts. In a glucose-rich medium, *P. kudriavezii* species was about two times more efficient than the *S. cerevisiae* CEN.PK122-2N in producing the three different volatiles compounds, whereas *C. lusitaniae* exhibited roughly the same capacity as *S. cerevisiae*. In a glucose mineral medium, the amount of fusel alcohols was, as expected, extremely low, indicating that the production of these higher alcohols in a glucose rich medium was likely the result of the bioconversion of excess of branched-chain

and aromatic amino acids present in this medium by the catabolic Ehrlich pathway (Hazelwood et al., 2008). Accordingly, the supplementation of the mineral medium with linear, branched and aromatic amino acids (leucine, isoleucine, valine and phenylalanine) boosted the production of these fusel alcohols, with a profile that was again yeast-dependent. Unlike *H. uvarum*, which was unable to grow in this medium, *C. lusitaniae* species showed a higher capacity than *S. cerevisiae* in the formation of the three flavours. Like on YEPD medium, *K. ohmeri* had the highest production of 3-MB but the lowest formation of 2-PE, when cultivated on SMM-AA. On the other hand, the flavours profile of *P. kudriavezii* on a synthetic medium supplemented with these amino acids was remarkably singular since the formation of 2-PE was strongly stimulated upon addition of phenylalanine, reaching a concentration that was 5 times higher than in *S. cerevisiae*, while levels 2-MB and 3-MB were significantly not different from those measured in YEPD.

This result suggest that the bioconversion of phenyl-lanine into 2-PE by the catabolic Erlich pathway is likely more efficient in *P. kudriavezii* than in *S. cerevisiae*. Thus, it would be worth to compare the enzymatic capacities of the Erlich pathway between these two yeast species, and in particular the amino transaminase and decarboxylase that play a key role in this pathway (Hazelwood et al., 2008). Moreover, the higher level of 2-PE produced by *P. kudriavezii* corroborated with the greater resistance of this yeast species to this aromatic compound. Finally, the corresponding esters of the higher alcohols (2-MB, 3-MB and 2-PE) were below detection in all of the five yeast species cultivated in SMM supplemented with amino acids (data not shown).

Conclusions

In this work, we isolated nine yeast species from traditional Algerian food and dairy products. Our analysis focused on five of these yeast isolates revealed important physiological and metabolic properties. In particular, it was found that *P. kudriavezii* was remarkably able to grow at higher temperature (>40°C), and much more tolerant to ethanol and osmotic pressure and was a better producer of 2-phenylethanol than *S. cerevisiae*. All these physiological traits may place this yeast species as an attractive alternative yeast platform in the field of bio-based production from renewable sources.

ACKNOWLEDGEMENTS

This work was supported in part by Region Midi Pyrénées (France) under Grant No. 09003813 and was carried out in the frame of COST Action FA0907 *BIOFLAVOUR* (www.bioflavour.insa-toulouse.fr) under the EU's Seventh Framework Programme for Research (FP7). M.A.R. is

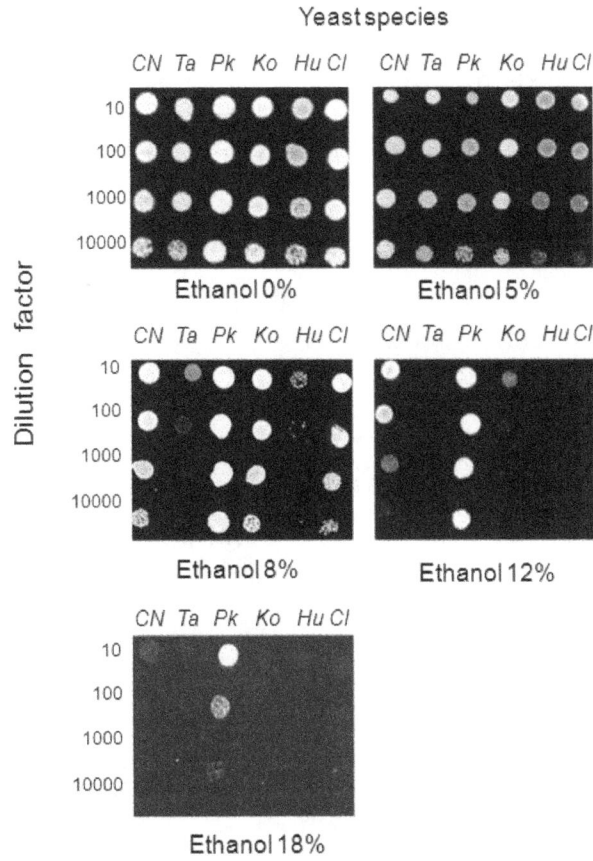

Figure 2. Effect of ethanol on the growth of yeast isolates on YEPD agar plates. Growth was scored after 2 days at 30°C. Abbreviations for yeast isolates are: *C.N* stands for *S cerevisiae* CEN.PK122-2N; Ta for *Trichosporon asahii*; Pk for *Pichia kudriavezii*; Ko for *Kodamaea ohmeri*; Hu for *Hanseniaspora uvarum* and Cl for *Clavispora lusitaniae*.

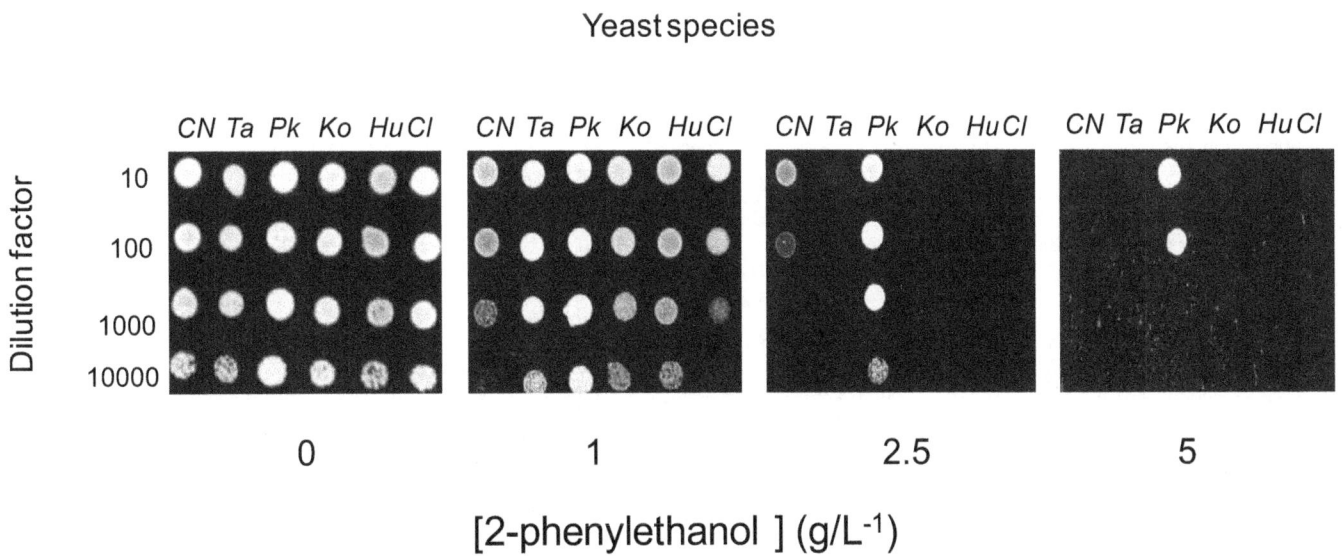

Figure 3. Effect of 2-phenyethanol on the growth of yeast isolates on YEPD agar plates. Growth was scored after 2 days at 30°C. Abbreviations are as in Figure 2.

Table 3. Flavours production by the different yeast isolates.

strain	YEPD (µmol/g dry mass)			SMM (µmol/g dry mass)			SMM + amino acids* (µmol/g dry mass)		
	2-MB	3-MB	2-PE	2-MB	3-MB	2PE	2-MB	3-MB	2-PE
S. cerevisiae CEN.PK122-2N	90	270	85	15	116	30	750	900	580
C. lusitaniae	80	137	60	25	108	60	880	1000	1600
P.kudriavezii	238	408	306	30	238	100	238	680	2857
H. uvarum[$]	82	578	85	-	-	-	-	-	-
K.ohmeri	288	638	20	85	320	20	800	2244	192
T. asahii	bd	bd	bd	bd	bd	bd	bd	bd	bd

The yeast isolates were cultivated on glucose-rich (YEPD) medium, glucose synthetic mineral medium (SMM) or supplemented with aromatic and branched chain amino acids (SMM + amino acids). Samples for analysis of flavours was taken at the end of growth when glucose was consumed. *Amino acids added to SMM were leucine, valine, isoleucine and phenylalanine at 1% (w/v); [$]No growth on SMM. Values are the mean of two independent experiments, with a standard deviation that was less than 15% between the two experiments. Abbreviations are: 2-MB; 2-methylbutanol or amyl alcohol; 3-MB; 3-methyl butanol or isoamyl alcohol; 2-PE, 2-phenylethanol; bd = below detection.

grateful to the L.B.R.A.P. for financial support during the work in Toulouse. The authors also thank all colleagues from the laboratory team for scientific and technical support during this work.

REFERENCES

Al Sweih N, Khan ZU, Ahmad S, Devarajan L, Khan S, Joseph L, Chandy R (2011). *Kodamaea ohmeri* as an emerging pathogen: a case report and review of the literature. Med. Mycol. 49:766-770.

Altschul SF, Madden TL, Schaffer AA, Zhang J, Zhang Z, Miller W, Lipman DJ (1997). Gapped BLAST and PSI-BLAST: a new generation of protein database search programs. Nucleic Acids Res. 25:3389-3402.

Barth G, Gaillardin C (1997). Physiology and genetics of the dimorphic fungus *Yarrowia lipolytica*. FEMS Microbiol. Rev. 19:219-237

Berger RG (2009). Biotechnology of flavours-the next generation. Biotechnol. Lett. 31:1651-1659.

Clemente-Jimenez J, Mingorance-Cazorla L, Martinez-Rodriguez S, Heras-Viazquez F, Rodriguez-Vico F (2004). Molecular characterization and oenological properties of wine yeasts isolated during spontaneous fermentation of six varieties of grap must. Food Microbiol. 21:149-155.

Ding J, Huang X, Zhang L, Zhao N, Yang D, Zhang K (2009). Tolerance and stress response to ethanol in the yeast Saccharomyces cerevisiae. Appl. Microbiol. Biotechnol. 85:253-263.

Ebright JR, Fairfax MR, Vazquez JA (2001). *Trichosporon asahii*, a non-Candida yeast that caused fatal septic shock in a patient without cancer or neutropenia. Clin. Infect. Dis. 33:E28-E30.

Fleet GH (2007). Yeasts in foods and beverages: impact on product quality and safety. Curr. Opin. Biotechnol. 18:170-175

Fortman JL, Chhabra S, Mukhopadhyay A, Chou H, Lee TS, Steen E, Keasling JD (2008). Biofuel alternatives to ethanol: pumping the microbial well. Trends. Biotechnol. 26:375-381.

Gallardo JC, Souza CS, Cicarelli RM, Oliveira KF, Morais MR, Laluce C (2011). Enrichment of a continuous culture of Saccharomyces cerevisiae with the yeast *Issatchenkia orientalis* in the production of ethanol at increasing temperatures. J. Ind. Microbiol. Biotechnol. 38:405-414.

Gargeya IB, Pruitt WR, Simmons RB, Meyer SA, Ahearn DG (1990). Occurrence of *Clavispora lusitaniae*, the teleomorph of *Candida lusitaniae*, among clinical isolates. J. Clin. Microbiol. 28:2224-2227.

Hazelwood LA, Daran JM, van Maris AJ, Pronk JT, Dickinson JR (2008). The Ehrlich pathway for fusel alcohol production: a century of research on *Saccharomyces cerevisiae* metabolism. Appl. Environ. Microbiol. 74:2259-2266.

Hua D, Xu P (2011). Recent advances in biotechnological production of 2-phenylethanol. Biotechnol. Adv. 29:654-660.

Kirby J, Keasling JD (2008). Metabolic engineering of microorganisms for isoprenoid production. Nat. Prod. Rep. 25:656-661

Kurtzman CP, Robnett CJ (1997). Identification of clinically important ascomycetous yeasts based on nucleotide divergence in the 5' end of the large-subunit (26S) ribosomal DNA gene. J. Clin. Microbiol. 35:1216-1223.

Kurtzman CP, Robnett CJ (1998). Identification and phylogeny of ascomycetous yeasts from analysis of nuclear large subunit (26S) ribosomal DNA partial sequences. Antonie Van Leeuwenhoek. 73:331-371

Kwon YJ, Wang F, Liu CZ (2011). Deep-bed solid state fermentation of sweet sorghum stalk to ethanol by thermotolerant Issatchenkia orientalis IPE 100. Bioresour. Technol. 102:11262-11265.

Lachance MA, Daniel HM, Meyer W, Prasad GS, Gautam SP, Boundy-Mills K (2003). The D1/D2 domain of the large-subunit rDNA of the yeast species *Clavispora lusitaniae* is unusually polymorphic. FEMS Yeast Res. 4:253-258.

Lachance MA, Starmer WT, Rosa CA, Bowles JM, Barker JS, Janzen DH (2001). Biogeography of the yeasts of ephemeral flowers and their insects. FEMS Yeast Res. 1:1-8.

Lloyd D, Morrell S, Carlsen HN, Degn H, James PE, Rowlands CC (1993). Effects of growth with ethanol on fermentation and membrane fluidity of *Saccharomyces cerevisiae*. Yeast. 9:825-833.

Lomascolo A, Stentelaire C, Asther M, Lesage-Meessen L (1999). Basidiomycetes as new biotechnological tools to generate natural aromatic flavours for the food industry. Trends Biotechnol. 17:282-289.

Martorell P, Stratford M, Steels H, Fernandez-Espinar MT, Querol A (2007). Physiological characterization of spoilage strains of Zygosaccharomyces bailii and Zygosaccharomyces rouxii isolated from high sugar environments. Int J Food Microbiol 114:234-242.

Merico A, Capitanio D, Vigentini I, Ranzi BM, Compagno C (2003). Aerobic sugar metabolism in the spoilage yeast Zygosaccharomyces bailii. FEMS Yeast Res. 4:277-283.

Octave S, Thomas D (2009). Biorefinery: Toward an industrial metabolism. Biochimie. 91:659-664.

Okuma Y, Endo A, Iwasaki H, Ito Y, Goto S (1986). Isolation and properties of ethanol-using yeasts with acid and ethanol tolerance. J. Ferment Technol. 64:379-382.

Silver S, Wendt L (1967). Mechanism of action of phenethyl alcohol: breakdown of the cellular permeability barrier. J. Bacteriol. 93:560-566.

Stanley D, Bandara A, Fraser S, Chambers PJ, Stanley GA (2010). The ethanol stress response and ethanol tolerance of *Saccharomyces cerevisiae*. J. Appl. Microbiol. 109:13-24.

van Dijken JP, Bauer J, Brambilla L, Duboc P, Francois JM, Gancedo C, Giuseppin ML, Heijnen JJ, Hoare M, Lange HC, Madden EA, Niederberger P, Nielsen J, Parrou JL, Petit T, Porro D, Reuss M, van Riel N, Rizzi M, Steensma HY, Verrips CT, Vindelov J, Pronk JT (2000). An interlaboratory comparison of physiological and genetic properties of four *Saccharomyces cerevisiae* strains. Enzyme Microb. Technol. 26:706-714.

Verduyn C, Postma E, Scheffers WA, Van DJ (1992). Effect of benzoic acid on metabolic fluxes in yeasts: a continuous- culture study on the regulation of respiration and alcoholic fermentation. Yeast. 8:501-517.

Yuangsaard N, Yongmanitchai W, Yamada M, Limtong S (2013). Selection and characterization of a newly isolated thermotolerant *Pichia kudriavzevii* strain for ethanol production at high temperature from cassava starch hydrolysate. Antonie Van Leeuwenhoek. 103:577-588.

22

Yeasts in marine and estuarine environments

Kathiresan Kandasamy*, Nabeel M. Alikunhi and Manivannan Subramanian

Centre of Advanced Study in Marine Biology, Annamalai University,
Parangipettai: 608502, Tamil Nadu, India.

Yeasts and other fungi are prevalent in marine and estuarine ecosystems where they play an important role in the food web. Marine yeasts are unique in performing fermentations under high salt concentrations. The mechanism underlying the high salt tolerance involves the ability to accumulate high concentrations of sodium without becoming intoxicated, and the exclusion of excessive sodium from the cytoplasm. Overall, the yeasts play major roles in fermentation, enzyme technology, pollution control, micro sensors, and in some medicinal and medical applications.

Key words: Marine yeast, deep sea, estuarine, mangrove, association.

INTRODUCTION

Yeasts are unicellular micro-fungi, capable of self perpetuating their populations in terrestrial and aquatic environments (Kurtzman and Fell, 1998). A key characteristic is the ability to ferment sugars for ethanol production. They live as saprophytes on plant or animal materials, where they preferentially catabolize sugars but can also utilize polyols, alcohols, organic acids, and amino acids as carbon and sources of energy (Spencer and Spencer, 1997). To promote efficient decomposition of substrates, many yeasts produce filaments or pseudo-hyphae and also produce hydrolytic enzymes. Research on yeasts has played a major role in the development of a number of modern scientific disciplines and much work is being carried out in studying their physiology, metabolism, genetics, and molecular biology and developing new applications for industry and medicine (Barnett, and Barnett, 2011). Although, a large number of studies about terrestrial and aquatic yeasts are available there are only few reports about marine and estuarine yeasts, and hence is the need for this review.

DISTRIBUTION OF MARINE YEASTS

Most studies on yeasts in estuaries and near-shore seawater was performed in Europe and north America

*Corresponding author. E- mail: kathirsum@rediffmail.com.

(Fell et al., 1960; Roth et al., 1962; Fell and Van Uden, 1963; Van Uden and Castelo-Granco, 1963; Taysi and Van Uden, 1964; Norkrans, 1966; Ahearn et al., 1968; Van Uden and Fell, 1968; Hoppe, 1972; Ahearn, 1973; Barnett and Pankhurst, 1974; Buck, 1975).

Marine and estuarine habitats

Marine yeasts display a high salt tolerance and the ability to perform fermentation. In general, yeast cell numbers decrease with increasing salt concentration and total organic carbon in the estuarine environment (Urano et al., 2001). Due to sewage pollution and terrestrial run-off in this environment, some species of yeasts are more prevalent in estuaries, as compared to open seas (Lazarus and Koburger, 1974). Yeast and other fungi are prevalent in salt marsh and mangrove ecosystems where they play an important role in the detritus food web of the coastal environment (Mayers et al., 1975; Hyde, 2002).

Yeasts in estuarine waters vary widely both in number and species. The most frequently isolated genera of yeasts are *Debaryomyces*, *Candida*, *Rhodotorula*, *Cryptococcus* and *Kloeckera*. While studying the yeast flora of the Suwannee River estuary in Florida, Lazarus and Koburger (1974) obtained highest yeast densities in low saline areas, and highest species diversity in the sewage-polluted waters in the estuaries. However, no ascosporogenous yeasts have been isolated from the areas of low salinity (Lazarus and Koburger, 1974). The researchers from University of Miami have isolated one ascomycetous yeast, *Lachancea meyersii* from the mangroves of the Bahamas (Fell et al., 2004).

Offshore and deep-sea environments

Only a few studies on yeasts from oceanic regions have been published in the last decades. This may be due to the high costs involved in offshore and oceanic sampling (Fell, 1976). Among the ascomycetous yeasts, the halotolerant species *Debaryomyces hansenii* is a typical ubiquitous species in oceanic regions as well as in other aquatic environments. Among the basidiomycetous yeasts, some species of *Cryptococcus*, *Rhodotorula*, *Sporobolomyces* and their teleomorphs are widespread across various oceanic regions. Generally, basidiomycetous yeasts often account for the majority of the total yeast population in oligotrophic oceanic water. *Candida* species also occur, but at lower frequencies than in the inshore or polluted freshwater regions. Some of the *Candida* species are only evident in the oceanic regions around Antarctica along with psychrophilic species such as *Leucosporidium* spp. and *Sympodiomyces parvus*.

They are probably autochthonous marine species (Lachance and Starmer 1998). *Metschnikowia* species are known to be associated with seawater, freshwater, algae, invertebrates and fish. Phylogenetic relationship analysis shows that *M. australis*, *M. bicuspidata var. bicuspidata*, *M. bicuspidate var. chathamia*, *M. krissii* and *M. zobellii*, prevalent in marine environments are monophyletic. However, the less prevalent aquatic species such as *M. reukaufii* and *M. pulcherrima* are phylogenetically distant (Mendonça-Hagler et al., 1993).. The latter two are usually found to associate with natural substrates of terrestrial origin such as flowers, fruits and insects. The monophyly of the marine species suggests that their divergence has evolved in the course of association with marine environments.

The ubiquitous species in various marine habitats are usually regarded as allochthonous, as many basidiomycetous types of yeast are often found to associate with the phyllosphere of terrestrial plants and their marine prevalence is believed to be due to run-off from the phylloplane (Hagler and Ahearn, 1987; Lachance and Starmer, 1998). The yeasts of the ballistosporogenous genera - Sporobolomyces and Bullera - and their teleomorphs are typical inhabitants of the phylloplane. The yeasts of the genera - Sporobolomyces and Bullera – are the most commonly encountered in the Pacific Ocean off Mexico (Hernandez-Saavedra et al., 1992). Interestingly, the frequencies of occurrence of the yeasts increase with increasing distance from the coastline and increasing depth of coastal sea. The yeasts of ballistosporogenous genera are also present in benthic invertebrates collected from deep-sea floors in the Pacific Ocean off Japan (Nagahama et al., 2001b). These facts indicate that ballistosporogenous yeasts are not effluents from terrestrial plant foliages but are indigenous to the sea.

Basidiomycetous types of yeasts are present in the seawater of the Atlantic Ocean off Faro in the south of Portugal (Gadanho et al., 2003). *Rhodosporidium babjevae* and *Rhodosporidium diobovatum* (the two possible species previously identified as *Rhodosporidium glutinis*, and *Sakaguchia dacryoides*) and *Pseudozyma aphidis* (ustilaginomycetous yeast) are the most frequently occurring yeasts among the basidiomycetous yeasts (Gadanho et al., 2003).

Yeast-like cells are reportedly abundant in deep-sea sediment around the Pacific Ocean. The most frequently surveyed site is around a cold seep at a depth of about 880 to 1,200 m near Hatsushima Island, Sagami Bay. Other less frequently surveyed sites include Suruga Bay (380 to 2,500 m), the Japan Trench (4,500 to 7,500 m) and Iheya Ridge (990 to 1,400 m). The sites surveyed only once include Kagoshima Bay, 220 to 260 m; the Mariana Trench, about 11,000 m; the Palau–Yap Trench, 3,700 to 6,500 m; and the Manus Basin, 1,600 to 1,900 m. The Iheya Ridge and the Manus Basin are biologically fertile spots owing to the hydrothermal vent ecosystem (Alongi 1992, Nagahama et al., 2001a, b, 2003a, b).

The species which occur most frequently in the above sites are *Rhodosporidium sphaerocarpum*, *Williopsis saturnus* and *Candida pseudolambica*, but their distribution is limited mostly to the sediments of Suruga Bay and Kagoshima Bay. *D. hansenii* occurs only in the sediments of Sagami Bay and Suruga Bay, although it is known to be the most common ascomycete in marine waters (Hagler and Ahearn 1987). Almost all ascomycetous yeasts have been isolated from sediments, with the exception of *Kloeckera nonfermentans*, which is common to both sediments and benthic invertebrates, specifically in Sagami and Suruga Bay. In contrast, *Rhodosporidium diobovatum* and *Rhodotorula mucilaginosa* are widely prevalent in the various locations and sources.

The frequency of occurrence of each corresponding phylogenetic taxon is obviously different according to the source and geographical origin. The ascomycetous yeasts constitute the majority of the total yeast population in the sediments of Sagami Bay, Suruga Bay and Kagoshima Bay, and these sites are relatively inshore (5 to 20 km) near urban and industrial areas and where the sea floors are affected by human activity. Species in the *Erythrobasidium* clade have been isolated mostly from the benthic invertebrates, and the initial isolates from the sediments of the Manus Basin are considered to give clues about the hydrothermal ecosystems. Many of these species belong to the *Occultifur* lineage although some are novel species yet to be classified (Nagahama et al., 2001a, 2003a).

The association with animals is probably favourable for yeasts, owing to the abundance of nutrients (Hagler and Ahearn, 1987). However, the reasons why the number of species associated with animals is low is yet to be known. Hymenomycetous species, mostly assigned to the genus Cryptococcus, are localized in the Japan Trench, Sagami Bay and Suruga Bay, and the genus does not appear farther southwest. Species of Sporidiobolales are present at all of the sites.

Marine yeasts are believed to have physiological adaptations but are not scientifically validated. In general, yeasts from both terrestrial and marine origins are moderately pressure-tolerant. However, the response of yeasts to elevated hydrostatic pressure has not been properly studied (ZoBell and Johnson, 1949; Yamasato et al., 1974).

The carotenogenic basidiomycetous yeasts such as *Rhodotorula* and *Rhodosporidium* are psychro-tolerant and pressure-tolerant (Davenport, 1980). *Rhodotorula* species grown at 20 MPa (equivalent to 2,000 m depth) are not significantly different as compared to those grown at 0.1 MPa; however, growth is reduced to 20 to 30% when the species is grown at 40 MPa (Lorenz and Molitoris, 1997).

The yeasts isolated from seafloors deeper than 4,000 m do not grow well under hydrostatic pressures corresponding to the sources at which they have been collected (2 to 4°C, > 40 MPa).

This may be due to the specifications of compressed incubation system, which allows sharp pressure changes and insufficient oxygen supply. Psychrophilic strains have not been found so far in the deep sea, but many isolates are psychrotolerant growing well at < 4°C (Lorenz and Molitoris, 1997).

Hypersaline habitats

Yeasts occur in hypersaline habitats world-wide (Butinar et al., 2005) and include *Rhodosporidium sphaerocarpus*, *R. babjevae*, *Rhodotorula larynges*, *Trichosporon mucoides*, *Candida parapsilosis C. glabrata*, *Pichia guilliermondii*, *Debaryomyces hansenii*, *Trimmatostroma* and *Yarrowia lipolytica*. Interestingly ascomycetous yeast, *Metschnikowia bicuspidate* is known to be a parasite of the brine shrimp and it occurs as a free-living form from the Great Salt Lake brine.

Antarctic habitats

The first *Candida*-like *Leucosporidium* species was isolated in the 1960's from Antarctic soil and seawater (Di Menna, 1960; Sinclair and Stokes, 1965; Fell et al., 1969; Watson and Arthur, 1976; Ray et al., 1992). *Leucosporidium antarcticum* is endemic to Antarctica. This yeast species can weakly utilize both sucrose and maltose, and is extremely sensitive to temperatures above 20°C.

Marine plant-associated yeasts

Yeasts are epiphytic on seaweeds, abundant on Chlorophytes and Rhodophytes, but of low abundance on Phaeophytes due to the release of growth-inhibitory phenolics from the brown seaweeds (Raja Seshadri and Sieburth, 1971).

The yeasts also associate with phytoplankton (Kriss and Novozhilova, 1954) and decaying seaweeds (Bunt, 1955; Suchiro and Tomiyasu, 1962; Van Uden and Castelo Granco, 1963). However, no specific association has been established for yeasts with marine algae and seagrasses (Roth et al., 1962).

Yeast communities of polluted estuary and mangrove ecosystems in subtropical marine environments are extremely diverse. Yeasts are prevalent in salt marshes or mangrove ecosystems where the yeasts play an important role in the detrital food web and they are food source for some marine invertebrates including zooplankton (Meyers et al., 1975). *L. meyersii* sp. nov. (type strain NRRL Y 27269, CBS 8951, ML 3925) is described from 18 strains collected from mangrove habitats in the northern Bahamas Islands.

This species is homothallic, producing spherical ascospores in asci that become deliquescent, and is delineated from other ascomycetous yeasts by sequence analysis of the D1/D2 domains of the large subunit ribosomal DNA.

The species can be distinguished from other members of the genus *Lachancea* by lack of growth on galactose and by growth on maltose.

This new species is named in honor of Professor Samuel P. Meyers in recognition of his pioneering research with marine fungi (Fell et al., 2004). *Candida intermedia*, *D. hansenii*, *Issatchenkia occidentalis* (*Candida sorbosa*), *Pichia guillier- mondii* and *Pichia membranifaciens* (*Candida valida*) are the ubiquitous ascomycetous species at the Sepetiba Bay, Japan (De Araujo et al., 1995; Soares et al., 1997). The identity of the yeast community in the subtropical mangrove ecosystem is unclear, owing to phenotypic characterization yielding ambiguous taxonomic results.

Yeast species, *Kluyveromyces aestuarii* is associated with detritus-feeding invertebrates and sediments within mangrove areas (De Araujo et al., 1995; Soares et al., 1997). The aquatic strains of *Kluyveromyces lactis* are isolated from rhizosphere sediments of the marine marsh lands (Naumova et al., 2004; *hansenii* Meyer et al., 1971; De Araujo et al., 1995; Soares et al., 1997).

Plant-associated yeasts on bromeliads in mangrove areas are distinct from those typical of polluted areas, and comprise a larger number of species and isolates with basidiomycetous affinities (Hagler et al., 1993). Two yeast species *Kluyveromyces lactis* and *Pichia spartinae* are prevalent in the outer- or intra-culm (fistulous stalk) cells and tissues of the saltmarsh grass, *Spartina alterniflora* (Buchan et al., 2002).

Human-associated marine yeasts

Yeasts can cause infection in humans. *Candida albicans* causes candidasis, resulting in vaginal infections and also diaper rash and thrush of the mouth and throat. *Debaryomyces hansenii* is generally considered a non-pathogenic yeast species; however, it is associated with one case of bone infection and is identified in several clinical isolates associated with bone infection, fever and chronic bronchitis (Wong et al., 1982; Nishikawa et al., 1996).

Human pathogenic yeasts can be found in coastal areas. For example, *C. albicans* is an obligate saprophyte of warm-blooded animals, occurring rarely in host-free environments and surviving in nature for only short periods outside of animals. *C. albicans* with sparse filamentation and weak fermentation has been reported to occur at the surface micro-layer of the North Sea, but not in subsurface waters. It is found sporadically in marine and fresh waters and is common in faeces and raw sewage. Human pathogenic yeasts enter and aggregate in the bivalve mollusks due to the filter feeding mechanism of these animals (Buck et al., 1977). *C. parapsilosis, Candida tropicalis* and *Torulopsis glabrata* are the human-associated yeasts most frequently isolated from bivalve shellfish (oysters and mussels) collected from estuarine areas (Dabrowa et al., 1964; Kobayashi et al., 1953; Buck et al., 1977). These pathogenic yeast are selectively inhibited in sewage filtrates by water soluble substances produced by bacterial strains of *Bacillus* (Coleman et al., 1975).

Water temperature and pollution are important factors that influence the distribution of human-associated yeasts. Temperatures lower than that of the human host may dictate the abundance of intestinal yeasts. Incubation of cultures at 37°C eliminated many saprophytic types of yeast but encouraged human-associated yeast. The samples closest to sources of domestic pollution have the greatest abundance and survival of *C. albicans* in seawater (Dzawachiszwili et al., 1964; Madri et al., 1966; Madri, 1968; Ahearn, 1973). The *C. albicans* population is greatest during colder months in the heavily polluted waters. The pumping rates of bivalves are minimal at low water temperatures. This does not kill yeasts, but the slow rate of pumping may account for the survival of human-associated yeast and other yeasts in bivalves in the winter (Tripp, 1960; Galtsoff, 1964; Buck et al., 1977). In oysters, internal phagocytosis and migration are the main processes by which yeast cells (*Saccharomyces cerevisiae*) are removed.

When raw shellfish containing pathogenic yeasts are consumed, human health can be affected. However, the infective dosages of the yeasts are still not known. Individuals who repeatedly handle contaminated shellfish with cut or damaged hands are most at risk from yeast infections. Potentially pathogenic microorganisms are therefore, a serious consideration in the assessment of water and shellfish quality of near-shore recreational areas (Buck et al., 1977).

CULTURAL CHARACTERISTICS OF MARINE YEASTS

In the marine environment, bacteria are usually more numerous than yeasts. For the selective isolation of yeasts from the environment, bacterial inhibitors such as chloramphenicol, chlorotetracycline and streptomycin are generally used in the culture medium either alone or in combination. Antibiotics are used at concentrations up to 50 times greater than that required for bacterial inhibition (Ahearn et al., 1968; Richards and Elliott, 1966). The indiscriminate use of antibiotics to suppress bacteria may inhibit some yeast. The low pH of many media may inhibit acid-sensitive yeasts growing in alkaline seawater. Thus, the media commonly used for the detection and enumeration of marine yeasts have their limitations (Meyer et al., 1967; Ahearn et al., 1968; Van Uden and Fell, 1968). Employing temperature - gradient gel electrophoresis, Gadanho and Sampaio (2004) have studied yeast diversity in the estuary of the Tagus River, Portugal. This molecular detection method is carried out directly from water samples in parallel with cultivation of the yeasts using enrichment media. The number of species detected after enrichment is higher than the number of taxa found using the direct detection method. The most common species detected in marine environments is *D. hansenii,* an ascomycetous yeast (Hagler and Ahearn, 1987), probably because of its broad salinity tolerance and ability to utilize a wide range of carbon sources (Yadav and Loper, 1999).

PHYSIOLOGICAL ADAPTATIONS OF MARINE YEASTS

Yeasts from seawater are of two types; obligate and facultative. The obligate marine yeasts originate from the marine and inhabit the seawater throughout their lives. The facultative marine yeasts originate from other environments such as rivers, soils, woods, or the surface of animals and are transported to the marine environment. The obligate yeasts have inherently high NaCl tolerance as well as fermentative activities under high salt conditions (Urano et al., 1998). The facultative marine yeasts have weak salt tolerance acquiring high NaCl tolerance gradually over long periods. Repeated cultivation of weak salt tolerant yeasts in NaCl-rich media transforms them to high salt tolerant organisms (Urano et al., 2001).

Microorganisms differ in their tolerance to osmotic stress, but in general yeasts and fungi are more tolerant than bacteria (Brown, 1978). Among yeasts, strains of *D. hansenii* and *S. rouxii* are highly osmotolerant and

Table 1. Potential of marine yeasts for industrial processes and biotechnology.

High value products/Application	Yeast	Source
Pollution degradation or algae blooms controlling yeasts	*Candida, Rhodotorula, Torulopsis, Hanseniaspora, Debaryomyces,* and *Trichosporon*	Hagler and Hagler, 1981
Gycerol kinase	*D. hansenii*	Nilsson and Adler, 1990
Biotransformation of aromatic polycyclic hydrocarbons	*Trichosporon penicillatum*	Ronald and Shiaris ,1993
Membranes - surfactants for pharmaceuticals	*C. bombicola*	Shepherd et al., 1995; Guilmanov et al., 2002
Convert prawn shell waste into microbial biomass protein	*Candida* species	Rhishipal and Rosamma Philip, 1998
Organic acids and amino acids-regulating the acidity of the fermented product, and also provides lipolytic and proteolytic activity contributing to flavour development	*Debaryomyces hansenii*	Urano et al., 1998
Superoxide dismutases-anti-inflammatory activities	*D. hanseii*	Gonzalez and Ochoa, 1999
Superoxide dismutase	*Saccharomyces cerevisiae*	Hernandez Saavedra and Ochoa, 1999
Microbialsensor-rapid measurements of bio-degradable substances.	*Arxula adeninivorans*	Tag et al., 2000
Glucoamylase gene Glycerol, compatible solutes	*C. magnolia*	Wartmann and Kunze, 2000; Rothschild and Mancinelli, 2001; Sahoo and Agarwal, 2001
Lipids- liposomes for drug delivery and cosmetic packaging		Cavicchioli and Torsten, 2000
Waste transformation and degradation	*C. utilis*	Cavicchioli and Torsten, 2000; Zheng et al., 2005
Hydrocarbon degradation	*Yarrowia lipolytica*	Oswal et al., 2002
Prolyl aminopeptidase (PAP)- role in meat fermentation	*D. hansenii*	Bolumar et al., 2003
Carotene-food colouring	*Rhodotorula mucilaginosa, Arxula adeninivorans*	Libkind et al., 2004
Viable cells- bioremediation of TNT polluted marine environments	*Y. lipolytica*	Jain et al., 2004
α glucosidases- facilitating assimilation of β - fructofuranosides and α glucopyranosides	*Leucosporidium antarcticum*	Turkiewics et al., 2005
Immunostimulant	*Fenneropenaeus indicus*	Sajeevan et al., 2006
Microorganism-useful to improve the final quality of fermented sausages	*D. hansenii*	Bolumar and Sanz, 2006
Protease	*Aureobasidium pullulans*	Chi et al., 2007
Inulinase	*Cryptococcus aureus*	Sheng et al., 2007
Reducing post harvest decay of tomatoes caused by *Alternaria alternate*	*Rhodosporidium paludigenum*	Wang et al., 2008

Table 1. Continued.

Silver nanoparticles	*Candida albicans, C. tropicals, Debaryomyces hansenii, Geotrichum sp., Pichia capsulata, Pichia fermentans, Pichia salicaria, Rhodotorula minuta, Cryptococcus dimennae and Yarrowia lipolylica*	Manivannan et al., 2010
Bio-ethanol production	*Candida albicans, C. tropicals, Debaryomyces hansenii, Geotrichum sp., Pichia capsulata, Pichia fermentans, Pichia salicaria, Rhodotorula minuta, Cryptococcus dimennae and Yarrowia lipolylica*	Kathiresan et al., 2011

capable of growth in media containing up to about 4 M NaCl (Onishi, 1963; Norkrans, 1966). *S. cerevisiae* is limited by NaCl concentrations above 1.7 M (Onishi, 1963). When *D. hansenii* is subjected to increased NaCl stress, intracellular K^+ decreases and intracellular Na^+ increases (Norkrans, 1968). However, the total salt level in the cells is not sufficient to balance the water potential of the medium; this is why additional osmotically active solutes such as polyols accumulate intracellularly when exposed to osmotic stress (Brown and Simpson, 1972; Gustafsson and Norkrans, 1976; Brown, 1978; Adler et al., 1985). Tolerance for a sudden osmotic dehydration is also better in cells having an increased amount of intracellular polyols (Adler and Gustafsson, 1980). Two polyols are produced and accumulated in *D. hansenii*; glycerol, which is the major internal solute in exponentially growing cells, and arabinitol, which predominates in stationary-phase cells (Adler and Gustafsson, 1980). A positive correlation exists between internal glycerol level and salinity of the surrounding medium (Adler et al., 1985; Andre et al., 1988). Glycerol is the major osmoticum, as its concentration may reach molar levels under strongly saline conditions (Gustafsson and Norkrans, 1976). The enzymes that control glycerol catabolism are glycerol kinase and mitochondrial glycerol 3-phosphate dehydrogenase (Gancedo et al., 1968; Sprague and Cronan, 1977; Adler et al., 1985). In yeasts lacking glycerol kinase, the presence of an NAD-dependent glycerol dehydrogenase and a dihydroxyacetone kinase is an alternative pathway (Babel and Hofmann, 1982; May et al., 1982).

Among several marine yeasts, *D. hansenii* accumulates high amounts of Na^+, and in this yeast, Na^+ is not more toxic than K^+ (Ross and Morris, 1962; Norkrans, I966; Prista, 1997). Besides Na^+, glycerol plays a role as a compatible solute for a glycerol/Na^+ symporter with homeosmotic function in this yeast species (Lages et al., I999; Lucas et al., I990). Increased transport activities might be needed in addition to the maintenance of a high osmotic pressure within the cell. In addition, existence of sodium efflux process may also be involved in saline tolerance. The mechanism for extrusion of Na^+ across the plasma membrane might be carried out via the function of Na^+-ATPase or Na^+/H^+ antiporters (Ramos, I999).

POTENTIAL OF MARINE YEASTS

Yeasts are used in many industrial processes, such as the production of alcoholic beverages, food, fodder yeasts and for the synthesis of various metabolic products. The last category includes enzymes, vitamins, polysaccharides, carotenoids, polyhydric alcohols, lipids, glycolipids, citric acid, ethanol and compounds synthesized by the introduction of recombinant DNA into yeasts. Some of these products are produced commercially, while others are potentially valuable in biotechnology. Some uses of marine yeasts in the food, beverage and fermentation industries are shown in Table 1.

CONCLUSION

Even after five decades of research, the potential of marine yeasts to contribute to biotechnological applications has not been fully realized or exploited. By virtue of their occurrence in extreme environmental conditions, the marine yeasts have superior qualities over their terrestrial counterparts with regards to salt tolerance, enzyme production, biosynthetic potential, pollution abatement, and ethanol and other fermentative processes, and hence deserve further investigation. Although much work has been carried out on molecular aspects of yeasts, such efforts for marine yeasts are lacking. There are no proper culture collections for marine yeasts. Only a few marine habitats have been investigated for yeast species and many additional species await discovery. Based on the fact that yeasts of terrestrial origin are widely used in traditional and modern biotechnology, the exploration for new species of marine origin should lead to additional novel technologies.

ACKNOWLEDGEMENTS

The authors are thankful to the authorities of Annamalai University for providing facilities and the Ministry of Environment and Forests, Govt. of India, New Delhi for financial support.

REFERENCES

Adler L, Blomberg A, Nilsson A (1985). Glycerol metabolism and osmoregulation in the salt-tolerant yeast *Debaryomyces hansenii*. J. Bacteriol. 162:300-306.

Adler L, Gustafsson L (1980). Polyhydric alcohol production and intracellular amino acid pool in relation to halotolerance of the yeast *Debaryomyces hansenii*. Arch. Microbiol. 124:123-130.

Ahearn DG (1973). Effects of environmental stress on aquatic yeast populations, pp:. 433-439. In L. H. Stevenson and R. R. Colwell (ed.). Estuarine Microbiology Ecology.

Ahearn DG (1973). Effects of environmental stress on aquatic yeast populations, pp:. 433-439. In L. H. Stevenson and R. R. Colwell (ed.). Estuarine Microbial Ecology..

Ahearn DG, Roth Jr FJ, Meyers SP (1968). Ecology and characterization of yeasts from aquatic regions of South Florida. Mar. Biol. 1: 291-308.

Alongi DM (1992). Bathymetric patterns of deep-sea benthic communities from bathyal to abyssal depths in the western South Pacific (Solomon and Coral Seas). Deep Sea Res. 39: 549-565.

Andre L, Nilsson A, Adler L (1988). The role of glycerol in osmotolerance of the yeast *Debaryomyces hansenii*. J. Gen. Microbiol. 134: 669-677.

Babel W, Hofmann KH (1982). The relation between the assimilation of methanol and glycerol in yeasts. Arch. Microbiol. 132:179-184.

Barnett JA, Barnett L (2011). Yeast Research: a Historical Overview. American Society for Microbiology Press, 392.

Barnett JA, Pankhurst RJ (1974). A new key to the yeasts. American Elsevier, New York.

Bolumar T, Sanz Y (2006). Sensory improvement of dry fermented sausage by the addition of cell free extracts from *Dabaryomyces hansenii* and *Lactobacillus sakei*. Meat Sci. 72:457-466.

Bolumar T, Sanz Y, Aristoy MC,Toldra F (2003). Purification and Characterization of a prolyl aminopeptidase from *Debaryomyces hansenii*. Appl. Environ. Microbiol. 69:227-232.

Brown AD, Simpson JR (1972). Water relations of sugar-tolerant yeasts: the role of intracellular polyols. J. Gen. Microbiol. 72:589-591.

Brown AD (1978). Compatible solutes and extreme water stress in eukaryotic microorganisms. Microb. Physiol. 17:181-242.

Buchan A, Newell SY, Moreta JI, Moran MA (2002). Analysis of internal transcribed spacer (ITS) regions of rRNA genes in fungal communities in a southeastern U.S. salt marsh. Microb. Ecol. 43:329-340.

Buck JD (1975). Distribution of aquatic yeasts-effect of incubation temperature and chloramphenicol concentration on isolation. Mycopathologia 56:73-79.

Buck JD, Bubucis PM, Combs TJ (1977). Occurrence of Human-Associated Yeasts in Bivalve Shellfish from Long Island Sound. Appl. Environ. Microbiol. 33:370-378.

Bunt JS (1955). The importance of bacteria and other microorganisms in the sea water at MacQuarie Island. Aust. J. Mar. Fresh. Res. 6:60-65.

Burke RM, Jennings DH (1990). Effect of growth characteristics of the marine yeast *Debaryomyces hansenii* in batch and continuous culture under carbon and potassium limitation. Mycol. Res. 94:378-388.

Butinar L, Santos S, Spencer-Martins I, Oren A, Gunde N (2005). Yeast diversity in hyper saline habitats FEMS Microbiol. Lett. 244:229-234.

Cavicchioli R, Torsten T (2000). Extremophiles. In: Lederberg J (ed) Encyclopedia of microbiology, vol 2, 2nd edn. Academic, San Diego, pp. 317-337.

Chand GP, Eckert JW (1996). Studies on transformation of *Candida cephila* and *Debaryomyces hansenii* with plasmids. Phytopathology 86:S34.

Chi Z, Ma C, Wang P, Li HF (2007). Optimization of medium and cultivation conditions for alkaline protease production by the marine yeast *Aureobasidium pullulans*. Bioresour. Technol. 98:534-538.

Coleman A, Cook WL, Ahearn DG (1975). Abstr. Annu. Meet. Am. Soc. Microbiol. N20:187.

Dabrowa N, Landau JW, Newcomer VD, Plunkett OA (1964). A survey of tide-washed areas of Southern California for fungi potentially pathogenic to man. Mycopathol. Mycol. Appl. 24:137-150.

Davenport RR (1980). Cold-tolerant yeasts and yeast-like organisms. In: Davenport RR (ed) Biology and activity of yeasts. Academic, London, UK, pp. 215-230.

De Araujo FV, Soares CA, Hagler AN, Mendonça-Hagler LC (1995). Ascomycetous yeast communities of marine invertebrates in a southeast Brazilian mangrove ecosystem. Antonie van Leeuwenhoek 68:91-99.

Di Menna, ME (1960). Yeast from Antarctica. J. Gen. Microbiol. 7:295-300.

Dzawachiszwili N, Landau JW, Newcomer VD, Plunkett OA (1964). The effect of sea water and sodium chloride on the growth of fungi pathogenic to man. J. Invest. Dermatol. 43:103-109.

Fedorak PM, Semple KM, Westlake DWS (1984). Oil degrading capabilities of yeasts and fungi isolated from coastal marine environment. Can. J. Microbiol. 30:565-571.

Fell JW, Kurtzman CP (1990). Nucleotide sequence analysis of the large subunit rRNA for identification of marine occurring yeasts. Curr. Microbiol. 21:295-300.

Fell JW, Van Uden N (1963). Yeasts in marine environments, pp 329-340. In C. H. Oppenheimer [ed.], Symposium on marine microbiology. Charles C Thomas, Publisher, Springfield, Ill.

Fell JW (1976). Yeasts in oceanic regions. In: Jones EBG (ed) Recent advances in aquatic mycology. Elec, London, pp. 93-124.

Fell JW, Statzell-Tallman A, Luit MJ, Kurtzman CP (1992). Partial rRNA sequences in marine yeasts-a model for identification of marine eukaryotes. Mol. Mar. Biol. 1:175-186.

Fell JW, Tallman AS, Kurtzman CP (2004). *Lachencea meyersii* sp. nov., an ascosporogenous yeast from mangrove region in the Bahama Islands. Stud. Mycol. 50:359-363.

Fell JW, Ahearn DG, Meyers SP, Roth Jr FJ (1960). Isolation of yeasts from Biscayne Bay, Florida and adjacent benthic areas. Limnol. Oceanogr. 4:366-371.

Fell JW, Statzell JW, Hunter IL, Phaff HJ (1969). *Leucosporidium gen. nov.*, the heterobasidiomycetous stage of several yeasts of the genus *Candida*. Antonie Van Leeuwenhoek. 35:433-462.

Fukumaki T, Inoue A, Moriya K (1994). Isolation of marine yeast that degrades hydrocarbon in the presence of organic solvent. Biosci. Biotechnol. Biochem. 58:1784-1788.

Gadanho M, Sampaio JP (2004). Application of temperature gradient gel electrophoresis to the study of yeast   diversity in the estuary of the Tagus river, Portugal. FEMS Yeast Res. 5:253-261.

Gadanho M, Almeida JM,Sampaio JP (2003). Assessment of yeast diversity in a marine environment in the south of Portugal by microsatellite-primed PCR. Antonie van Leeuwenhoek 84:217-227.

Gadd, GM, Edwards SW (1986). Heavy-metal-induced flavin production by *Debaryomyces hansenii* and possible connections with ion metabolism. Trans. Br. Mycol. Soc. 87:533-542.

Galtsoff PS (1964). The American oyster, *Crassostrea virginica* Gmelin. Fish. Bull. 64:1-480.

Gancedo C, Gancedo JM, Sols A (1968). Glycerol metabolism in yeasts. Pathways of utilization and production. Eur. J. Biochem. 5:165-172.

González AG, Ochoa JL (1999). Anti-Inflammatory activity of *Debaryomyces hansenii* Cu, Zn-SOD. Arch. Med. Res. 1:69-73.

Govind NS, McNally KL, Trench RK (1992). Isolation and sequence analysis of the small subunit ribosomal RNA gene from the euryhaline yeast *Debaryomyces hansenii*. Curr. Gen. 22:191-195.

Guilmanov V, Ballistreri A, Impallomeni G,Gross RA (2002). Oxygen transfer rate and sophorose lipid production by *Candida bombicola*. Biotechnol. Bioeng. 77:489-494.

Gustafsson, L, Norkrans B (1976). On the mechanism of salt tolerance. Production of glycerol and heat during growth of *Debaryomyces hansenii*. Arch. Microbiol. 110:177-183.

Hagler AN, Ahearn DG (1987). Ecology of aquatic yeasts. In: Rose A.H, Harrison J.S (ed) The yeasts, vol 2, Yeasts and the environment. Academic, London, pp. 181-205.

Hagler AN, Hagler LCM (1981). Yeasts from marine and estuarine waters with different levels of pollution in the state of Rio de Janeiro, Brazil. Appl. Environ. Microbiol. 41:173-178.

Hagler AN, Rosa Morais PB, Hagler LCM, Franco GMO, Araujo FV, Soares CAG (1993). Yeasts and coliform bacteria of water accumulated in bromeliads of mangrove and sand dune ecosystems of southeast Brazil. Can. J. Microbiol. 39:973-977.

Hernandez-Saavedra, NY, Hernandez-Saavedr D, Ochoa JL (1992). Distribution of *Sporobolomyces* (Kluyver et van Niel) Genus in the Western Coast of Baja California Sur, Mexico. Syst. Appl. Microbiol. 15:319-322.

Hernandez-Saavedra, NY, Ochoa JL, Vazquez- Dulhalt R (1994). Effect of salinity on the growth of the marine yeast *Rhodotorula rubra*. Microbiology 80:99-106.

Hernendez Savedra, NY, Ochoa JL (1999). Copper zinc superoxide dismutase from the marine yeast *Debaryomyces hansenii*. Yeast 15:657-668.

Hirayama K (1992). Part VI. Physiology in growth. In: Jap. Soc.Fisheries Sci. (ed.). The rotifer *Brauchionus plicatilis*-biology and mass culture, pp. 52-68. (In Japanese).

Hyde KD (2002). Fungi in marine environments. Fungal Diversity Press, Hong Kong. Kimura, T., K. Hayashi and I. Sugahara, 1985. Studies on C1- compounds-utilizing yeasts from coastal water and sediments. Bull. Fac. Fish. Mie Univ. 12:61-67.

Jain MR, Zinjarde SS, Deobagkar DD, Deobagkar DN (2004). 2,4,6-Trinitrotoluene transformation by a tropical marine yeast, *Yarrowia lipolytica* NCIM 3589. Mar. Poll. Bull. 49:783-788.

Kathiresan K, Saravanakumar K, Senthilraja P (2011). Bio-ethanol production by marine yeasts isolated from coastal mangrove sediment. Int. Multidiscip. Res. J. 1:19-24.

Kobayashi Y, Tsubaki K, Soneda M (1953). Marine yeasts isolated from little-neck clam. Bull. Nat. Sci. Mus. 33:47-52.

Kriss AE, Novozhilova MI (1954). Are yeast organisms inhabitants of seas and oceans? Mikrobiologia. 23:669-683.

Lachance MA, Starmer WT (1998). Ecology and Yeasts. In: Kurtzman CP, Fell JW (eds) The yeasts a taxonomic study, 4th edn. Elsevier, Amsterdam, pp. 21-30.

Lages F, Silva-Grac M, Lucas C (1999). Active glycerol uptake is a mechanism underlying halotolerance in yeasts: a study of 42 species. Microbiol. 145:2577-2585.

Lazarus CR, Koburger JA (1974). Identification of yeasts from the Suwannee River Florida estuary. Appl. Environ. Microbiol. 27:1108-1111.

Li Z, Obita H, Kamishima S, Fukuda S, Kakita H, Kobayashi Y, Higashihara T (1995). Improvement of immobilization conditions for biodegradation of floating oil by a bio-system, co-immobilizing marine

oil-degrading yeast *Candida* sp. and nutrients. Seibutsu Kogaku Kaishi, 73:295-299 (in Japanese).

Libkind D, Brizzio S, van Broock M (2004). *Rhodotorula mucilaginosa*, a carotenoid producing yeast strain from a Patagonian high-altitude lake. Folia Microbiol. 49:19-25.

Lorenz R, Molitoris HP (1997). Cultivation of fungi under simulated deep sea conditions. Mycol. Res. 101:1355-1365.

Lucas C, Da-Costa M, Van-Uden N (1990). Osmoregulatory active sodium-glycerol co-transport in the halotolerant yeast *Debaryomyces hansenii*. Yeast 6:187-191.

Madri P (1968). Factors influencing growth and morphology of *Candida albicans* in a marine environment. Bot. Mar. 11:31-35.

Madri P, Claus D, Moss EE (1966). Infectivity of pathogenic fungi in a simulated marine environment. Rev. Biol. 5:371-381.

Manivannan S, Alikunhi NM, Kandasamy K (2010). *In vitro* Synthesis of Silver Nanoparticles by Marine Yeasts from Coastal Mangrove Sediment. Adv. Sci. Lett. 3:1-6.

Marija V, Goran M, Ivanka P (1993). Capability for degradation of crude oil hydrocarbons by sea water yeasts and bacteria from Kvarner Bay. Period. Biol. 94:169-177.

May JW, Marshall JH,Sloan J (1982). Glycerol utilization by *Schizosaccharomyces pombe*: phosphorylation of dihydroxyacetone by a specific kinase as the second step. J. Gen. Microbiol. 128:1763-1766.

Mayers SP, Ahearn DG, Alexander S, Cook W (1975). *Pichia spartinae*, a dominant yeast of the *Spartina* salt marsh. Dev. Ind. Microbiol. 16:262-267.

Mendonça-Hagler LC, Hagler AN, Kurtzman CP (1993). Phylogeny of *Metschnikowia* species estimated from partial rRNA sequences. Int. J. Syst. Bacteriol. 43:368-373.

Meyers SP, Ahearn DG, Miles P (1971). Characterization of yeasts in Baratara Bay. La St. Univ. Coastal Stud. Bull. 6:7-15.

Meyers SP, Ahearn DG, Gunkel W,Roth Jr, FJ (1967). Yeasts from the North Sea. Mar. Biol. 1:118-123.

Morita K, Usami R, Horikoshi K (1994). Marine killer yeasts isolated from deep sea and their properties. J. Mar. Biotechnol. 2:135-138.

Morris EO (1968). Yeasts of marine origin. Oceanogr. Mar. Biol. Annu. Rev. 6:201-230.

Nagahama T, Hamamoto M, Nakase T, Takami H, Horikoshi K (2001a). Distribution and identification of red yeasts in deep-sea environments around the northwest Pacific Ocean. Antonie van Leeuwenhoek 80:101-110.

Nagahama T, Hamamoto M, Nakase T, Horikoshi K (2001b). *Rhodotorula lamellibrachii* sp. nov., a new yeast species from a tubeworm collected at the deep-sea floor in Sagami bay and its phylogenetic analysis. Antonie van Leeuwenhoek 80:317-323.

Nagahama T, Hamamoto M, Nakase T, Horikoshi K (2003a). *Rhodotorula benthica* sp. nov. and *Rhodotorula calyptogenae* sp. nov., novel yeast species from animals collected from the deep-sea floor, and *Rhodotorula lysiniphila* sp. nov., which is related phylogenetically. Int. J. Syst. Evol. Microbiol. 53:897-903.

Nagahama T, Hamamoto M, Nakase T, Takaki Y, Horikoshi K (2003b). *Cryptococcus surugaensis* sp. nov., a novel yeast species from sediment collected on the deep-sea floor of Suruga Bay. Int. J. Syst. Evol. Microbiol. 53:2095-2098.

Naumova ES, Sukhotina NN, Naumov GI (2004). Molecular-genetic differentiation of the dairy yeast *Kluyveromyces lactis* and its closest wild relatives. FEMS Yeast Res. 5:263-269.

Nilsson A, Adler L (1990). Purification and characterization of glycerol-3-phosphate dehydrogenase (NAD⁺) in the salt-tolerant yeast *Debaryomyces hansenii*. Biochim. Biophy. Acta. 16:180-185.

Nishikawa A, Tomomatsu H, Sugita T, Ikeda R and Shinoda T (1996). Taxonomic position of clinical isolates of *Candida famata*. J. Med. Vet. Mycol. 34:411-419.

Norkrans B (1966). Studies on marine occurring yeasts: growth related to pH, NaCl concentration and temperature. Arch. Microbiol. 54:374-392.

Norkrans B (1968). Studies on marine occurring yeasts: respiration, fermentation and salt tolerance. Arch. Microbiol. 62:358-372.

Onishi H (1963). Osmophilic yeasts. Adv. Food Res. 12:53-94.

Oswal N, Sarma PM, Zinjarde SS, Pant A (2002). Palm oil mill effluent treatment by tropical marine yeast. Bioresour. Technol. 85:35-37.

Prista C, Almagro A, Loureiro-Dias MC, Ramos J (1997). Physiological basis for the highs tolerance of *Debaryomyces hansenii*. Appl. Environ. Microbiol. 63:4005-4009.

Ramos J (1999). Contrasting salt tolerance mechanism in *Saccharomyces cerevisiae* and *Debaryomyces hansenii*, pp. 377-390. In, S.G.Pandalai(ed) in recent research developments in Microbiology. Research Signpost Publication, Trivandrum, India.

Ranu G (1994). Emulsifying activity of hydrocarbonoclastic marine yeast. Nutr. Bioact. Subst. Aquat. Org. Pap. Symp. pp. 276-285.

Ray MK, Uma DK, Seshu KG, Shivai S (1992). Extra cellular protease from the Antarctic yeast *Candida humicola*. Appl. Environ. Microbiol. 58:1918-1923.

Rhishipal R, Philip R (1998). Selection of marine yeasts for the generation of single cell protein from prawn-shell waste. Bioresour. Technol. 65: 255-256.

Ricaurte ML, Govind NS (1999). Construction of plasmid vectors and transformation of marine yeast *Debaryomyces hansenii*. Mar. Biotechnol. 1:15-19.

Richards M, Elliott FR (1966). Inhibition of yeast growth by streptomycin. Nature 209:536.

Ronald MA, Shiaris MP (1993). Biotransformation of Polycyclic Aromatic Hydrocarbons by Yeasts Isolated from Coastal Sediments. Appl. Environ. Microbiol. 59:1613-1618.

Ross SS, Morris EO (1962). Effect of sodium chloride on the growth of certain yeasts of marine origin. J. Sci. Food Agr. 13:467-475.

Roth FJ, Ahearn DG, Fell JW, Meyers SP, Meyers SA (1962). Ecology and taxonomy of yeasts isolated from various marine substrates. Limnol. Oceanogr. 7:178-185.

Rothschild LJ, Mancinelli RL (2001). Life in extreme environments. Nature 409:1092-1101.

Sahoo DK, Agarwal GP (2001). An investigation on glycerol biosynthesis by an osmophilic yeast in a bioreactor. Process Biochem. 36:839-846.

Sajeevan TP, Philip R, Bright Singh IS (2006). Immunostimulatory effect of a marine yeast *Candida sake* S165 in *Fenneropenaeus indicus*. Aquaculture 257:150-155.

Seshadri R, Sieburth JM (1971). Cultural Estimation of Yeasts on Seaweeds,. Appl. Microbiol. 22:507-512.

Sheng J, Chi Z, Li J, Gao L, Gong F, (2007). Inulinase production by the marine yeast *Cryptococcus aureus* G7a and inulin hydrolysis by the crude inulinase. Proc. Biochem. 42:805-811.

Shepherd R, Rockey J, Sutherland IW, Roller S (1995) Novel bioemulsifiers from microorganisms for use in foods. J. Biotechnol. 40:207-217.

Sinclair NA, Stokes JL (1965). Obligatory psychrophilic yeast from the Polar Regions. Can. J. Microbiol. 11:259-269.

Soares CAG, Maury M, Pagnocca EC, Araujo EV, Mendonca-Hagler LC, Hagler AN, (1997). Ascomycetous yeasts from tropical intertidal dark mud of southeast Brazilian estuaries. J. Gen. Appl. Microbiol. 43:265-272.

Sprague GF, Cronan JE (1977). Isolation and characterization of Saccharomyces cerevisiae mutants defective in glycerol catabolism. J. Bacteriol. 129:1335-1342.

Suchiro S, Tomiyasu Y (1962). Studies on the marine yeasts. V. yeasts isolated from sea weed. J. Fac. Agric. 12:163-169.

Tag K, Lehmann M, Chan C, Renneberg R, Riedel K, Kunze G (2000). Measurement of biodegradable substances with a mycelia-sensor based on the salt tolerant yeast *Arxula adeniniIorans* LS3. Sens. Actuat. B. 67:142-148.

Taysi I, Van-Uden N (1964). Occurrence and population densities of yeast species in an estuarine-marine area. Limnol. Oceanogr. 9:42-45.

Tripp MR (1960). Mechanisms of removal of injected microorganisms from the American Oyster, *Crassostrea virginica* (Gmelin). Biol. Bull. 119:273-282.

Turkiewicz M, Pazgier M, Donachie SP, Kalinowska H (2005). Invertase and _-glucosidase production by the endemic Antarctic marine yeast *Leucosporidium antarcticum*. Pol. Polar Res. 26:125-136.

Urano N, Hirai H, Ishida M, Kimura S (1998). Characterization of ethanol-producing marine yeasts isolated from coastal water. Fish. Sci. 64:633-637.

Urano N, Yamazaki M, Ueno R (2001). Distribution of Halotolerant and/or Fermentative Yeasts in Aquatic Environments. J. Tokyo Univ. Fish. 87:23-29.

Van Uden N, Fell JW (1968). Marine yeasts, pp. 167-201. In M. R. Droop and E. J. F. Wood (ed.), Advances in microbiology of the sea, vol. 1. Academic Press Inc., New York.

Van Uden N, Castelo-Granco R (1963). Distribution and population densities of yeast species in Pacific water, air animals and kelp off Southern California. Limnol. Oceanogr. 8:323-329.

Wartmann T, Kunze G (2000). Genetic transformation and biotechnological application of the yeast *Arxula adeninivorans*. Appl. Microbiol. Biotechnol. 54:619-624.

Watson K, Arthur H (1976). *Leucosporidium* yeasts: obligate psychrophiles which alter mem- brane-lipid and cytochrome composition with temperature. J. Gen. Microbiol. 97: 11--18.

Wong B, Kiehn TE, Edwards F, Bernard EM, Marcove RC, De Haven E, Armstrong D (1982). Bone infection caused by *Debaryomyces hansenii* in a normal host: a case report. J. Clin. Microbiol. 16:545-548.

Yadav JS, Loper JC (1999). Multiple P450 alk (cytochrome P450 alkane hydroxylase) genr from the halotolerant yeast *Dabaryomyces hansenii*. Gene. 226:139-146.

Yamasato K, Goto S, Ohwada K, Okuno D, Araki H, Iizaka H (1974). Yeasts from the Pacific Ocean. J. Gen. Appl. Microbiol. 20:289-307.

Yifei W, Bao Y, Shen D, Feng W, Yu T, Zhang J, Zheng XD (2008). Biocontrol of *Alternaria alternata* on cherry tomato fruit by use of marine yeast *Rhodosporidium paludigenum* Fell & Tallman. Int. J. Food Microbiol. 123:234-239.

Zheng S, Yang M, Yang Z (2005). Biomass production of yeast isolate from salad oil manufacturing wastewater. Bioresour. Technol. 96:1183-1187.

ZoBell CE, Johnson FH (1949). The influence of hydrostatic pressure on the growth and viability of terrestrial and marine bacteria. J. Bacteriol. 57:179-189.

23

Some new rust fungi (Uredinales) from Fairy Meadows, Northern Areas, Pakistan

N.S. AFSHAN[1]*, A. N. KHALID[2] AND A. R. NIAZI[2]

[1]Centre for Undergraduate Studies, University of the Punjab Quaid-e-Azam Campus, Lahore, 54590, Pakistan.
[2]Department of Botany, University of the Punjab Quaid-e-Azam Campus, Lahore, 54590, Pakistan.

Puccinia opizii on *Carex curta* **is reported here as a new record for Pakistan. Furthermore,** *Coleosporium lycopi, Cronartium ribicola, Melampsora epitea, Peridermium thomsonii, Phragmidium rosae-moschatae, Puccinia carthami, Puccinia chrysanthemi, Puccinia circaeae, Puccinia graminis, Puccinia komarovii* **and** *Uromyces trifolii* **are additions to the rust flora of Fairy Meadows, Northern Areas, Pakistan.**

Key words: Byal camp, *carex curta*, fairy meadows, Hunza, Nanga Parbat, Northern areas.

INTRODUCTION

This paper is a continuation of the enumeration of rust fungi of Fairy Meadows, Pakistan. Previously, about 73 species of rust fungi have been reported from Northern Pakistan (Afshan et al., 2011), with only 22 taxa from Fairy Meadows, including one species each of *Aecidium, Chrysomyxa, Cronartium, Hyalopsora, Melampsora* and *Pucciniastrum*, two species of *Uromyces* and fourteen species of *Puccinia*. During surveys of the rust flora of Fairy Meadows, sixteen plants infected with rust fungi were collected. Among these, *Puccinia opizii* on *Carex curta* is a new record for Pakistan. In addition, *Coleosporium lycopi* on *Campanula benthamii, Cronartium ribicola* on *Ribes orientale, Melampsora epitea* on *Salix flabellaris, S. hastata* and *S. tetrasperma, Puccinia carthami* on *Centaurea calcitrapa, P. chrysanthemi* on *Artemisia brevifolia, A. dracunculus* and *A. maritima, P. circaeae* on *Circaea alpina, P. komarovii* on *Impatiens brachycentra* and *Uromyces trifolii* on *Trifolium resupinatum* are new records for Fairy Meadows, Northern Areas, Pakistan. The uredinial stage (II) of *Phragmidium rosae-moschatae* on *Rosa webbiana* and aecidial stage (I) of *Puccinia graminis* on *Berberisvulgaris* are also additions to the rust flora of this area of Pakistan. Although *Peridermium thomsonii* on *Picea smithiana* has previously been reported from Fairy Meadows, it is re-described here to illustrate important morphological features. With these additions, the number of rust fungi known from Fairy Meadows, Pakistan is raised to 31.

MATERIALS AND METHODS

Freehand sections of infected tissue and spores were mounted in lactophenol and gently heated to boiling. The preparations were observed under a NIKON YS 100 microscope and photographed with a digipro-Labomed and JSM5910 scanning electron microscope. Drawings of spores and paraphyses were made using a camera lucida (Ernst Leitz Wetzlar, Germany). Spore measurements were made with the use of an ocular micrometer. At least 25 spores were measured for each spore stage. In addition to comparisons using light microscopy, images were obtained of the rust spores using a scanning electron microscope (SEM). The rusted specimens have been deposited in the herbarium of the Botany Department at the University of the Punjab, Lahore (LAH).

Enumeration of taxa

Puccinia opizii Bubák, *Zentbl. Bakt. ParasitKde*, Abt. II 9: 925 (1902) (Figure 1) ≡ *Puccinia dioicae* var. *opizii* (Bubák) U. Braun, *Feddes Repert. Spec. Nov. Regni Veg.* 93(3-4): 264 (1982).

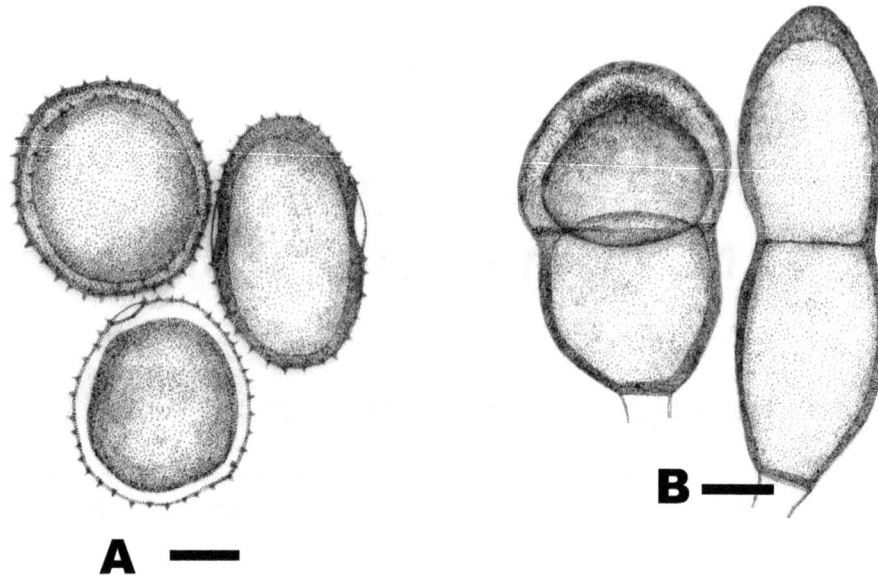

Figure 1. Lucida drawings of *Puccinia opizii*. (A) Echinulated urediniospores; (B) Teliospores. Scale bar for A = 2 cm, B = 8 μm & C = 9 μm.

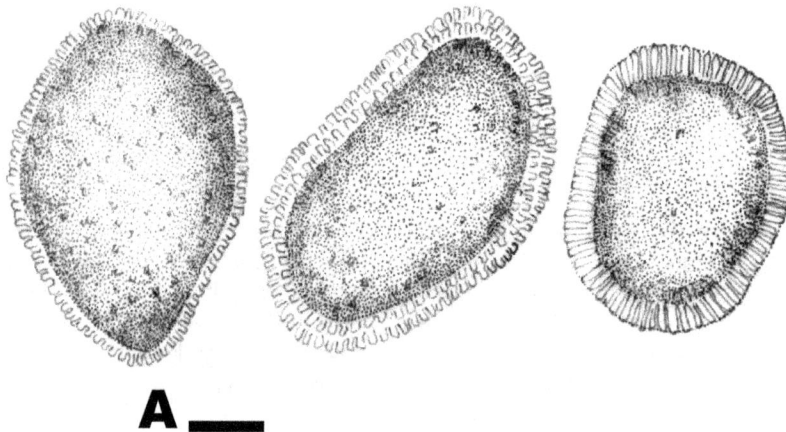

Figure 2. Lucida drawing of urediniospores of *Coleosporium lycopi*. Scale Bar = 4 μm. *Cronartium ribicola* J.C. Fisch., *Hedwigia* 11: 182 (1872).

Spermogonia and aecia unknown. Uredinia hypophyllous, on leaves and culms, covered by epidermis, yellowish brown to dark brown, 0.09 to 0.2 × 0.1 to 3.0 mm. Urediniospores globose to ellipsoid, 17 to 22 × 21 to 24 μm; wall up to 2 μm thick, pale brownto cinnamon brown, finely echinulate; germ pores 1 to 2, equatorial; pedicel hyaline, short. Telia mostly hypophyllous, sometimes amphigenous, intermixed with uredinia, dark brown to blackish brown, 0.2 to 0.4 × 0.2 to 0.5 mm. Teliospores clavate to oblong, constricted at the septum, 1523 × 37–49 μm (mean 19.9 × 44.2 μm); wall 2–3 μm thick, chestnut brown to golden brown, smooth; apex mostly rounded, 8–12 μm thick; germ pores obscure; pedicel hyaline to light brown, thin walled, 7–10 × 60–70 μm.

Material examined

On *Carex curta* Gooden, with II and III stages, Pakistan, Northern

On *Carex curta* Gooden, with II and III stages, Pakistan, Northern Areas, Fairy Meadows, 3036 m a. s. l., 12 Aug 2007, N. S. Afshan, NSA 72. (LAH NSA 1074).

Puccinia opizii is a new record for Pakistan. Rust fungi previously reported on *Carex* spp. from Pakistan include *Puccinia caricina* DC. and *P. caricis-filicinae* Barclay on *Carex filicina* Nees, *P. dioicae* Magnus and *P. pakistani* S. Ahmad on *Carex nubigena* D. Don, *P. bolleyana* Sacc. on *Carex flacca* Schreb., *P. subepidermalis* Afshan, Khalid and S.H. Iqbal on *Carex curta* Gooden, *P. caricis-kouriyamensis* on *Carex karoi* Freyn, *P. caricis-pocilliformis* on *Carex* sp. and *P. extensicola* var. *linosyridis-caricis* on *Carex divulsa* Stokes (Khalid and Saba, 2011).

Coleosporium lycopi Syd. & P. Syd., *Annls. mycol.* 11(5): 402 (1913) (Figure 2).

Spermogonia, aecia and telia not found. Uredinia on adaxial surface, cup- shaped, scattered, light yellow to yellowish orange,

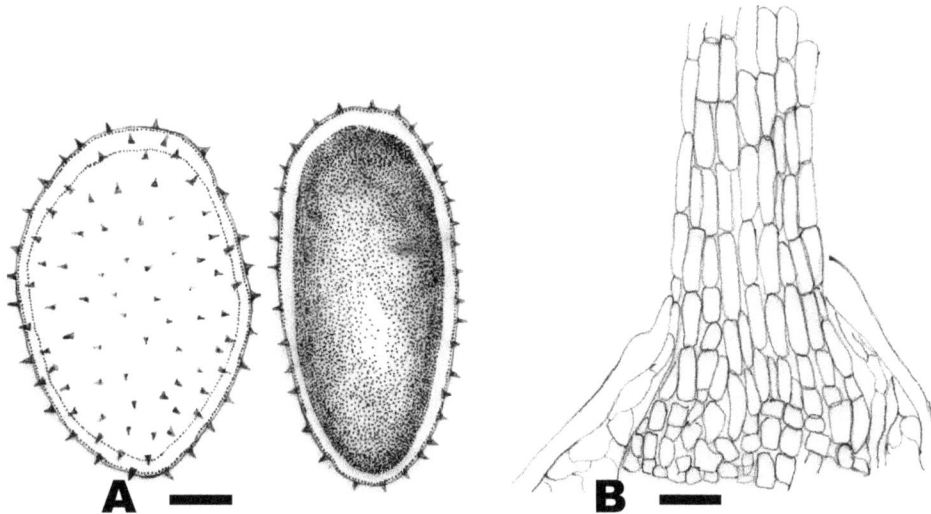

Figure 3. Lucida drawing of *Cronartium ribicola*. (A) Urediniospores; (B) Telium containing teliospores. Scale Bar = 6 µm.

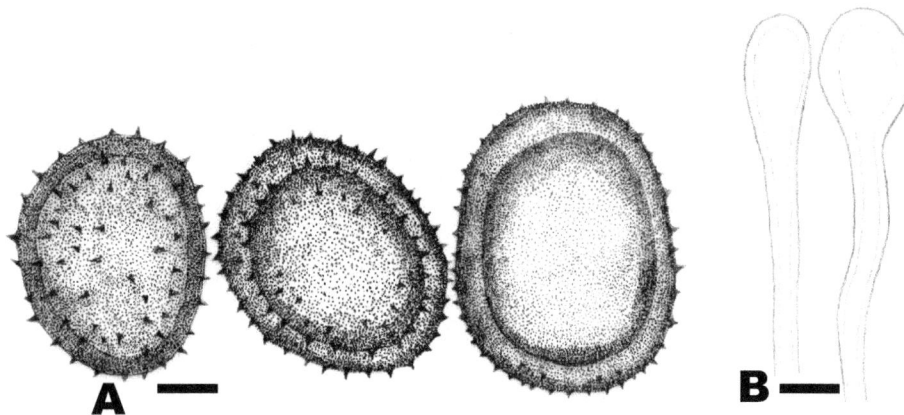

Figure 4. Lucida drawings of *Melampsora epitea*. (A) Urediniospores; (B). Capitate paraphyses. Scale Bar for D = 8 µm and E = 14 µm.

pulverulent. Urediniospores globose to subglobose, ovoid or ellipsoid, 15 to 20 × 19 to 25 µm; wall thickness 1.5 to 2 µm, annulate to verrucose, hyaline to light yellow with orange granules;germ pores obscure.

Material examined

On *Campanula benthamii* Wall., with II stage, Pakistan, Northern Areas, Fairy Meadows, Byal Camp (Nanga Parbat base camp), at 3036 m a. s. l., 12 Aug, 2007, N.S.Afshan, NSA G 39. (LAH NSA 1010).

Coleosporium lycopi is a new record from the Northern Areas, Fairy Meadows and Byal Camp. *Cronartium ribicola* J.C. Fisch., *Hedwigia* 11: 182 (1872) (Figure 3) Uredinia hypophyllous, forming groups, yellow, surrounded by a delicate peridium, opening by a central pore. Urediniospores ellipsoid to obovoid, sharply echinulate, yellow to yellowish orange, 12 to 19 × 20 to 27 µm, wall hyaline, (1.5 to) 2 to 3 µm thick. Telia hypophyllous, crowded, mostly along veins of leaf, arising in Uredinia, orange to yellowish

brown, producing columns of teliospores up to 3 mm long and 122 to 156 µm thick. Teliospores ellipsoid to broadly ellipsoid or cylindric, 10 to 20 × (27 to) 30 to 75 µm, wall hyaline, smooth, (1.5 to) 2 to 3 µm thick.

Material examined

On *Ribes orientale* Desf., with II and III stages, Pakistan, Northern Areas, Fairy Meadows, 3036 m a. s. l., 12 Aug 2007, N.S. Afshan and A.N. Khalid, NSA G 20. (LAH NSA 1011).

Cronartium ribicola has been reported on *Ribes emodense* Rehder, *R. aff. himalensis* Royle ex Decne and *R. aff. orientale* Desf. from Kaghan valley, Hazara and Naran by Malik and Khan (1944), Ahmad (1956a, b) and Ono (1992). It is a new record for Fairy Meadows. *Melampsora epitea* Thüm., *Mittheil. Ver. Österr.* 2: 38 and 40 (1879) var. *epitea* (Figure 4) Telia not found. Uredinia hypophyllous, mostly scattered, sometimes in the form of groups, yellow to orangish yellow.

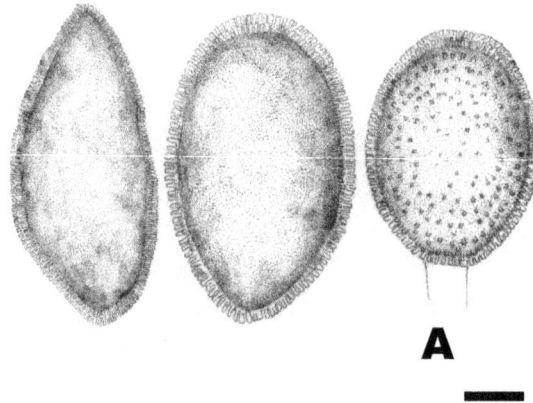

Figure 5. Lucida drawing of aeciospores of *Peridermium thomsonii*. Scale Bar = 10 μm.

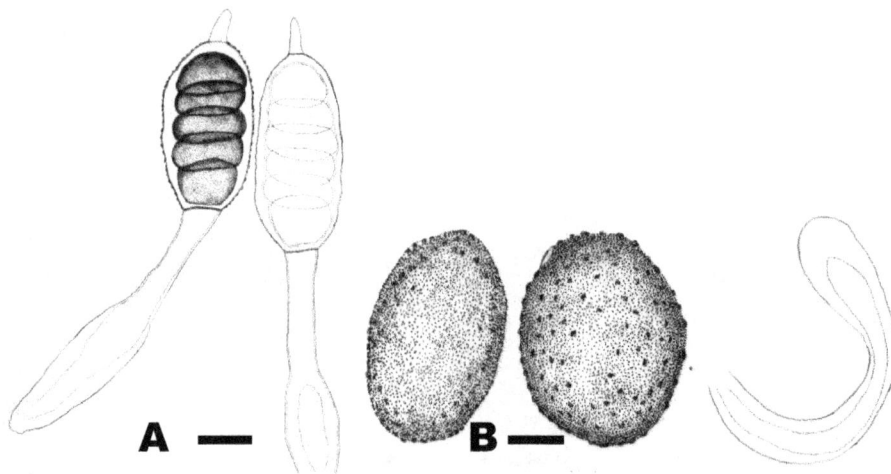

Figure 6. Lucida drawings of *Phragmidium rosae-moschatae.* (A) Teliospores; (B) Urediniospores and paraphysis. Scale Bar for A = 32 μm and B = 8 μm.

Urediniospores globose to obovoid or broadly ellipsoid; 18 to 21 × 21 to 24 μm; wall 1.5 to 2 μm thick, echinulate, hyaline to yellowish brown; germ pores obscure. Paraphyses abundant, capitate, hyaline to light yellow, 16 to 19 μm wide at apex, 7 to 8 μm thick at base and up to 70 μm long.

Material examined

On *Salix flabellaris* N. J. Andress, *S. hastata* L., *S. tetrasperma* Roxb., with II stage, Pakistan, Northern Areas, Fairy Meadows, 3036 m a. s. l., 12 Aug 2007, N. S. Afshan, NSA G 40. (LAH NSA 1018).

Melampsora epitea has been reported on *Salix* sp., *S. tetrasperma* Roxb., *S. hastata* L. and *S. flabellaris* N. J. Anderson from Kaghan valley, Swat, Choa Saiden Shah, Naltar, Byal camp and Bashu Jungle by Sultan (2005). *Melampsora epitea* is first time reported from Fairy Meadows.

Peridermium thomsonii Berk., Indian Forester, 3: 94 (1852) (Figure 5) subcuticular, conspicuous, 0.2 to 0.24 × 0.1 to 0.2 mm, hemispherical. Aecia on cone scales and needles; on needles,

mostly epiphyllous, causing hypertrophy, peridermioid, cylindric to flat; on scales amphigenous, causing destruction of seed, peridia firm; peridial cells ellipsoid, hyaline; aecidium 0.3 to 0.4 × 0.3 to 0.5 mm; aeciospores 23 to 32 × 28 to 42 μm, wall 1 to 1.5 μm thick excluding verrucae, densely verrucose, hyaline with yellow contents.

Material examined

On *Picea smithiana* (Wall.) Boiss., with 0 & I stage, Pakistan, Northern Areas, Fairy Meadows and Byal camp, 3036 m a. s. l., 12-13 Aug, 2007, N.S. Afshan & A.N. Khalid, NSA # G 15. (LAH NSA 1021).

Peridermium thomsonii has previously been reported on *Picea smithiana* from Kaghan valley and Fairy Meadows by Ahmad (1956a, b) and Sultan (2005).

Phragmidium rosae-moschatae Dietel, *Hedwigia* 44: 132 (1905) (Figure 6) Spermogonia and aecia not found. Uredinia hypophyllous, scattered, minute. Urediniospores globose to subglobose or broadly ellipsoid, 16 to 21 × 17 to 24 μm; wall yellow

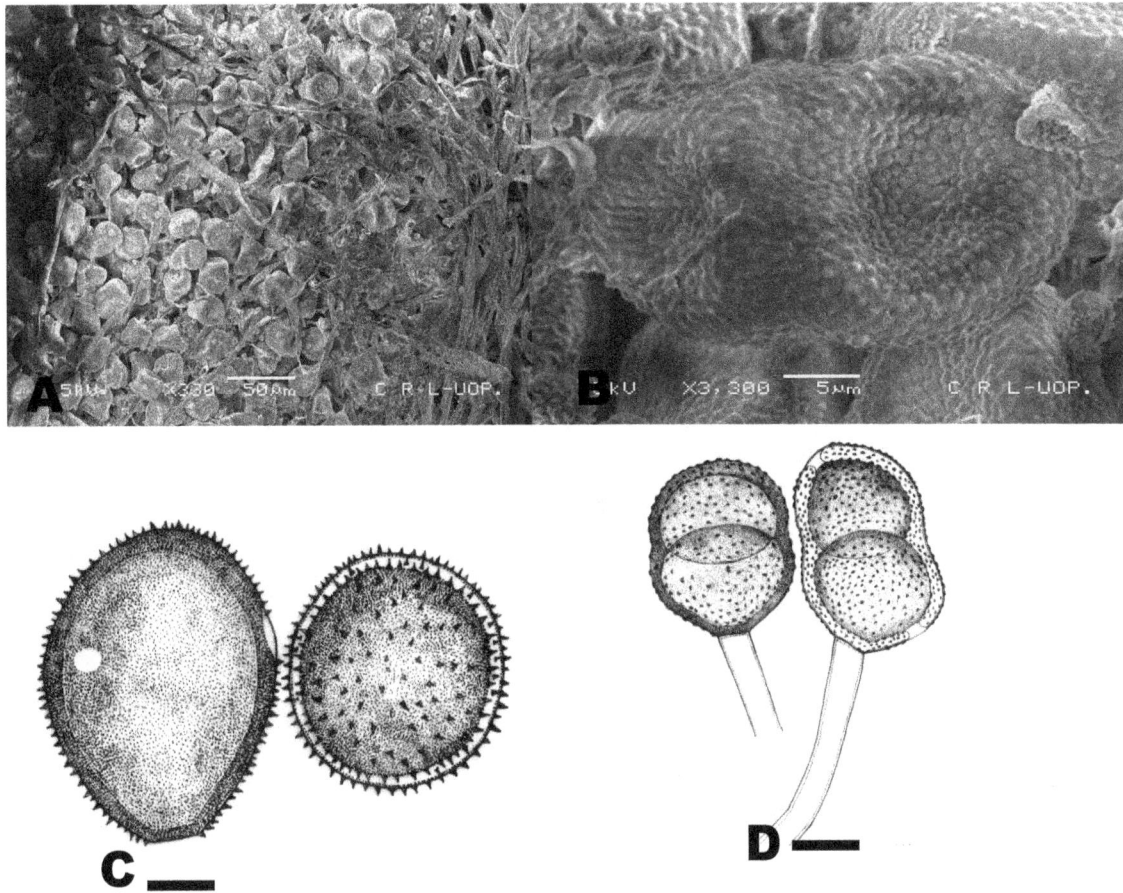

Figure 7. *Puccinia carthami* (A) SEM photograph of a telium. (B) SEM photograph of a teliospore showing verrucose wall ornamentation. (C) Lucida drawings of Urediniospores; (D) Teliospores. Scale bar for C = 7 μm and D = 12 μm.

to light brown, 2–2.5 μm thick, verrucose. Telia hypophyllous, scattered, minute, black. Teliospores 5 to 7 (to 8) celled, cylindric, 28 to 35 × (55 to) 63 to 84 μm excluding apiculus (mean 32.4 × 69.9 μm); apiculus 10 to 20 μm high; wall dark brown to chestnut brown, 2 to 3 μm thick, verrucose; pedicel hyaline to light yellow, 8 to 10 × 94 to 110 μm, thickened to 20 μm at the base.

Material examined

On *Rosa webbiana* Wall., with II and III stages, Pakistan, Northern Areas, Fairy Meadows, 3036 m a. s. l., 12 Aug 2007, N.S. Afshan, NSA G 41. (LAH NSA 1025).

Phragmidium rosae-moschatae has been reported on *Rosa brunonii* Lindl. (= *R. moschata*), *R. lacerans* Boiss. & Buhse, *R. webbiana* Wallich ex Royle, and *R. centifolia* from Quetta, Murree, Peshawar, Tarnab, Swat, Kaghan valley and Tatu-Fairy Meadows by Malik and Khan (1944), Ahmad (1956a, b), Malik et al. (1968), Malik and Virk (1968), Ono and Kakishima (1992), Ono (1992), Kakishima et al. (1993a, b) and Sultan (2005).

Sultan (2005) reported only the telial stage of *Phragmidium rosae-moschatae* from Fairy Meadows, thus discovering of the uredinial stage of *Phragmidium rosae-moschatae* is new from Fairy Meadows.

Puccinia carthami Corda, *Icon. fung.* (Prague) 4: 15 (1840) (Figure 7) = *Puccinia calcitrapae* var. *centaureae* (DC.) Cummins, *Mycotaxon* 5(2): 402 (1977)=*Puccinia centaureae* DC., in de Candolle & Lamarck, *Fl. franç.*, Edn 3 (Paris) 5/6: 59 (1815)

Spermogonia and aecia unknown. Uredinia amphigenous, scattered, soon naked, pulverulent, brown, 0.1 to 0.2 × 0.2 to 0.4 mm. Urediniospores globose to broadly obovoid or broadly ellipsoid, 22 to 28 × 25 to 30 μm (mean 25.2 × 28.3 μm); wall 1.5 to 2 (to 3) μm thick, golden brown to cinnamon brown, echinulate; germ pores 2–4, equatorial, with slight or no caps; pedicel hyaline, short, 6 to 8 × 10 to 15 μm.

Telia amphigenous, dark brown to blackish brown, 0.08 to 0.1 × 0.2 to 0.8 mm. Teliospores ellipsoid to broadly ellipsoid, rounded at both ends, not or slightly constricted at septum, 23 to 29 × 30 to 41 μm (25.8 × 36.7 μm); wall 2 to 3 μm thick, verrucose, chestnut brown; germ pore 1 per cell, pore of upper cell apical or usually depressed, pore of the lower cell near septum, with slight or no caps; pedicel hyaline, deciduous, 4 to 6 × 47 to 120 μm.

Material examined

On *Centaurea calcitrapa* L., with II + III stages, Pakistan, Northern Areas, Fairy Meadows, 3036 m a. s. l., 12 Aug, 2007, N.S. Afshan,

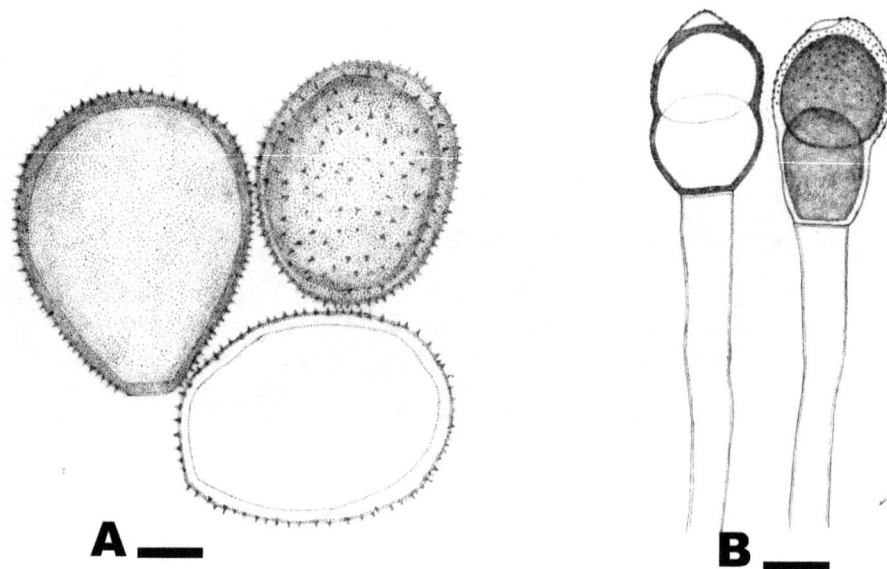

Figure 8. Lucida drawings of *Puccinia chrysanthemi*. (A) Lucida drawings of Urediniospores;
(B) Teliospores. Scale bar for A = 7 μm and B = 10 μm.

NSA G 03. (LAH NSA 1036).*Puccinia carthami* has previously been reported on *Carthamus oxyacantha* M. B. and *C. tinctorius* L. from Faisalabad, Swat and Lahore by Ahmad (1956a, b) and Khan and Kamal (1968); as *P. calcitrapae* var. *centaureae* (= *P. centaureae*) on *Centaurea bruguieriana* (DC.) Bornm. ex Rech. f., *C. candolleana* and *Schischkinia albispina* (Bunge.) Iljm from Peshawar; as *P. carduorum* Jacky on *Carduus edelbergii* Rech. f. from Kaghan valley and Swat by Ahmad (1956a, b); as *P. cirsii* DC. On *Cirsium argyracanthum* DC., *C. wallichii* DC. and *Cnicus* sp. from Swat, Changla gali (NWFP) and Kaghan Valley by Ahmad (1956a, b), Ono (1992) and Kakishima et al. (1993b).

Puccinia carthami is a new record for Fairy Meadows, Northern Areas.

Puccinia chrysanthemi Roze, *Bull. Soc. mycol. Fr.* 16: 92 (1900) (Figure 8) = *Puccinia absinthii* DC., *Encycl. Méth. Bot.* 8: 245 (1806) Spermogonia and aecia unknown. Uredinia amphigenous, light brown, scattered, 0.09 to 0.2 × 0.2 to 1.0 mm. Urediniospores ovoid or obovoid to ellipsoid, light yellow to pale brown, 20 to 24 × 26 to 34 μm; germ pores 1 to 2, equatorial, without a papilla; echinulate, wall 1.5 to 2 μm thick at sides; pedicel hyaline, short, not persistent. Telia amphigenous, on leaves, small pustules, roundish, dark brown to black, scattered, 0.09 to 0.1 × 0.3 to 1.0 mm. Teliospores ellipsoid to broadly ellipsoid or obovoid, not or slightly constricted at septum, attenuated towards base, 19 to 24 × (39 to) 51 to 55 μm, chestnut brown, wall 3 to 3.5 μm thick at sides, 4 to 8 μm thick apically, apex conical or rounded, verruculose at apex, smooth at lower side; germ pores 2, apical or subapical in distal cells and close to septum in proximal cells, with hyaline papilla; pedicel hyaline, persistent, 8 to 12 × 105 to 142 μm.

Material examined

On *Artemisia brevifolia* Wall., with II + III stages, Pakistan, Northern Areas, Fairy Meadows, at 3036 m a.s.l., 12 August, 2007, N.S. Afshan, NSA # G 47. (LAH NSA 1037); On *A. dracunculus* L., with II stage, Pakistan, Northern Areas, Fairy Meadows, at 3036 m a.s.l.,

13 August, 2007, N.S. Afshan & A.N. Khalid, NSA # G 47A; On *A. maritima* L., with II + III stages, Pakistan, Northern Areas, Fairy Meadows, at 3036 m a.s.l., 12 August, 2007, N.S. Afshan, NSA # G 47B.

Puccinia chrysanthemi has previously been reported on *A. persica* Boiss. and *A. parviflora* Buch.-Ham. ex Roxb. from Quetta, Chitral, and NWFP., on *A. dubia* Wall. from Swat and on *A. dracunculus* from Kaghan valley (Ahmad et al., 1997). *P. chrysanthemi* is a new record for Fairy Meadows.

Puccinia circaeae Pers., *Roemer's Neues Magazin für die Botanik*: 119 (1794) (Figure 9)
=*Leptopuccinia circaeae* (Pers.) Syd., (1922).= *Micropuccinia circaeae* (Pers.) Arthur and Jacks., (1921). Spermogonia, aecia and uredinia not found. Telia hypophyllous, on yellow spots, scattered or circinate, brown, 0.2 to 0.4 × 0.2 to 0.6 mm. Teliospores fusoid or oblong, slightly constricted at the septum, attenuated towards base; wall 1.5 to 2 μm thick, smooth, yellowish brown; 10 to 13 × 28 to 43 μm (mean 11.85 × 35.73 μm); apex rounded or conical, 4 to 10 μm thick; pedicel hyaline, persistent, 5 to 7 × 40 to 50 μm.

Material examined

On *Circaea alpina* L. (= *Circaea pricei* Hayata), with III stage, Pakistan, Northern Areas, Fairy Meadows, at 3036 m a.s.l., 12 August, 2007, N.S. Afshan, NSA # G 14. (LAH NSA 1038).

Puccinia circaeae has been reported on *Circaea alpina* from Changla Gali and Kaghan valley (Ahmad et al., 1997). *Puccinia circaeae* is reported for the first time from Fairy Meadows.

Puccinia graminis Pers., *Roemer's Neues Magazin für die Botanik*: 119 (1794) (Figure 9). Plate 55, Figure. A–F) Spermogonia, uredinia and telia not found. Aecia adaxial, cup shaped, light yellow to yellowish orange spots with brown margins, mostly grouped, sometimes scattered, 0.3 to 0.5 × 0.4 to 0.6 mm. Aeciospores globose to subglobose or ovoid, catenulate, 17 to 20 × 20 to 26 μm;

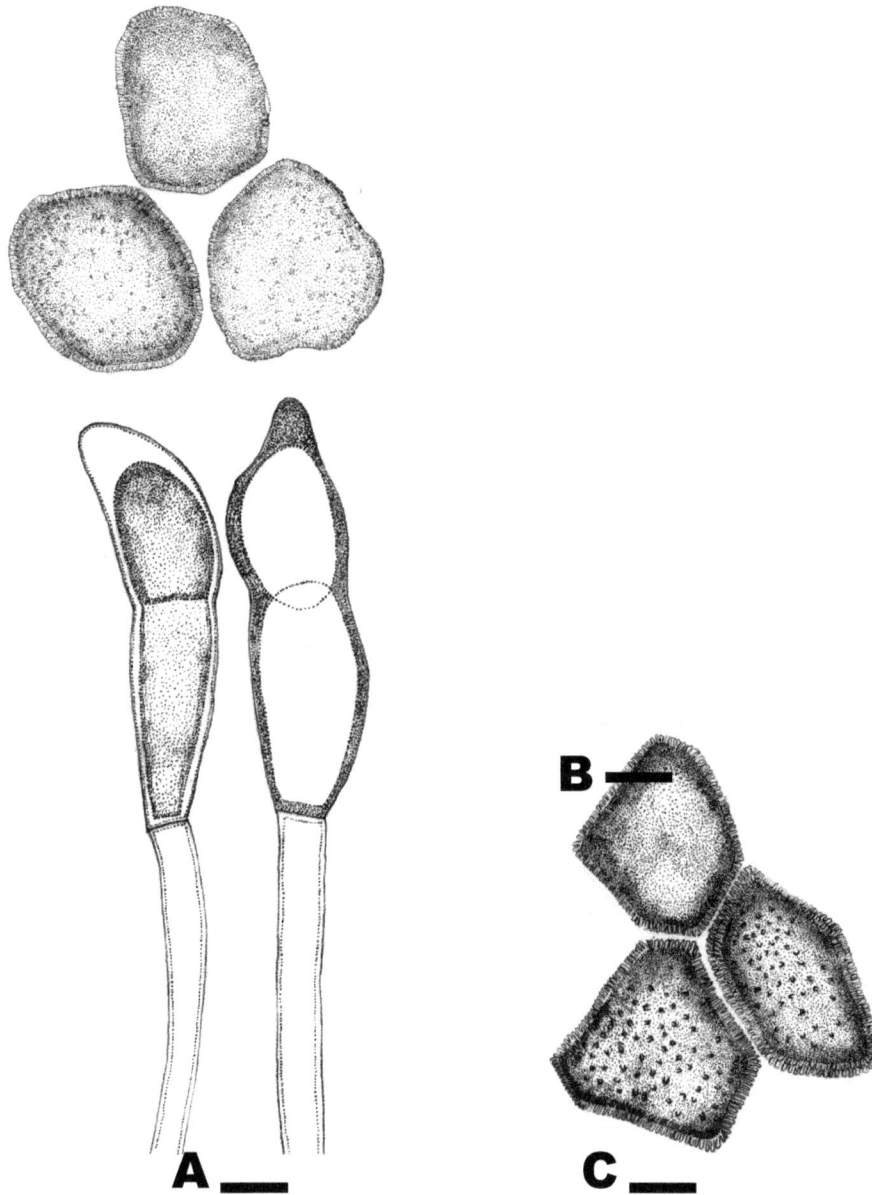

Figure 9. (A) Lucida drawings of Teliospores of *Puccinia circaeae.* (B) Lucida drawing of aeciospores of *Puccinia graminis*; (C) Peridial cells of *Puccinia graminis*. Scale bar for A = 12 µm, B and C = 10 µm.

wall 1–1.5 µm thick, hyaline to light yellow, verrucose. Peridial cell rhomboidal to irregular in shape, hyaline to light yellow, 20 to 28 × 25 to 30 µm

Material examined

On *Berberis vulgaris* Aitch., with I stage, Pakistan, Northern Areas, Fairy Meadows, at 3036 m a. s. l., 12 August, 2007. N.S. Afshan, NSA # G 42. (LAH NSA 1057).

Aecidial stage (I) of *Puccinia graminis* has been reported on *Berberis vulgaris* from Naltar by Sultan (2005) but it is being reported first time from Fairy Meadows.

Puccinia komarovii Tranzschel, *Annls mycol.* 34: 59 (1936).

(Figure 10) Spermogonia and aecia unknown. Uredinia hypophyllous, scattered or circinate, pulverulent, cinnamon brown, 0.1–0.2 × 0.2–0.4 mm. Urediniospores subglobose to obovoid or ellipsoid, 15–20 × 22–29µm; wall 1.5–2 µm thick, pale brown to cinnamon brown, sparsely echinulate; with a single apical germ pore, apex 3-4 µm thick; pedicel hyaline, short, 7–8 × 15–20 µm. Telia similar, chestnut brown to blackish brown, 0.08-0.1 × 0.2-0.8 mm.

Teliospores ellipsoid ovate or subclavate, rounded at both ends or sometimes attenuated downwards, not or slightly constricted at the septum, 18–24 (–27) × (28–) 31–38 µm; wall 2–3 µm thick, smooth, cinnamon brown to chestnut brown; apex 4–8 µm thick, with a hyaline papilla, pale in color; germ pore 1 per cell, pore of upper cell apical or sub apical, of the lower at septum; pedicel

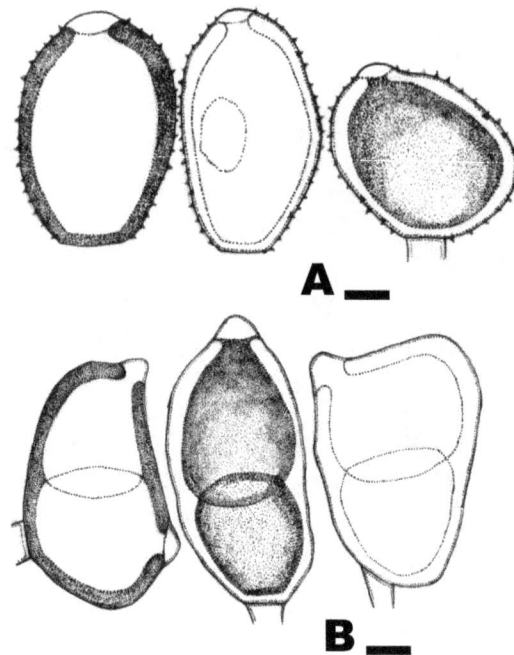

Figure 10. Lucida drawings of *Puccinia komarovii.* (A) Urediniospores showing germ pores; (B) Teliospores. Scale bar for A = 8 µm and B = 10 µm.

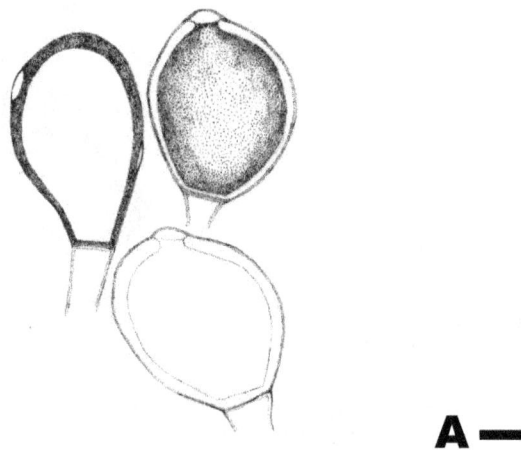

Figure 11. Lucida drawings of teliospores of *Uromyces trifolii.* Scale bar = 7 µm.

short, hyaline, deciduous, 5-7 × 24-30 µm.

Material examined

On *Impatiens brachycentra* G. M. Schulze & Launert, with II & III stages, Pakistan, Northern Areas, Fairy Meadows, at 3036 m a.s.l., 12 August, 2007. N.S. Afshan, NSA # G09. (LAH NSA 1009). *Puccinia komarovii* has previously been reported on *Impatiens* sp. from Poonch and Changla Gali (Ahmad, 1956b). It is a new record for Fairy Meadows.

Uromyces trifolii (R. Hedw.) Lév., Annls Sci. Nat., Bot., sér. 3 8 : 371 (1847) (Figure 11)

=*Uromyces flectens* Lagerh., *Svensk bot. Tidskr.* 3: 36 (1909)
=*Uromyces nerviphilus* (Grognot) Hotson, *Publ. Puget Sound Biol. Sta. Univ. Wash.* 4: 368 (1925).

Spermogonia, aecia and uredinia not found. Telia mostly hypophyllous, sometimes epiphyllous, rounded, scattered, first covered by the epidermis, then naked, pulverulent, surrounded by the ruptured epidermis, dark brown to blackish brown, 0.08 to 0.2 × 0.1 to 0.3 mm. Teliospores globose to subglobose or ellipsoid to obovoid, 15 to 24 × 21 to 30 μm (mean 19.58 × 26.02 μm), the apex rounded with minute hyaline papilla; wall 1 to 2 μm thick, brown to chestnut brown, smooth or with minute scattered warts; apex 3 to 4 μm thick; germ pore 1, obscure; pedicel hyaline, 4 to 8 μm wide and up to 24 μm long.

Material examined

On *Trifolium resupinatum* L., with III stage, Pakistan, Northern Areas, Fairy Meadows, at 3036 m a. s. l., 12 August, 2007, N.S. Afshan & A.N. Khalid, NSA # G81. (LAH NSA 1112). *Uromyces trifolii* is a new record from Fairy Meadows. It has previously been reported on leaves of *Trifolium repens* L., *T. resupinatum* L. and *T. alexandrianum* L. from Quetta, Peshawar, Sargodha, Sangla Hill and Kaghan valley by Malik and Khan (1944), Ahmad (1956a, b), Malik et al. (1968), Khan and Kamal (1968) and Malik and Virk (1968).

DISCUSSION

Fairy Meadows located at the base of Nanga Parbat, which, at 8126 m, is the 9th highest mountain in the world and second in Pakistan afer K2. Te Fairy Meadows are lush green alpine pastures situated in the middle of a pine forest at an altitude of 3306 m. The pine forests skirting Fairy Meadows are one of the virgin forests in the North of Pakistan and home to a number of species of wild flowers, birds, and wildlife (Singh et al., 2004). Te altitudinal range of the Fairy Meadows vegetation belt is defined as montane belt. Although the montane belt on Fairy Meadows/Nanga Parbat is by far the richest in species number and potential differentiation of vegetation types, it is floristically depauperate as compared to the outer Himalayan slopes (Troll, 1939). Being rich in plant diversity, these forests harbor a large number of rust fungi that are obligate parasites of plants from which they obtain nutrients, and on which they reproduce and complete their life cycles. Although the fungal flora of Pakistan has been explored by several workers in the past, a very important group of fungi has largely been neglected resulting in the paucity of literature and very fragmentary knowledge of these fungi, particularly in Fairy Meadows, Pakistan. Keeping in view these facts, the present study was undertaken to explore and assess the diversity and distribution of rust fungi along with their respective host plants in this floristically rich area. During a recent survey of the rust fungi in the Northern Areas of Pakistan, specifically Fairy Meadows, sixteen species of rusts were encountered. Previously, 22 species of rust fungi have been reported from this area (Afshan et al. 2011). This work has raised the number of reported rust fungi from Fairy Meadows to 31.

REFERENCES

Afshan NS, Khalid AN, Niazi AR, Iqbal SH (2011). New records of Uredinales from Fairy Meadows. Pakistan. Mycotaxon 115:203-213.

Ahmad S (1956a). Uredinales of West Pakistan. Biologia 2(1):29-101.

Ahmad S (1956b). Fungi of Pakistan. Biological Society of Pakistan, Lahore Monograph 1:1-126.

Ahmad S, Iqbal SH, Khalid AN (1997). Fungi of Pakistan. Nabiza Printing Press, Karachi, Pakistan.

Kakishima M, Izumi O, Ono Y (1993a). Rust Fungi (Uredinales) of Pakistan collected in 1991. Cryptogamic Flora Pak. 2:169-179.

Kakishima M, Izumi O, Ono Y (1993b). Graminicolous Rust Fungi (Uredinales) from Pakistan. Cryptogamic Flora Pak. 2:181-186.

Khalid AN, Saba M (2011). *Puccinia cortusae* (Basidiomycota; Uredinales) on *Cortusa brotheri* (Primulaceae), new to southern Asia (Fairy Meadows, Pakistan). Mycotaxon 117:317-320.

Khan SA, Kamal M (1968). The fungi of South West Pakistan. Part 1. Pak. J. Sci. Ind. Res.11:61-80.

Malik SA, Khan MA (1944). Parasitic fungi of the North-West Frontier Province. Ind. J. Agric. Sci. 13:522-527.

Malik SA, Virk (1968). Contribution to the knowledge of parasitic fungi of Quetta-Kalat Region. Biologia 14:27-35.

Malik SA, Javaid MT, Ahmad M (1968). Uredinales of Quetta-Kalat region of Pakistan. Biologia 14:37-46.

Ono Y (1992). Uredinales collected in the Kaghan Valley, Pakistan. Cryptogamic flora Pak. 1:217-240.

Ono Y, Kakishima M (1992). Uredinales collected in the Swat Valley, Pakistan. Cryptogamic flora Pak. 1:197-216.

Singh S, Brown L, Bennett-Jones O, Mock J, O'Neil K, Yasmeen S (2004). Pakistan and the Karakoram Highway. Lonely Planet.

Sultan MA (2005). Taxonomic study of rust flora of Northern Areas of Pakistan. Ph.D. Thesis, University of the Punjab. Lahore, Pakistan.

Troll C (1939). Das Pflanzenkleid des Nanga Parbat. Begleitworte zur Vegetationskarte der Nanga Parbat-Gruppe (Nordwest-Himalaja), 1: 50,000. Wissenschafliche Verö?entlichungen des Deutschen Museums für Länderkunde zu Leipzig NF 7:149-193.

24

The surface display of phytase on yeast cell and activity assay of the displayed protein

Shumin Yao*, Jing Huang and Jingli Tan

College of Life Science, Qufu Normal University, Qufu 273165, People's Republic of China.

Phytase is a new-style enzyme used in animal feed additive. It can increase phosphorus availability, decrease environmental phosphorus pollution and improve the performance of animals. Phytase gene was cloned by reverse transcription-polymerase chain reaction (RT-PCR) using first strand cDNA as template after reverse transcripting *Aspergillus niger* total RNA. The phyA gene was cloned into plasmid pYD1 which allows regulated expression, secretion and detection of expressed proteins on the surface of *Saccharomyces cerevisiae* cells using immunofluorescence. The construct was propagated in *Escherichia coli* DH5α and then was transformed into the yeast strain EBY100. After induction, the phytase activity was measured every 12 h. The results indicated that the activity of the fusion protein reached the highest level after being induced with 2.0% galactose for 48 h. This enzyme had pH optima (pH 7) and its optimum temperature was about 65°C.

Key words: Yeast surface display, *Aspergillus niger*, phytase.

INTRODUCTION

Phytate (myoinositol hexakisphosphate or myoinositol 1, 2, 3, 4, 5, 6-hexakis dihydrogen phosphate), which is also known as phytic acid, is a form of phosphate storage in plants such as soybeans, cottonseeds, and other legumes and cereals. Therefore, phytase can be incorporated into commercial poultry, swine, and fish diets and has a wide range of applications in animal and human nutrition as it can reduce phosphorus the excretion of monogastric animals by replacing inorganic phosphates in the animal diet. Phytase A (PhyA) from *Aspergillus niger* is known to be the most active enzyme, and is frequently used in animal feeds to improve the availability of phosphorous and minerals because it has two pH optima (2 to 2.5 and 5 to5.5) and a temperature optimum between 55 and 60°C (Han et al., 1999, Lim et al., 2001).

The low optimum pH and high thermo-stability of phytase provide the additional advantage of being able to smoothly pass through the stomach acid and the heat denaturation (60 to 80°C) during feed pelleting, respectively (Wyss et al., 1999). It combines the advantages of both a prokaryotic system, such as high expression levels and easy scale-up, as well as a eukaryotic system to conduct most of the post-translational modification. The yeast-based expression system is unique compared to other expression systems. Moreover, the yeast is non-conventional hemiascomycetous yeast and is able to grow on hydrophobic substrates; it has been used in several industrial processes and is non-pathogenic (GRAS, generally regarded as safe) (Madzak et al., 2004). Consequently, it is used in livestock feeds for fish, poultry and fur-bearing animals, and as a food supplement for consumption by humans.

Recently, a number of heterologous proteins of varied size have been displayed on yeast cell surface using a genetic engineering technique (Colby et al., 2004). The yeast cell-surface display allows peptides and proteins to be displayed on the surface of yeast cells by fusing them with the anchoring motifs. The characteristics of anchoring motif, displayed protein and host cell, and fusion method all affect the efficiency of surface display of proteins. The yeast cell surface display has many potential applications, including live vaccine development, peptide library screening, bioconversion using whole cell biocatalyst and bioadsorption (Kondo et

*Corresponding author. E-mail: yaoshumin299@163.com.

al., 2004). Yeast surface display used in combination with directed evolution is a robust platform for protein engineering, allowing the discrimination of subtle phenotype differences using fluorescence activating cell sorter (FACS), immunofluorescence, and the characterization of protein kinetics and thermal stability measured directly on the cell surface. Yeast surface display also confers a eukaryotic expression bias, resulting in posttranslational assembly such as that mediated by foldases and chaperones, and posttranslational modification such as glycosylation.

A large variety of proteins have been successfully engineered using yeast surface display. In addition to protein engineering, more recent applications of this platform include enzyme engineering (Antipov et al., 2008), epitope mapping (Chao et al., 2006), and cell panning for the discovery of novel surface receptors (Wang et al., 2007). In this study, we attempted to achieve the expression of PhyA from A. niger on the cell surface of S. cerevisiae to apply it as a new candidate for dietary yeast supplementation and as a whole cell bio-catalyst that can hydrolyze the phytate in animal feed and waste, respectively.

MATERIALS AND METHODS

Experimental strains

The Escherichia coli strain used in this study was DH5α[F− endA1 hsdR17(rK−/mK+) supE44 thi-1λ− recA1 gyr96ΔlacU169(φ80lacZΔM15)] kept in this laboratory. A. niger 424-1 was used for the cloning of phyA and S. cerevisiae EBY100(MATa ura 3-52 trp 1 leu2Δ1 his3Δ200 pep4:HIS3 prb1Δ1.6R can1 GAL) was used as the recipient cell for phytase production. Plasmid pYD1 for phyA production in EBY100 was purchased from Invitrogen. S. cerevisiae EBY100 was maintained in a YPD medium (1% yeast extract, 2% peptone, and 2% dextrose), and a tryptophan-deficient selective medium (0.67% yeast nitrogen base without amino acids, and 1.0% ammonium sulfate, 2.0% glucose, 0.01% leucine, 1.5% agar) was used at 30°C to screen the transformants.

Cloning of phyA genes from A. niger 424-1

Genomic RNA from A. niger 424-1 was prepared according to the methods described by Tritol manual. According to PrimeScript™ 1st Strand cDNA Synthesis Kit instructions on reverse transcription, the first strand cDNA. PhyA ds cDNA was amplified by one pair of primer (P1: 5'-CGGAATTCATGGGCAGTCCCAGACTCGAG-3'; P2:5'-CGGCGGCCGCCTAAGCAAAACACTCC-3') designed according to the phyA gene sequence (XM_001401676) in A. niger and multiple cloning sites on pYD1. The phyA genes were amplified from the genomic RNA with the pair of primers by PCR with PrimeSTAR™ HS DNA Polymerase, respectively. The PCR procedure was as follows: initial denaturation at 94°C for 5 min, followed by 30 cycles of 45 s at 94°C, 40 s at 55°C, 2 min at 72°C, followed by additional 10 min at 72°C. The resultant 1404 bp PCR products were digested with EcoR I and Not I and cloned into pYD1 vector, respectively. The resultant plasmids were named as pYD-phyA. These plasmids were transformed into E. coli DH5α and phyA genes on the plasmids purified from the transformants were sequenced, respectively.

Transformation of yeast strain

The competent cells of the yeast strain EBY100 were prepared and pYD1- phyA, and pYD1 were transformed into the competent cells according to the manufacturer instructions of pYD1 Yeast Display Vector Kit (Invitrogen), respectively. The transformants were grown on the selective medium which contained 0.67% YNB without amino acids and 1.0% ammonium sulfate, 2.0% glucose, 0.01% leucine, 1.5% agar at 30°C for 48 h. The derived transformants containing pYD1- phyA and pYD1 were confirmed by colony PCR technique and named EBY100/pYD1-phyA, and EBY100/pYD1, respectively.

phyA genes expression and bioassay of surface displayed protein

EBY100/pYD1- phyA display on the yeast cells were carried out according to the manufacturer's instruction of pYD1 Yeast Display Vector Kit (Invitrogen). The cells were induced by growing the cells in YNB-CAA medium containing 2.0% galactose for 0, 12, 24, 36 and 48 h, respectively. The cell cultured over a 48-h time period (0, 12, 24, 36, and 48 h) was assayed to determine the optimal induction time for maximum display. The staining of displayed proteins on the yeast cells was performed according to the methods described in pYD1 Yeast Display Vector Kit (Invitrogen). The stained cells with fluorescence were suspended in 40 μL of phosphate buffered saline (PBS) buffer and observed under microscope with ultraviolet (UV) light, and the percentage of the cells displaying phyA were calculated. A volume of the yeast cells equivalent to 2.0 OD 600 nm units were collected and washed with PBS buffer by centrifugation after EBY100/pYD1-phyA, EBY100(negative control), EBY100/PYD1 (positive control) were induced in YNB-CAA medium containing 2.0% galactose for 0, 12, 24, 36 and 48 h.

Phytase assay

In brief, 1 ml of the cell culture was centrifuged at 5,000×g for 10 min. The supernatant obtained was used as the crude extracellular phytase preparation. The phytase activity was assayed as follows: 0.8 ml of sodium phytate solution (5.0 mM sodium phytate in 0.2 M sodium acetate pH 5.0) was pre-incubated at 65°C for 5 min, and 0.2 ml of the crude extracellular phytase preparation was added and mixed well. The mixture was incubated at 65°C for 30 min and afterward, the reaction was stopped by the addition of 1.0 ml of 50.0 g/L trichloroacetic acid. The inorganic phosphate liberated was quantitatively determined by using the ammonium molybdate method (Chi et al., 1999) spectrophotometrically at 700 nm. One unit of phytase activity was defined as the amount of enzyme causing the release of 1.0 μM of inorganic phosphate per minute under the assay conditions.

Effects of pH and temperature on phytase activity and stability

The effect of pH on the recombinant enzyme activity was determined by incubating the recombinant enzyme between pH 3.0 and 9.0 using the standard assay conditions. The buffers used were 0.1 M citric acid buffer(pH 3.0 to 6.0) and 0.1 M barbital sodium buffer (pH 7.0 to 10.0). The pH stability was tested via a 6-h pre-incubation of the recombinant enzyme in appropriate buffers that had the same ionic concentrations at different pH values ranging from 3.0 to 10.0 at 0°C. The remaining activities of phytase were

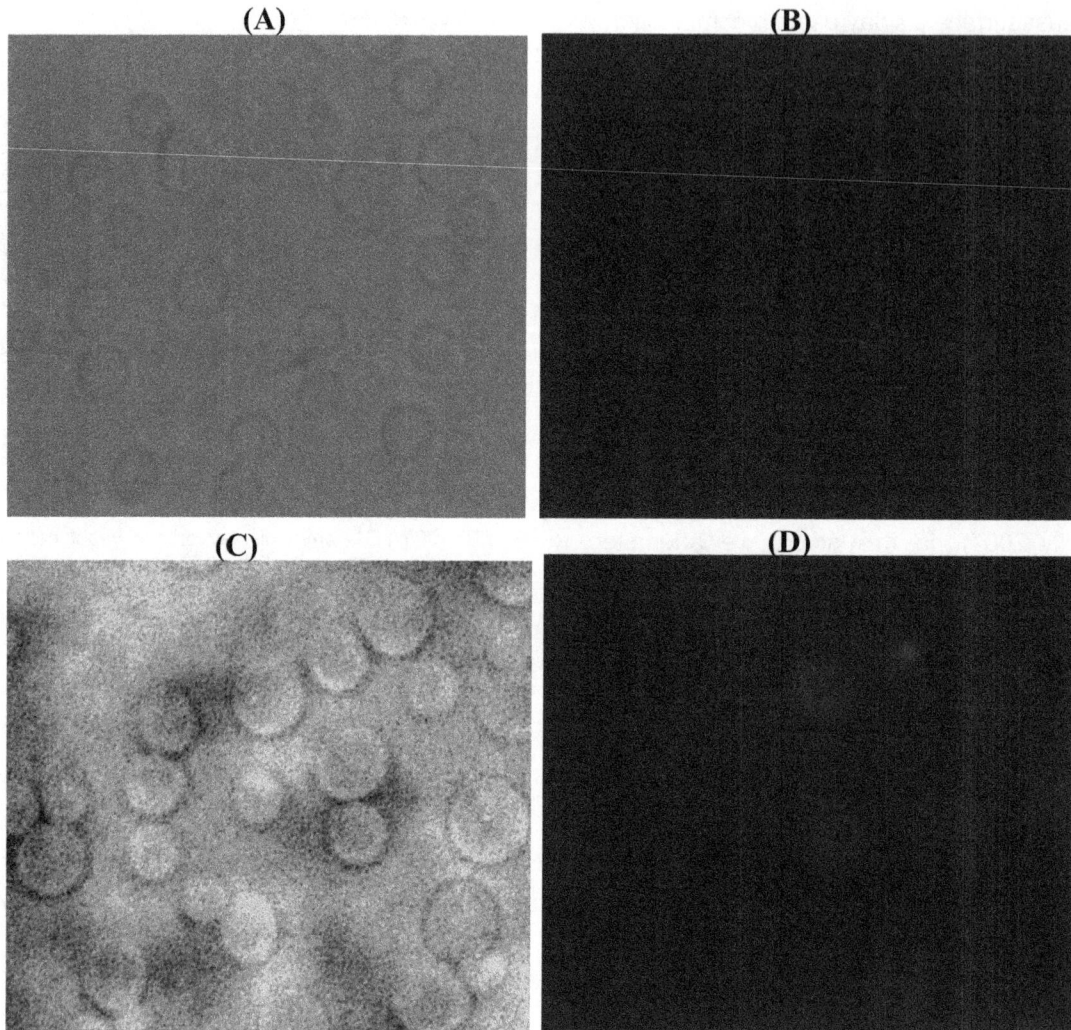

Figure 1. Detection of phyA on the cell surface by confocal laser scanning microscopy. (A-B) *S. cerevisiae* EBY100 under normal white light and the FITC filter (492 nm); (C-D) *S. cerevisiae* EBY100 carrying pYD1-phyA under normal white light and the FITC filter(492nm). Magnification: 100×10.

measured immediately after this treatment with the standard method as mentioned earlier.

The optimal temperature for activity of the enzyme was determined at 30, 35, 40, 45, 50, 55, 60, 65, 70, 75, 80 and 85°C in the same buffer as described in the preceding text. Temperature stability of the purified enzyme was tested by pre-incubating the enzyme at different temperatures (20, 30, 40, 50, 60, 65, 70, and 80°C) for 1 h; the residual activity was measured immediately as described in the preceding text. Here, a pre-incubated sample at 0°C was used as reference to calculate the residual activity.

Effects of different metal ions on phytase activity

To examine effects of different metal ions on phytase activity, an enzyme assay was performed for 1 h in the reaction mixture as described in the preceding text with various metal ions at final concentrations of 1.0 and 5.0 mM. The activity assayed in the absence of metal ions was defined as the control. The metal ions tested included zinc (Zn^{2+}), copper (Cu^{2+}), magnesium (Mg^{2+}), iron (Fe^{3+}), calcium (Ca^{2+}), potassium (K^+), manganese (Mn^{2+}), mercury

(Hg^{2+}), lithium (Li^+) ,Fe^{2+}, gold (Ag^+), sodium (Na^+), barium (Ba^{2+}), and cobalt (Co^{2+}).

RESULTS AND DISCUSSION

Display of phyA on yeast- cell surface

We selected the method of displaying phytase on the yeast cell surface because of its convenience and ease of handling enzyme with a highly homogeneous quality without purification. In order to obtain maximal displayed phytase on yeast cells, the transformed cells were induced by galactose for 0, 12, 24, 36 and 48 h. Immunofluorescence analysis showed that the largest amount of phytase was displayed on the yeast cells after induction for 48 h (Figure 1). Under this condition, about one-third of the yeast cells harboring pYD1-phyA had the

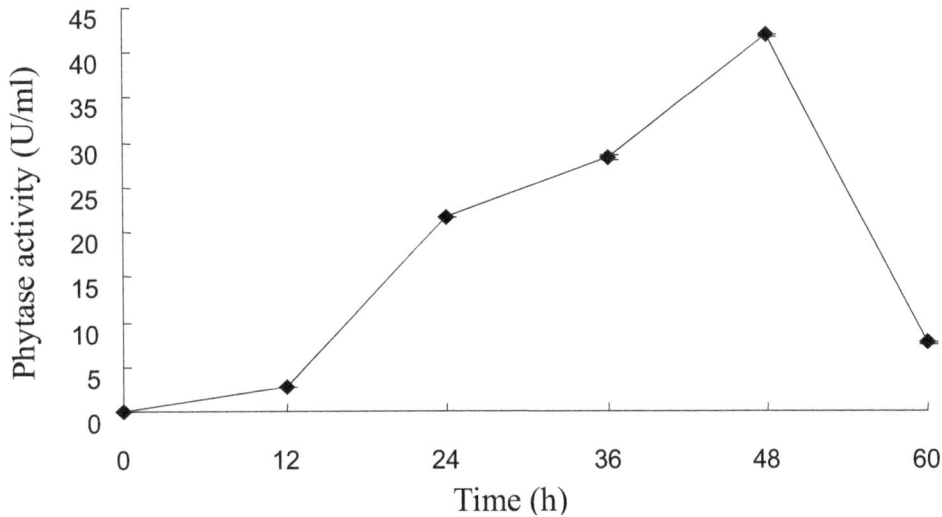

Figure 2. Time course of phytase production by the recombinant yeast. All the data are given as means ±SD, n = 3.

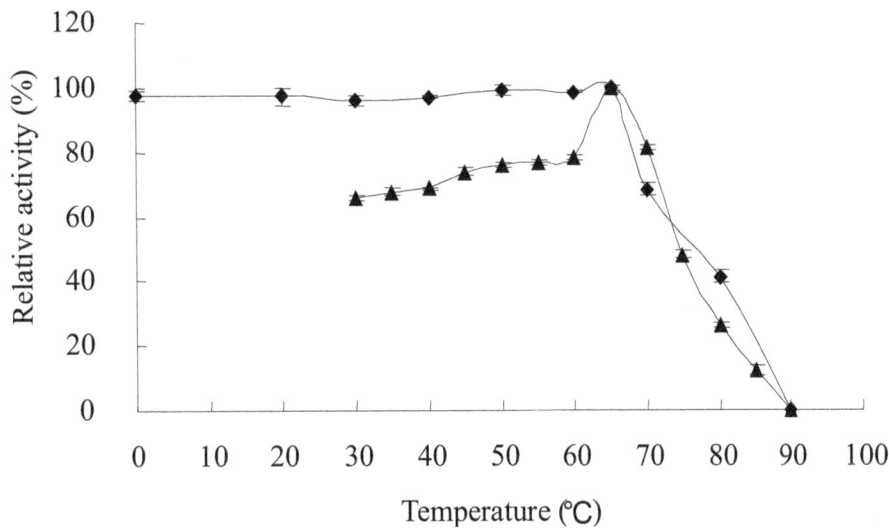

Figure 3. Effects of different temperature on activity (▲) and stability (◆) of the phytase. Temperature stability of the recombinant enzyme was tested by pre-incubating the enzyme at different temperatures (0, 20, 30, 40, 50, 60, 65, 70, 80, 90°C) for 1 h; the residual activity was measured immediately as earlier described. Here, pre-incubated sample at 0°C was used as reference to calculate the residual activity. Data are given as means ±SD, n = 3.

displayed phytase. It was found that most of the intense fluorescence as first localized in the small bud, then observed on the entire cell wall.

Induced expression of the recombinant yeast

After being induced, the phytase activity was measured every 12 h. The results indicated that the activity of the fusion protein reached the highest level after induced for 48 h. The results in Figure 2 indicate that 42.1 U/ml of

phytase activity could be reached within 48 h of the induction. These results demonstrate that the *recombinant* yeast strain could produce high yield of extracellular phytase.

Optimum temperature and thermal stability

The phytase activity measured as a function of temperature from 30 to 90°C shows that the activity was highest at 65°C (Figure 3). Thermostability is considered

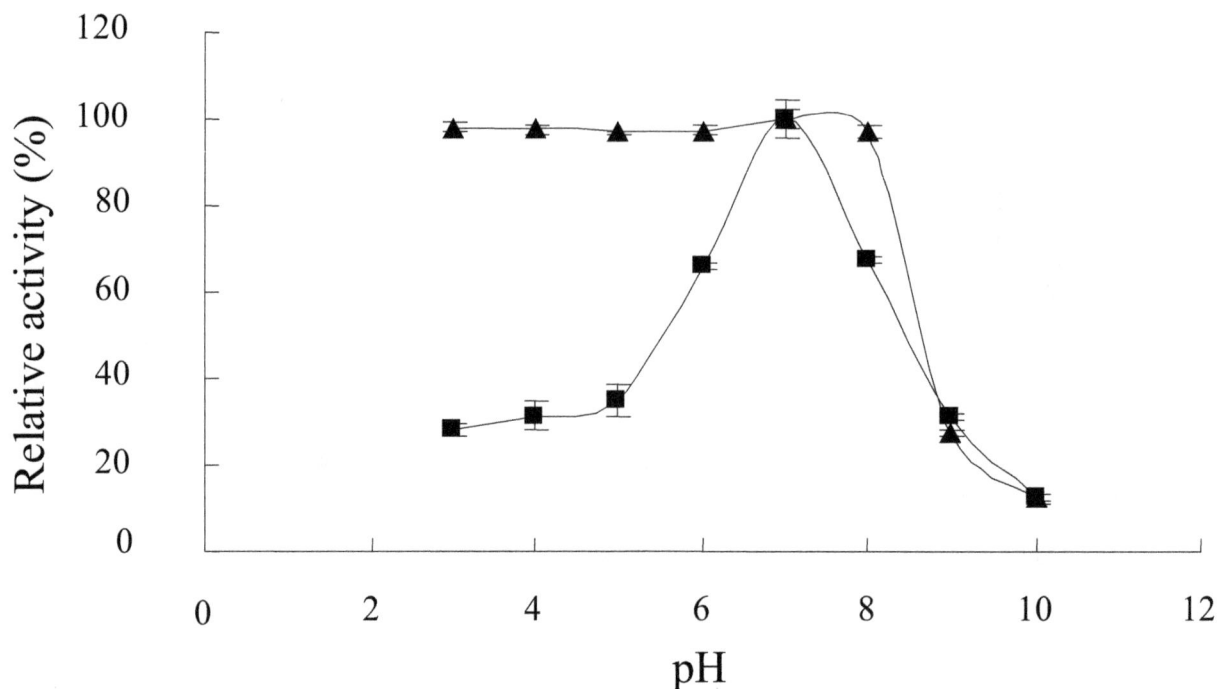

Figure 4. Effects of different pH on activity (■) and stability (▲) of the phytase. The pH stability was tested via a 6-h pre-incubation of the purified enzyme in appropriate buffers that had the same ionic concentrations at different pH values ranging from 3.0 to 10.0 at 4°C. The remaining activities of phytase were measured immediately after this treatment with the standard method as earlier described. The phytase activity of the finally concentrated elute without pre-incubation was regarded as 100%. Data are given as means ± SD, n = 3.

an important and useful criterion for industrial application of phytase. For example, thermostability is a prerequisite for the successful application of enzymes in animal feeds that are exposed to 60 to 90°C during the pelleting process. Therefore, thermostability was investigated by pre-incubating the enzyme in the same buffer as earlier described for 1 h, and the remaining activity was determined. As shown in Figure 3, the residual phytase activity still maintained 99.5% of the control after treatment at 65°C for 1 h, indicating that the enzyme was stable up to 65°C. Figure 3 also reveals that the recombinant enzyme was inactivated rapidly at temperatures higher than 65°C and was almost inactivated at 90°C within 1 h.

From these results, the phytase seemed to have considerable thermostability. For example, the phytase produced by *Saccharomyces castellii* exhibited an uncommon preference for high temperatures, with optimum activity at 77°C and thermostability up to 74°C (Segueilha et al., 1992; Pandey et al., 2001). This means that the thermostability of phytase from the recombinant yeast was not higher than that from *S. castellii*.

Optimum pH and pH stability

Phytase activity was measured at various pH values in buffers with the same ionic concentrations. The results

(Figure 4) show that the maximum activity was observed at pH 7.0. pH stability was tested via 6 h pre-incubation of the purified enzyme in appropriate buffers that had the same ionic concentrations at different pH values ranging from 3.0 to 10.0 at 0°C. The remaining activities of phytase were measured immediately after this treatment with the standard method as mentioned earlier. It can be seen from the results in Figure 4 that the activity profile of the enzyme was stable from pH 3.0 to pH 8.0, and greater than 97.0% the residual activity was maintained after treatment at pH from 3.0 to 8.0 and 0°C for 6 h. These results suggest that the enzyme was very stable in the pH range of 3.0 to 8.0, and at pH 9, the activity fell sharply.

Effects of different cations and enzyme inhibitors on the phytase activity of the recombinant yeast

Ca^{2+}, Na^+, Li^+, Ba^{2+} Cu^{2+} and Mg^{2+} (at concentrations of 5.0 mM) stimulated the phytase activity of the recombinant yeast. However, K^+, Mn^{2+}, Fe^{2+}, Li^+, Fe^{3+}, Ag^+, Co^{2+} and Zn^{2+} (at concentrations of 5.0 mM) acted as inhibitors in decreasing the phytase activity, with Fe^{3+} (at a concentration of 5.0 mM) showing the lowest level (15%) (Table 1). The results show that phytase activity have a certain relationship with some metal ions, the specific activation of the principles will be investigated further in

Table 1. Effect of different cations on the recombinant yeast phytase activity.

Metal ions	Final concentration in reaction mix (mM)	Relative phytase activity (%)
Ca^{2+}	1.0	92.8 ± 0.8
	5.0	104.6 ± 2.3
Na^+	1.0	92.2 ± 0.6
	5.0	98.3 ± 2.5
K^+	1.0	125.0 ± 1.2
	5.0	96.4 ± 1.7
Mn^{2+}	1.0	129.2 ± 1.0
	5.0	100.3 ± 1.5
Fe^{2+}	1.0	96.4 ± 0.2
	5.0	35.4 ± 2.5
Li^+	1.0	91.1 ± 1.8
	5.0	86.3 ± 1.6
Fe^{3+}	1.0	87.7 ± 1.1
	5.0	15.1 ± 0.4
Ag^+	1.0	139.2 ± 0.4
	5.0	96.0 ± 1.7
Cu^{2+}	1.0	89.8 ± 1.2
	5.0	101.2 ± 1.3
Co^{2+}	1.0	133.5 ± 2.8
	5.0	68.4 ± 1.6
Mg^{2+}	1.0	90.1 ± 0.8
	5.0	93.6 ± 1.7
Zn^{2+}	1.0	139.2 ± 0.5
	5.0	98.0 ± 1.1

Data are given as means ±SD, n = 3. The phytase activity in the absence of metal ions is regarded as 100%.

future studies.

Conclusion

Here, we explored the potential of a *S. cerevisiae* expression and display system that allows proteins of interest to be targeted to the yeast surface. The system not only displays single-subunit proteins, but also hetero-oligomeric multisubunits. In this study, we constructed a novel phytate-degrading yeast strain displaying the phytase on the cell surface. To our knowledge, this is the first report of the phytase expression on the plasmid pYD1, which will be applied as a dietary complement and whole cell bio-catalyst in animal foods and waste. This study also demonstrated the direct hydrolysis of phytate using the recombinant strain.

ACKNOWLEDGEMENTS

The authors would like to thank their advisors who made

them introduce the field of Molecular Biology into this research. They also acknowledge their families and friend who supported them all the time.

REFERENCES

Antipov E, Cho AE, Wittrup KD, Klibanov AM (2008). Highly L and D enantioselective variants of horseradish peroxidase discovered by an ultrahigh-throughput selection method. Proc. Natl. Acad. Sci. USA 105:17694-17699.

Chao G, Lau WL, Hackel BJ, Sazinsky SL, Lippow SM, Wittrup KD (2006). Isolating and engineering human antibodies using yeast surface display. Nat. Protoc. 1:755-768.

Chi, Z., Kohlwein, SD, Paltauf F (1999). Role of phosphatidylinositol (PI) in ethanol production and ethanol tolerance by a high ethanol producing yeast. J. Ind. Microbiol. Biotechnol. 22:58-63.

Colby DW, Kellogg BA, Graff CP, Yeung YA, Swers JS, Wittrup KD (2004). Engineering antibody affinity by yeast surface display. Methods Enzymol. 388:348-358.

Han Y, Wilson DB, Lei XG (1999). Expression of an *Aspergillus niger* phytase gene (phyA) in, Saccharomyces cerevisiae. Appl. Environ. Microbiol. 65:1915-1918.

Kondo A, Uda M (2004) Yeast cell-surface display applications of molecular display. Appl. Microbiol. Biotechnol. 64:28-40.

Lim YY, Park EH, Kim JH, Park SM, Jang HS, Park YJ, Yoon S, Yang MS, Kim DH (2001). Enhanced and targeted expression of fungal phytase in *Saccharomyces cerevisiae*. J. Microbiol. Biotechnol. 11:915-921.

Madzak C, Gaillardin C, Beckerich JM (2004). Heterologous protein expression and secretion in the non-conventional yeast Yarrowia lipolytica: a review. J. Biotechnol. 109:63-81.

Pandey A, Szakacs G, Soccol CR, Rodriguez-Leond JA, Soccol VT (2001). Production, purification and properties of microbial phytases. Bioresour. Technol. 77:203-214.

Segueilha L, Lambrechts C, Boze H, Moulin G, Galzy P (1992). Purification and properties of the phytase from *Schwanniomyces castellii*. J. Ferment. Bioeng. 74:7-11.

Wang XX, Cho YK, Shusta EV (2007). Mining a yeast library for brain endothelial cell-binding antibodies. Nat. Methods 4:143-145.

Wyss M, Pasamontes L, Friedlein A, Remy R, Tessier M, Kronenberger A, Middendorf A, Lehmann M, Schnoebelen L, Rothlisberger U, Kusznir E, Wahl G, Muller F, Lahm HW, Vogel K, van Loon AP (1999). Biophysical characterization of fungal phytases (myoinositol hexakisphosphate phosphohydrolases): molecular size, glycosylation pattern, and engineering of proteolytic resistance. Appl. Environ. Microbiol. 65:359-366.

25

Cultivation of *Pleurotus ostreatus* mushrooms on *Coffea arabica* and *Ficus sycomorus* leaves in Dilla University, Ethiopia

Fekadu Alemu

Department of Biology, College of Natural and Computational Sciences, Dilla University, P.O.Box., 419, Dilla, Ethiopia.

Coffea arabica and *Ficus sycomorus* leaves were assessed for supporting growth of *Pleurotus ostreatus*. Comparatively, sholla leaves were highly supportive and their fruiting bodies appeared late, while coffee leaves were also supportive and their fruiting bodies appeared early. Results of the cultivation of *Pleurotus* on coffee and sholla leaves showed that they have promising effect for sustainable development, food security and supply. This study therefore revealed that both coffee and sholla leaves can serve as good substrates for mushroom cultivation.

Key words: *Coffea arabica, Ficus sycomorus,* fruiting body, mushroom, *Pleurotus ostreatus,* Sholla, coffee.

INTRODUCTION

Mushroom cultivation is a potential biotechnological process where waste plant materials or negative value crop residues can be converted into valuable food. Mushroom has been studied for nutritional and medical purposes, and various potential antitumor and immune modulator substances, mainly polysaccharides have been identified (Zhang et al., 2007) for medical purposes. Mushrooms are consumed to prevent cancer and cardiac diseases, to improve blood circulation and to reduce cholesterol (Wasser and Weis, 1991). They are used for physical and emotional stress, asteoporosis, gastric ulcers and chronic hepatitis; for the improvement of the quality of life of patients with diabetes and especially for the stimulation of immunity (Menoli et al., 2004, Guterrez et al., 2004; Angeli et al., 2006; Choi et al., 2006; Grind et al., 2006).

Unlike in developed countries where mushroom food consumption is increasing (Kurtzman, 2005; Gregori et al., 2007; Neyrinck et al., 2009) especially in Ethiopia, eating of mushroom is very poor (Dawit, 1998). Information on nutritive value and sensory properties of edible oysters mushroom foods cultivated on agricultural residues in Ethiopia is limited. Such information is important to facilitate the population of mushroom cultivation, processing, marketing and consumptions.

Ficus sycomorus is commonly known as Sholla or Bamba in Amharic in Ethiopia. It is a large, semi-deciduous spreading savannah tree, up to 21 (max. 46) m; it is occasionally buttressed. Its leaves are broadly (ob) ovate or elliptic, the sub base is cordate, apex is rounded or obtuse, margin is entirely or slightly repand-dentate (2.5-13 (max. 21) x 2-10 (max. 16 cm) and is scabrous above; petiole is 1-5 cm long, with five to seven pairs of yellow lateral veins; lowest pair originates at the leaf base. It can be monocious or diecious (Berg and Corner, 2005).

Ficus is the Latin form of fig, derived from the Persian 'fica'. In Greek 'syka' means fig. The name of the species comes from the Greek 'sykamorea' (sycamore), used in the Gospel according to St. Luke; it was the tree that Jesus cursed because it was barren. Its leaves are said to be effective against jaundice, are antidote for snake

Figure 1. This shows during (sample) Coffee and Sholla leaves collection.

bite and also are a much-sought fodder with fairly high nutritive value (9% crude protein and 7 mJ/kg net energy dry matters); .they are valuable fodder in overstocked semi-arid areas where the trees occur naturally. Its fruits are eaten by livestock, wild animals and birds (Orwa et al., 2009).

Coffee has other several components including cellulose, mineral, sugars, lipids, tannin, and several poly phenols amino acids such as alanine, arginine, asparagines and cytosine found in coffee bean (Belitz et al., 2009; Grembecka et al., 2007; Santos and Oliveira, 2001). Additionally, coffee bean contains vitamin complex B, niacin and chologenic acid proportion that may vary from 7 to 12%, three to five times higher than caffein (Beliz et al., 2009; Lima, 2003; Trugo and Macrae, 1984).

The present study deals with the cultivation of *Pleurotus ostreatus* on some common and abundantly available wastes of coffee and sholla leaves; they are used for conversions in foods, which otherwise are left for natural degradation and also provide necessary information for their further utilization. Most of these studies focus on the higher yield and quality of fruiting bodies of *Pleurotus ostreatus* with respect to higher yield substrate.

MATERIALS AND METHODS

Origins of selected mushroom species

The selected mushroom species used in this study were pleurotus species (*Pleurotus ostreatus*). They were initially imported from where they were available and obtained from Addis Ababa University, Mycology laboratory. They were finally brought to Microbiology Laboratory, Dilla University.

Preparations of culture media

Culturing of *Pleurotus ostreatus* on malt extract agar

The mushroom species, pure culture of pleurotus species was maintained on potato dextrose Agar (oxoid) at 4°C. After few days the culture was again recultured on Malt Extract Agar at 28°C.

Spawn production

Spawn was prepared in spawn bottles; 4 kg of whole grain of sorghum was soaked in water over night to moisten it. The ratio of sorghum and water was 1:1. It was mixed with 160 g of calcium carbonate; 360 g of wheat bran sorghum grain was packed in 10 spawn bottles and sterilized in autoclave at 121°C for 30 min. After the sterilization, the spawn bottles were inoculated with actively growing mycelium of pleurotas from media growth and incubated at 28°C for mycelium growth without any light for 15-20 days until the mycelium full covered the sorghum grains.

Substrate collection

The basic organic raw materials used for cultivation were *Coffea arabica* (coffee leaf) and *Ficus sycomorus* (Sholla) leaf and other additive like cow dung. Coffee and sholla leaves were collected from Dilla University main campus area. Cow dung was collected from Dilla University main campus (Figure 1).

Preparation of substrate

Based on the treatment, about 4 kg of fresh coffee and sholla leaves were mixed thoroughly with 80 g cow dung of calcium

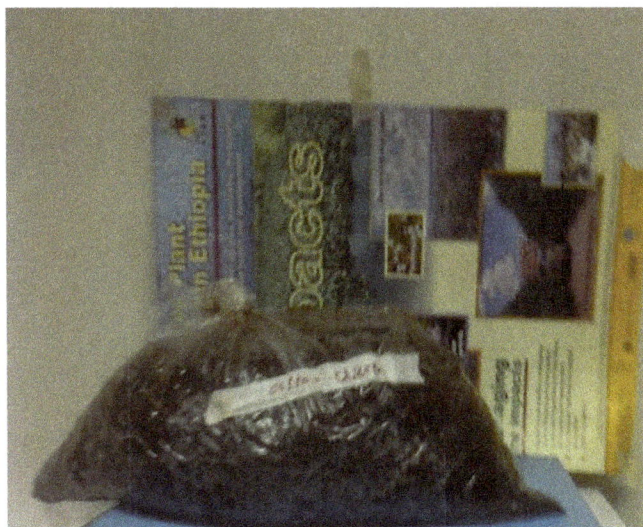

Figure 2. Substrate prepared after sterilization for cultivating of mushrooms.

carbon. Ash was also added. Water was added to moisten it based on the rule of thumb method given by Buswell (1984) (Figure 2).

Cultivation of *Pleurotus ostreatus*

After preparation of substrate, each substrate was steam sterilized at 121°C for 15 min in autoclave. Balloon bags were filled with sterilized substrate; multilayered technique was adapted for spawning 4 kg of sholla leaves plus cow dung filled in bags and 4 kg of coffee leaves plus cow dung filled in plastic bags. Then spawn was added to each substrate after inoculation was maintained at room temperature. With sufficient light, the balloon tubes were torn off following the running of the spawn. The materials of the fruiting bodies were evident within some days, after removal of balloon tube bags.

Data analysis

The data of actively mycelium growth during spawn making and formation of full morphology of pleurotus mushroom species and fruiting body were observed during cultivation on substrate.

RESULTS AND DISCUSSION

Pleurotus ostreatus was cultured on malt extract agar for seven days at 28°C and mycelium covered the medium as indicated in Figure 3.

Preparation of spawn

Full white mycelia invasion of *pleurotus ostreatus* was observed in sorghum after 20 days of incubation (Figure 4). It was ready to for inoculation on solid substrate.

In spawn preparation, the *pleurotus ostreatus* mycelium

gradual invaded the sorghum grains. Finally, the sorghum grain in the spawn bottles became completely covered by mycelium invasion; the growth of the mycelium is faster if it is kept in incubator under optimal temperature (Figure 4) within 15-20 days.

Cultivation of *Pleurotus ostreatus*

The formation of full morphology of fruiting bodies of *pleurotus ostreatus* was observed within 20 days.

Comparing the two lignocellulosic residues as substrates for the cultivation of *P. ostreatus* shows that sholla leaves led to the best growth of *P. ostreatus* as evidenced by the complete and heavy colonization of substrates forming a compact white mass of mycelium within 25 days of inoculation (Figure 6). Cultivation of the oyster mushroom on similar by-products has manifested variable levels of biological efficiency. These variations are mainly related to spawn rate, fungal species used and supplement added to the substrate (Mane et al., 2007).

The performances of the two substrates were also evident by their elevated fruiting body values on sholla leaves followed by coffee leaves (Figure 5). This might be due to the leaf content that contains some acids and phenols. Coffee waste is rich in anti-nutritional factors such as tannins. These substances have high capacity to bind proteins, making them unavailable to the organism and also act as enzyme inhibitors (Bressani, 1979; Mazzafera, 2002). Due to the presence of these anti-physiological/anti-nutritional factors, coffee waste is not considered as an adequate feed supplement for cattle and other livestock (Pandey et al., 2000).

In conclusion, mushroom cultivation involves having the wisdom and knowledge of growing fungi on waste plant materials that are not necessary consumed by humans. They may be converted to available foods. Commercial production of *P. ostreatus* mushrooms is largely determined by the availability and utilization of cheap by-products waste materials, which are agricultural wastes that are ideal and most promising substrates for cultivation. The substrates used in this study can be considered practical and economically feasible due to their availability throughout the year at no cost in large quantities. Utilization of these agro-wastes for the production of *P. ostreatus* mushrooms could be more economical, ecological and can improve food quality and similarly reduce food scarcity.

ACKNOWLEDGEMENTS

The authors greatly acknowledge the Departments of Biology, College of Natural, and Computational Sciences of Dilla University for the kind assistance in providing the laboratory facilities and all the required consumables and

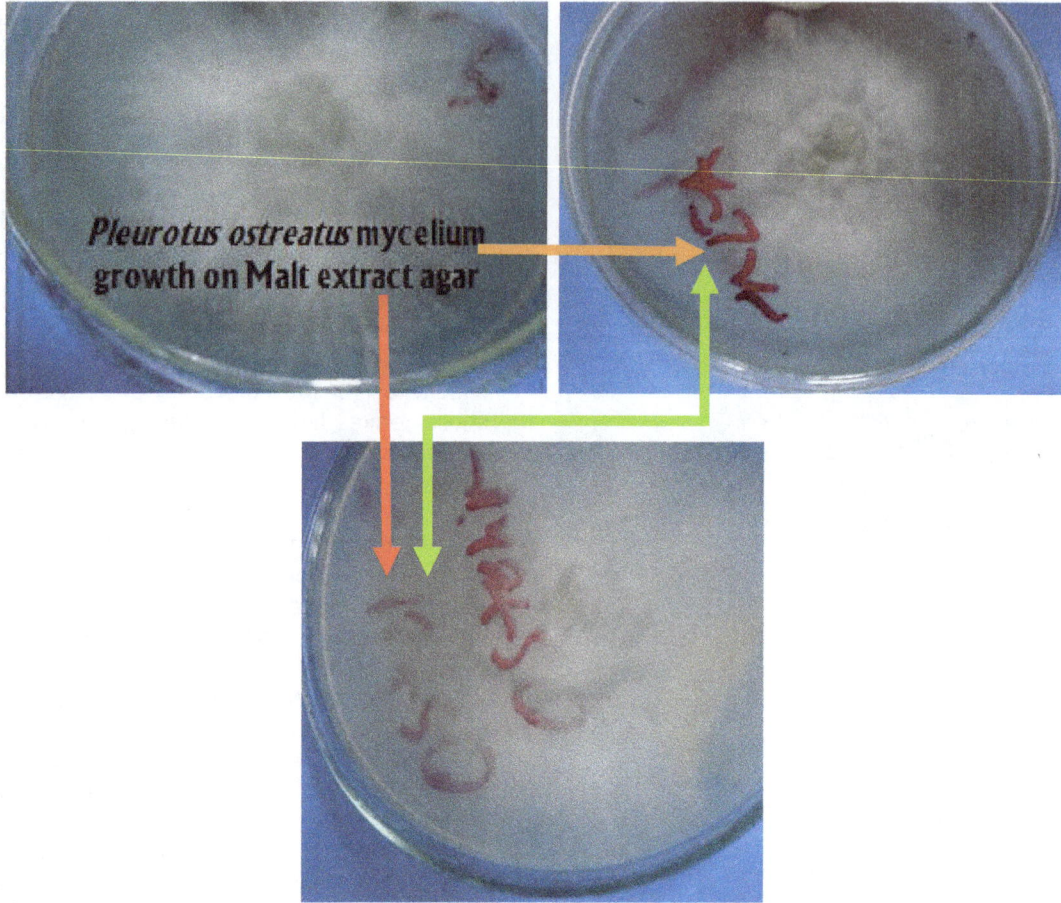

Figure 3. Culture of *Pleurotus ostreatus* on malt extract agar.

Figure 4. Full white Mycelium invasion of *Pleurotus ostreatus* was observed on Sorghum after 20 days of incubation.

Figure 5. The growing of fruiting body of *Pleurotus ostreatus* on coffee leaves substrate.

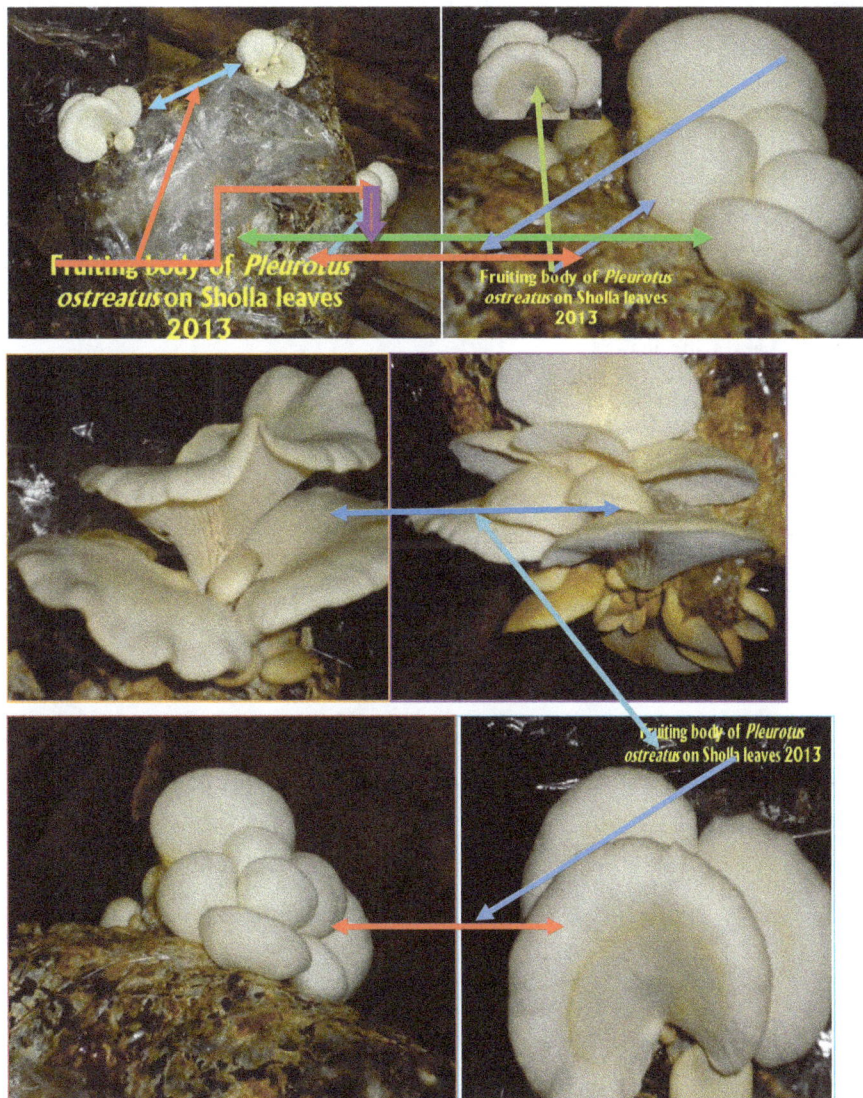

Figure 6. The growing of fruiting body of *Pleurotus ostreatus* on Sholla leaves substrate.

equipment during the whole period of this research work.

REFERENCES

Angeli JPF, Ribeiro LR, Gonzaga MLC, Soares SDA, Ricardo MPSN, T suboy MS (2006). Protective effect of glucan extracted from Agaricus blazeic against chemical induced DNA damaged in human lymphocytye. Cell Biol. Toxicol. 22:285-291.

Belitz HD, Grosch W, Schieberle P (2009). Coffee tea, cocoa in H-D Belitz W. Grosch and p. schiebherl (Eds). food chemistry 4th edn. Leigzing, Springer. Pp. 938-951.

Berg CC, Corner EJH (2005). Moracea Flora Malesiava 1(17):2-4. Birk Y, Bondi A, Gestetner B and Ishaya IA (1963). Thermostable Hemolytic Factors in Soyabeans. Nature. 197:1089-1090.

Bressani R (1979). Antiphysiological factors in coffee pulp. In: Coffee Pulp Composition,Technology and Utilization. (Braham, J.E. and Bressani, R. eds). IDRC Publisher. Canada. pp. 83-96.

Buswell JA (1984). Potential of spent mushroom substrate for Bioremediation purposes. Compost 2:31-35.

Choi YHGH, Chal OH, Kim HJ, Choi YH, Zhang X, Lim JM, Kim J-H, Lee MS, Han EH, Kim HT (2006). Inhibitory effects of Agaricus blazei on mast cell-mediated anaphylaxis-like reaction. Biol. Pharm. Bull. 29:1366-1371.

Dawit A (1998). *Mushroom cultivation; Apractical Approach.* Berhanea Selam printing Enterprise, Addis Ababa.

Gregori AS, Vagels M, Pohleven J (2007). Cultivation Technique and Medicinal propertie of pleurotous Spp. Food Technol. Biotechnol. 45:238-249.

Grembecka M, Malinoluska G, Szefer P (2007). Differentiation of market coffee and its infusion in view of their mineral composition. Sci Total Environ. 20:383(1-3):59-69.

Grind B, Hetland G, Johson E (2006). Effect on gene expression and viral load of medicinal extract from Agaricus blazei in patient with chronic hepatitis C infection. Int. Immuno Pharmacol. 6:1311-1314.

Guterrez ZR, Mantovani MS, Eira AF, Ribeiro IR, Jordao BQ (2004). Variation of the antimutagenicity effect of water extract of Agaricu blazei Murillaq in vitro. Toxicol. In Vitro. 8:301-309.

Kurtzman RHJr (2005). A review Mushrooms: sources for Modern Western Medicine. Micologia Aplicada International 17:21-33.

Lima DR (2003). Cafee Jaude, Manvalde formacologia clinicaTerapeuticae Toxicologia. Riodejanero: Meds: Editora.

Mane VP, Patil SS, Syed AA, Baig MMV (2007). Bioconversion of low quality lignocellulosic agricultural waste into edible protein by *Pleurotus sajorcaju* (Fr.) Singer. J Zhejiang Univ. Sci. B. 8:745-51.

Mazzafera P (2002). Degradation of caffeine by microorganism and potential use of decaffeinated coffee husk and pulp in animal feeding. Scientia Agricola. 59:815-821

Menoli RCRN, Montovan MS, Ribeiro LR, Speit G, Jorerdao BQ (2004). Anti mutagenic effect of mushroom Agaricau blazei murill extracts on V79 cells. Mutat Res. 496:5-13.

Neyrinck AM, Bindles LB, DeBacker F, Pachikian BD (2009). Dietor ysupplementation with chitosan derived from mushroom changes adipocytokine profile indiet induced obesce mice aphenomena linked to its lipid lowering action. Int. immune pharmaco. 9:767-773.

Orwa C, Mutua A, Kindt R, Jamnadass R, Simons A (2009). Agroforestree Database: a tree reference and selection guide version 4.0 (http://www.worldagroforestry.org/af/treedb/).

Pandey A, Soccol CR, Nigam P, Brand D, Mohan R, Roussos S (2000). Biotechnological potential of coffee pulp and coffee husk for bioprocesses. J. Biochem. Eng. 6:153-162.

Santos EJ, Olivera E (2001). Determination of mineral nutrient and Toxic Element in Brozilian soluble coffee.

Trugo LC, Macrae R (1984). Study of the effect of roasting on the chloronic acid composition of coffee using food chemistry. 15:369-382.

Wasser P, Weis AL (1991). Medicinal properties of substance occurring in higher Basidom mushrooms.curent perspective (review). Int. J. Med. Mushr. 1:31-62.

Zhang M, Cui SW, Cheung PLK,Wang Q (2007). Antitumor Polysaccharide from mushrooms. Areview on ther isolation process, structural characteristic and antitumor activity. Trends Food Sci. Technol. 39:14-19.

26

Mycotoxicological studies of an *Aspergillus oryzae* strain

A. Sosa[1], E. Kobashigawa[2], J. Galindo[1], R. Bocourt[1], C. H. Corassin[2] and C. A. F. Oliveira[2*]

[1]Instituto de Ciencia Animal, Apartado Postal 24, San José de las Lajas, La Habana, Cuba.
[2]Departamento de Engenharia de Alimentos, Faculdade de Zootecnia e Engenharia de Alimentos, Universidade de São Paulo.

The safety of *Aspergillus oryzae* used in industrial fermentations of food-grade products has long been recognized. However, production of fungal toxic secondary metabolites is strain-specific and environment-dependent. For these reasons, the present study aimed to conduct a mycotoxicological study of the strain H/6.28.1 of *A. oryzae* (from the collection of the Cuban Institute of Sugar Cane Derivatives Research), intended for use as microbial feed additive for ruminants. Analysis of mycotoxins in fungal culture extracts was carried out by thin layer chromatography and high-performance liquid chromatography. Results show that the strain under study does not produce detectable levels of aflatoxins B_1, B_2, G_1, G_2 and "ochratoxin". Cyclopiazonic acid was detected at level of 14.47 $\mu g \cdot ml^{-1}$, which has negligible toxicology significance for animal health. It is concluded that the *A. oryzae* evaluated could be used as a feed additive for ruminants.

Key words: Aflatoxins, ochratoxin A, cyclopiazonic acid, *A. oryzae*, feed additive.

INTRODUCTION

In the last decades, nutritionists have shown great interest in a wide variety of microorganisms that, when added in small quantities to the diet, improve the health and the productivity of animals (Kamra and Agarwal, 2004). This is particularly true for the conidial fungus *Aspergillus oryzae*, which have been shown to produce significant improvements in the performance of ruminants (Wiedmeier et al., 1987; Gómez-Alarcón et al., 1991; Humphry et al., 2002; Kim et al., 2006; Di Francia et al., 2008).

The safety of any microbial product intended for animal feed should be carefully evaluated. The primary issue in the safety evaluation of these microbial feed additives is its toxigenic potential, specifically the possible synthesis of toxins by a particular strain that can cause health disorders in the animals after ingestion through the diet (Pariza and Johnson, 2001). *A. oryzae* is considered as "generally recognized as safe" by the U.S. Food and Drug Administration, and is used in several fermented foods and enzyme preparations (Beuchat, 2001). However, mycotoxin production by some strains has been reported (Blumenthal, 2004). On the other hand, the Joint FAO/WHO Expert Committee on Food Additives (JECFA) recommends that enzyme preparations derived from fungal sources should be evaluated for those mycotoxins that are known to be produced by strains of the species used in the production of the enzyme preparation or related species (Food and Agriculture Organization, 2001). In this context, the aim of the present study was to determine the ability of an *A. oryzae* strain to produce some mycotoxins that could be produced by the species, before using it as a microbial feed additive in ruminant's diets.

MATERIALS AND METHODS

A strain of *A. oryzae* (H/6.28.1, belonging to the collection of the Cuban Institute of Sugar Cane Derivatives Research) was used

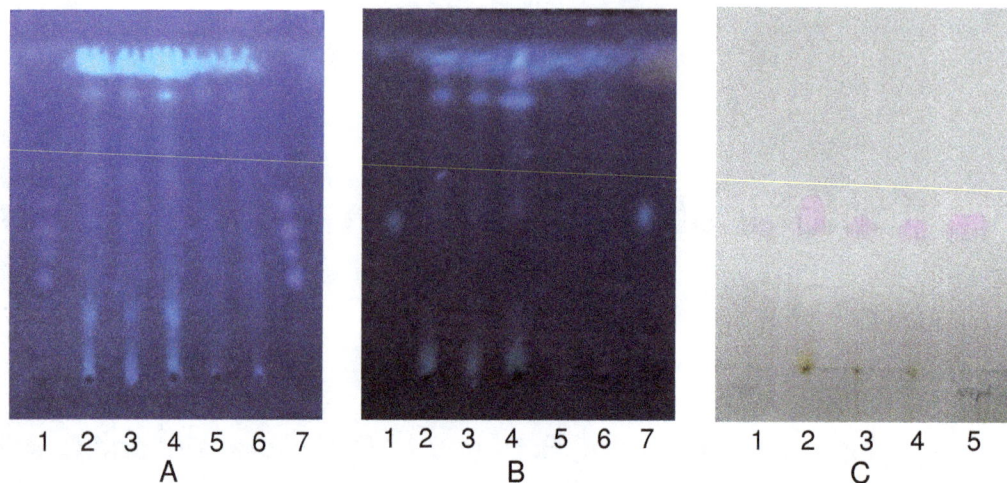

Figure 1. Thin layer chromatographic plates from samples of culture material of *Aspergillus oryzae* showing spots of: (A) aflatoxins B_1, B_2, G_1, and G_2 (spots 1 and 7: standards; spots 2-5: samples); **(B)** ochratoxin A (spots: 1 and 7: standards; spots 2-5, samples **(C)** cyclopiazonic acid (spots 1 and 5: standards; spots 2-4: samples).

The strain was previously isolated from soil and stored in Czapek agar (pH 6.8) at 4 °C. Confirmation of the strain identity was accomplished by DNAr digestion in the Spanish Collection of Type Cultures (CECT) from the University of Valencia.

The *A. oryzae* strain was grown for 10 days at 25 °C on Czapek yeast autolysate agar plates (Klich and Pitt, 1988) prepared from Czapek-Dox agar added with 5 gl^{-1} yeast extract (Difco). The combined agar and culture were then transferred to erlenmeyers of 250 ml and the mycotoxins were extracted with chloroform/methanol (2:1, v/v; 50 ml per plate) by shaking for 30 min. The mixture was filtered through anhydrous sodium sulphate. The filtrate was evaporated to dryness in a water bath at 40 °C under air stream (López-Díaz et al., 2001).

The dried extracts were dissolved in 1 ml chloroform and small aliquots (25 μl) were screened by spotting onto silica gel plates (Merck) along with standard solutions of the following mycotoxins under investigation: aflatoxins (AF) B_1, B_2, G_1, G_2, ochratoxin A (OA) and cyclopiazonic acid (CPA). The standards and the samples were spotted at 1.0 cm from the bottom of the silica gel plates, and spaced 1.0 cm between each one. In the case of aflatoxins, 5 and 10 μl of the standard solution containing 1.0 μg ml^{-1} of each AF were spotted. For OA, 10 and 15 μl of the standard solution (2.0 μg ml^{-1}) were spotted. The mobile phase consisted in a mixture of ether:methanol:water (96:3:1, v/v/v) (Soares and Rodrigues-Amaya, 1989), being the toxins detected by the presence of fluorescent spots of AF`s or OA under ultraviolet (UV) light (366 nm). For CPA, 50 and 100 μl of the standard solution (4.0 μg ml^{-1}) were spotted, and a mobile phase consisted of ethyl acetate:2-propanol:sodium hydroxide (50:15:10, v/v/v) was used (Lansden, 1986). When the chromatographic run for CPA finished, the silica gel plate was dried at 35 °C for 3 min and sprayed with Ehrlich's reagent (Lansden, 1986), for observation of CPA purple spots after about 4 min. The spots remained evident for approximately 7 min.

Quantification of the mycotoxins was accomplished in a Shimadzu 10VP liquid chromatographic system equipped with a 10 AXL fluorescence detector (excitation at 360 and 365 nm and emission above 440 and 450 nm, for aflatoxin and OA, respectively. A Phenomenex (Torrance, CA, USA) Sinergy C18 column (4.6 x 150 mm, 4 μm) and a Shimadzu Shim-Pack CLC G-ODS precolumn (4 x 10 mm) were used. A total of 20 μl of sample extracts were injected for determination of aflatoxin. The samples were previously derivatizated with 100 μl of acetic acid and 200 μl of n-hexane and filtered through a membrane (PTFE, 0.45 μm, Millex, Millipore). The mobile phase consisted on a mixture of acetonitrile:water:methanol (60:20:20, v/v/v) with a flow rate of 1.0 ml min^{-1}. Under these conditions, the retention times for aflatoxins were: G1: 4.275 min; B1: 5.136 min; G2: 7.278 min; B2: 9.420 min. Determination of OA determination was developed under the same condition, but the mobile phase consisted on a mixture of methanol-water-acetic acid (60:40:1, v/v/v) with a flow rate of 1.2 ml min^{-1}. A total of 20 μl of sample extracts, previously filtered, were injected. The retention time for OA was approximately 15.219 min. A calibration curve of CPA was prepared using standard CPA solutions at concentrations of 2.41; 4.82 and 9.63 μg·ml^{-1}. A UV visible SPD 10AVP detector, at 279 nm, was used. Chromatography conditions were as follows (Urano et al., 1992): 100% of methanol:water (7:3, v/v) (solvent A) during 5 min, later linear gradient from 100% of solvent A to 100% methanol:water (7:3, v/v) with 4 mM $ZnSO_4 x 7 H_2 O$ (solvent B) in 10 min, followed by 100% of solvent B for 5 min. The return of the gradient to 100% of solvent A was reached in 15 min. A total of 20 μl of sample extracts, previously filtered, were injected at a constant flow rate of 1.0 ml min^{-1}. Under these conditions, retention time for CPA was approximately 5.8 min.

RESULTS AND DISCUSSION

The thin layer chromatography revealed that the strain under study did not produce aflatoxins B_1, B_2, G_1 and G_2 and OA (Figures 1, A and B). These results were confirmed with the HPLC (Figures 2 and 3). The detection limit for each aflatoxin (B_1, B_2, G_1, G_2), OA and CPA were 0.5 ng·g^{-1}, 8.7 ng·g^{-1} and 5.0 μg·g^{-1} of culture material, respectively (Table 1).

The non-occurrence of aflatoxins in *A. oryzae* cultures was demonstrated by other authors (Murakami, 1971; Kusumoto et al., 1990). Early studies on the factors determining the aflatoxin production focused on physiological aspects of strains, including growth temperature, age of the culture and nutrient components in the medium (Reddy et al., 1979). The *A. oryzae* strain studied was

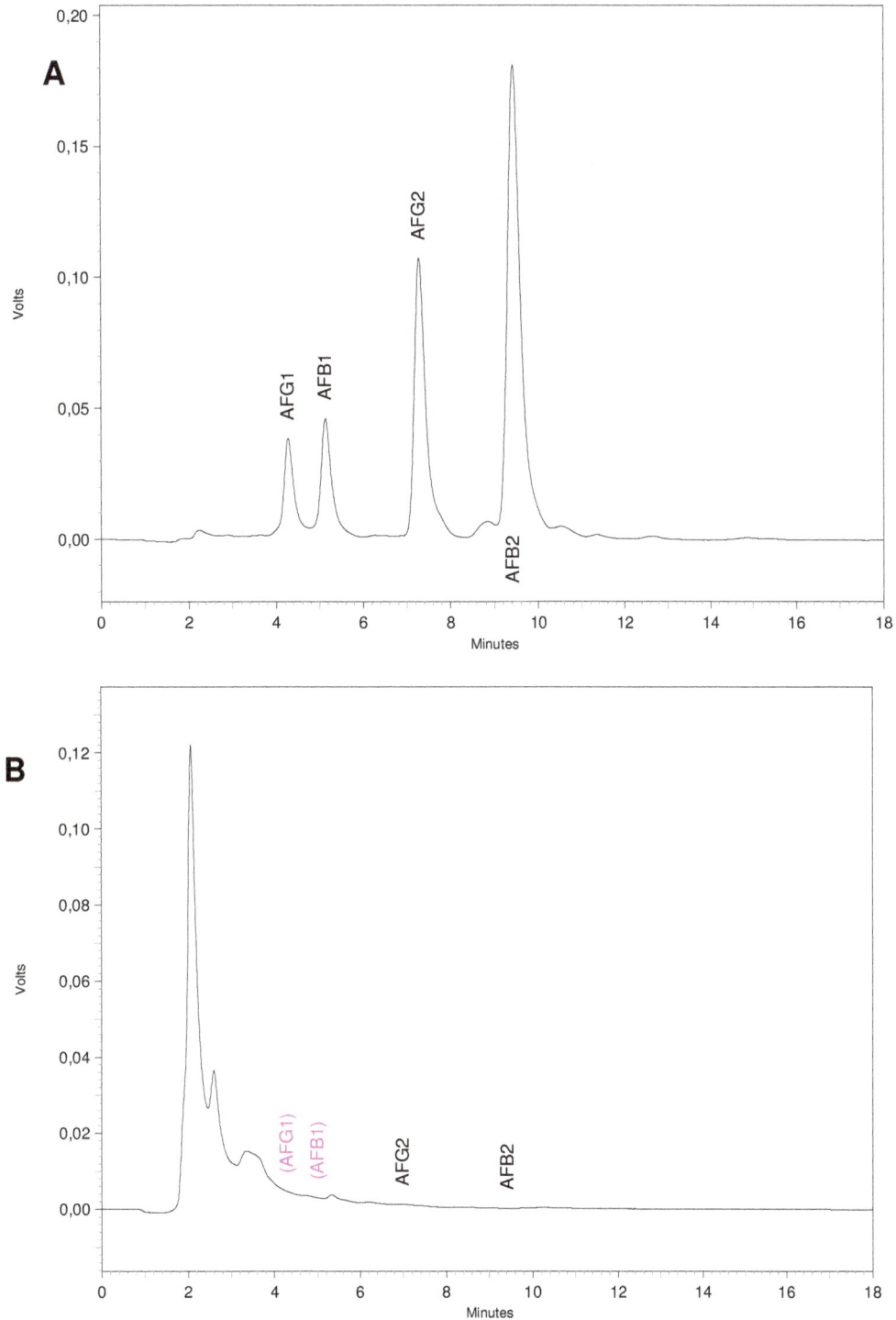

Figure 2. Chromatograms showing (A) Aflatoxins B_1, B_2, G_1, and G_2 standards (20.0 ng·ml^{-1} of each aflatoxin); (B) Sample of culture material of *Aspergillus oryzae*.

cultured under favorable conditions for the synthesis of aflatoxins (López-Díaz et al., 2001). Although in the present study no genetic approach was used, the absence of aflatoxin production by the strain evaluated could be related to its genetic characteristics. Previous work showed a number of mutations within the aflatoxin biosynthesis

gene homolog cluster in *A. oryzae* relative to the *A. flavus* sequence, including deletions, frameshift mutations, and base pair substitutions, which induce inactivation at the protein level and consequently impair the aflatoxin synthesis (Kiyota et al., 2011). Aflatoxin production in *A. oryzae* was reported by some authors (Adebajo, 1992; El-Kady

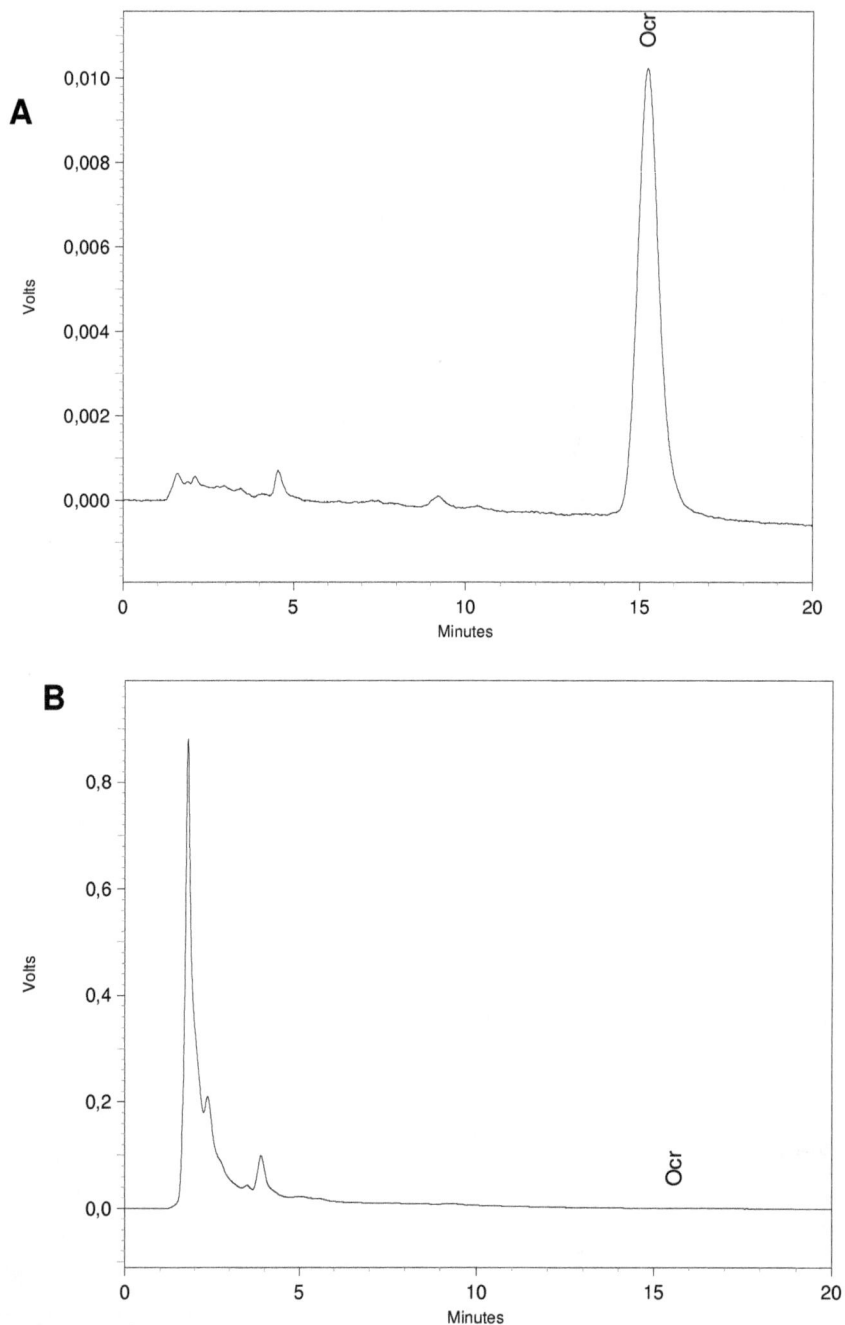

Figure 3. Chromatograms showing **(A)** Ochratoxin A standard (402.8 ng·ml[-1]); **(B)** sample of culture material of *Aspergillus oryzae*.

Table 1. Analysis of aflatoxins B_1, B_2, G_1, G_2, ochratoxin A and cyclopiazonic acid in the culture material of *Aspergillus oryzae* after 10 days of cultivation.

Mycotoxin	LOD	Concentration in the culture material [1]
Aflatoxins B_1, B_2, G_1, G_2 (ng·g[-1])	0.5	ND
Ochratoxin A (ng·g[-1])	8.7	ND
Cyclopiazonic acid (µg·g[-1])	5.0	14.5

[1]Mean value of 3 replicates analyzed; LOD, limit of detection; ND, Not detected in any sample analyzed.

Padrão CPA (2,41 ug/ml)

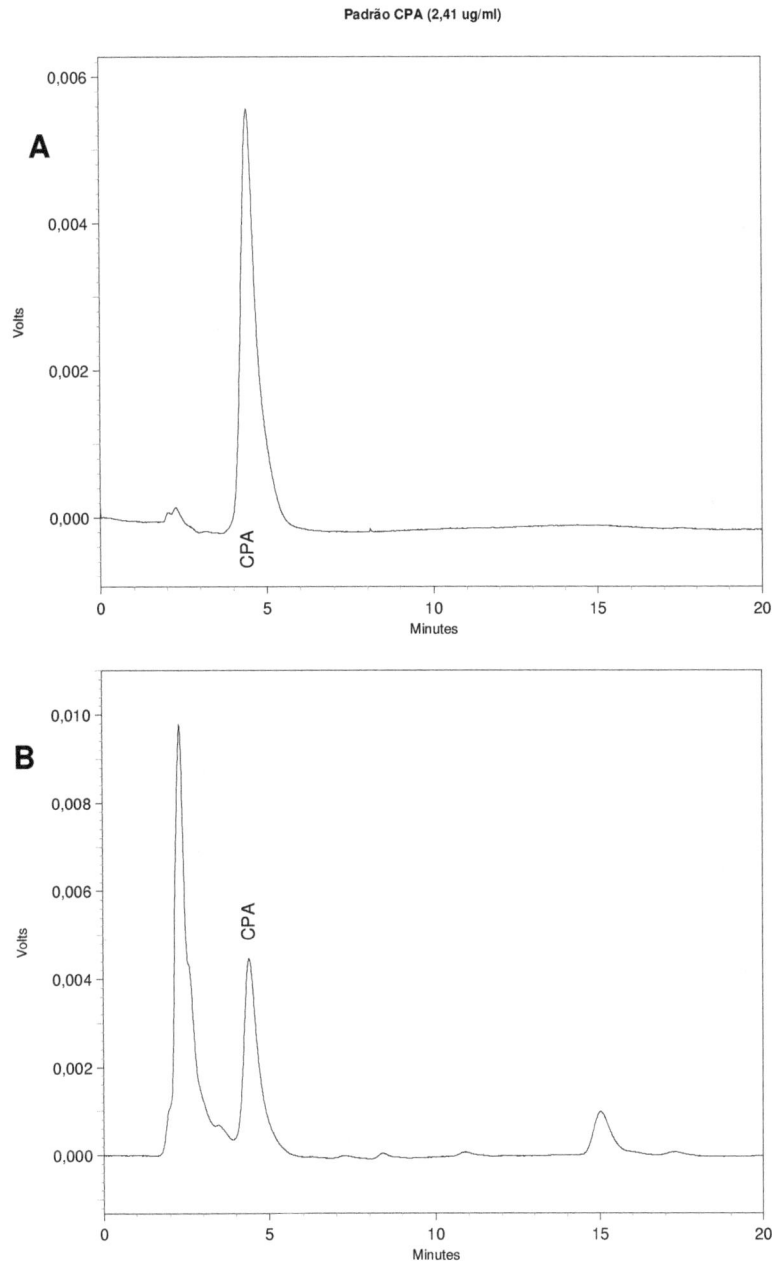

Figure 4. Chromatograms showing **(A)** Cyclopiazonic acid standard (2.41 μg·ml⁻¹). **(B)** Sample of culture material of *Aspergillus oryzae* containing 12.0 μg·g⁻¹

et al., 1994; Atalla et al., 2003). However, it is possible that these aflatoxin-producing strains were incorrectly identified as *A. oryzae* because of the close taxonomic relatedness between this fungus and other members of *A. flavus* group (Blumenthal, 2004).

The absence of OA in our study agree with those observed by Lane et al. (1997), who also didn't find this toxin in various batches of the enzyme preparations. Ochratoxins were first known to be produced by species from Section Circumdati (*A. orchraceus* group) (Cole and Cox, 1981). Moreover, the production of this toxin was repeatedly

reported in Section Nigri (Martínez-Culebras et al., 2009), which stress the importance of testing the ochratoxin production in *Aspergillus*-derived enzyme preparations. All samples showed spots corresponding to the CPA standard onto TLC plates (Figure 1C), which were thereafter confirmed and quantified by HPLC. The mean concentration of CPA in the samples was 14.5 μg·g⁻¹ (Figure 4). Production of CPA by *A. oryzae* was firstly reported by Orth (1977), and subsequently by other authors (Vinokurova et al., 2007). Although the molecular genetics of CPA biosynthesis have not been as well studied as for aflatoxins biosynthesis, some essential genes for the toxin pathway in the CPA-producing *A. oryzae* have been identified (Christensen et al., 2005; Tokuoka et al., 2008). It is presumed that some chromosomal deletion that affects aflatoxin production *A. oryzae* strains can also affects CPA production (Tokuoka et al., 2008; Chang et al., 2009). On the other hand, only a certain group of aflatoxin-nonproducing strains have the potential to produce CPA (Kusumoto et al., 2000; Tokuoka et al., 2008). According to the results obtained in our study, the evaluated strain seems to belong to this group. Production of CPA in *A. oryzae* was reported to start nearly 50 h after inoculation, reaching maximum levels in two weeks (Goto et al., 1987). Fondevila et al. (1990) reported that fermentation extracts of *A. oryzae* extracts remains in the rumen for less than 24 h. There-fore, when using *A. oryzae* as feed additive for ruminants, the short duration of the fungus in the rumen would not be enough to allow the production of secondary metabolites such as CPA and cause toxic effects.

Conclusions

The *A. oryzae* strain tested does not produce detectable levels of aflatoxins B_1, B_2, G_1 or G_2, and ochratoxin A. Although cyclopiazonic acid was detected in the fungal extracts during 10 days of cultivation, the amount produced is of negligible toxicology significance for animal health, taking into account the short period of the fungus in the rumen. The *A. oryzae* strain tested therefore could be used as a feed additive for ruminants.

REFERENCES

Adebajo LO (1992). Spoilage moulds and aflatoxins from poultry feeds. Food/Nahrung 36:523-529.

Atalla MM, Hassanein NM, El-Beih AA, Youssef YA (2003). Mycotoxin production in wheat grains by different *aspergillius* in relation to different relative humidities and storage periods. Food/Nahrung 47:6-10.

Beuchat LR (2001). Traditional fermented foods. In: Doyle MP, Beuchat LR, Montville TJ (Eds.), Food Microbiology Fundamentals and Frontiers, 2nd ed. ASM Press, Washington DC, pp.701-719.

Blumenthal CZ (2004). Production of toxic metabolites in *Aspergillus niger*, *Aspergillus oryzae*, and *Trichoderma reesei*: justification of mycotoxin testing in food grade enzyme preparations derived from the three fungi. Regul. Toxicol. Pharmacol. 39:214-228.

Chang PK, Horn BW, Dorner JW (2009). Clustered genes involved in cyclopiazonic acid production are next to the aflatoxin biosynthesis gene cluster in *Aspergillus flavus*. Fungal Gen. Biol. 46:176-182.

Christensen BE, Mollgaard H, Kaasgaard S, Lehmbeck J (2005). Methods for producing polypeptides in *Aspergillus* mutant cells. United States Patent 20050153397.

Cole RJ, Cox RH (1981). Handbook of Toxic Fungal Metabolites. Academic Press, New York.

Di Francia A, Masucci F, De Rosa G, Varricchio ML, Proto V (2008). Effects of *Aspergillus oryzae* extract and a *Saccharomyces cerevisiae* fermentation product on intake, body weight gain and digestibility in buffalo calves. Anim. Feed Sci. Technol. 140:67-77.

El-Kady I, El-Maraghy S, Zohri A (1994). Mycotoxin producing potential of some isolates of *Aspergillus flavus* and Eurotium groups from meat products. J. Microb. Res. 149:297-307.

Fondevila M, Newbold CJ, Hotten PM, Orskov ER (1990). A note on the effect of *Aspergillus oryzae* fermentation extract on the rumen fermentation of sheep given straw. Anim. Prod. 51:422-425.

Food and Agriculture Organization (2001). Summary and conclusions of the Joint FAO/WHO Expert Committee on Food Additives. FAO, Rome. (JECFA 57th meeting).

Gómez-Alarcón RA, Huber JT, Higginbotham GE, Wiersma F, Ammon D, Taylor B (1991). Influence of feeding *Aspergillus oryzae* fermentation extract on the milk yields, eating patterns and body temperatures of lactating cows. J. Anim. Sci. 69:1733-1740.

Goto T, Shinshi E, Tanaka K, Masaru M (1987). Production of cyclopiazonic acid by koji molds and possibility of cyclopiazonic acid contamination of Japanese fermented foods. Shokuhin Sogo Kenkyusho Kenkyu Hokoku 51:23-28.

Humphry JB, Coffey KP, Moyer JL, Brazle FK, Lomas LW (2002). Intake, digestion, and digestive characteristics of Neotyphodium coenophialum-infected and uninfected fescue by heifers offered hay diets supplemented with *Aspergillus oryzae* fermentation extract or laidlomycin propionate. J. Anim. Sci. 80:225-234.

Kamra DN, Agawal N (2004). Probiotics as feed additives for the ruminants. Indian Vet. Res. Inst. Izatnagar, India. 43:122-127.

Kim HS, Ahn BS, Chung SG, Moon YH, Ha JK, Seo IJ, Ahn BH, Lee SS (2006). Effect of yeast culture, fungal fermentation extract and non-ionic surfactant on performance of Holstein cows during transition period. Anim. Feed Sci. Technol. 126:23-29.

Kiyota T, Hamada R, Sakamoto K, Iwashita K, Yamada O, Mikami S (2011). Aflatoxin non-productivity of Aspergillus oryzae caused by loss of function in the aflJ gene product. J. Biosci. Bioeng. 111:512-517.

Klich MA, Pitt JI (1988). A laboratory guide to common Aspergillus species and their teleomorphs. CSIRO, North Ryde, NSW, Australia.

Kusumoto KI, Goto T, Manabe M (1990). Evaluation for conversion of sterigmatocystin to aflatoxin in koji-molds. Rep. Natl. Food Res. Inst. 54:14-17.

Kusumoto KI, Nogata Y, Ohta H (2000). Directed deletions in theaflatoxin biosynthesis gene homolog cluster of *Aspergillus oryzae*. Curr. Genet. 37:104-111.

Lane RW, Yamakoshi J, Kikuchi M, Mizusawa K, Henderson L, Smith M (1997). Safety evaluation of tannase enzyme preparation derived from *Aspergillus oryzae*. Food Chem. Toxicol. 35:207-212.

Lansden JA (1986). Determination of cyclopiazonic acid in peanuts and corn by thin layer chromatography. J. Assoc. Off. Anal. Chem. 69:964-966.

López-Díaz TM, Santos JA, García-López ML, Otero A (2001). Surface mycoflora of a Spanish fermented meat sausage and toxigenicity of Penicillium isolates. Int. J. Food Microb. 68:69-74.

Martínez-Culebras PV, Crespo-Sempere A, Sánchez-Hervás M, Elizaquivel P, Aznar R, Ramón D (2009). Molecular characterization of the black *Aspergillus* isolates responsible for ochratoxin A contamination in grapes and wine in relation to taxonomy of *Aspergillus* section Nigri. Int. J. Food Microb. 132:33-41.

Murakami H (1971). Classification of the koji mold. J. Gen. Appl. Microb. 17:281-309.

Orth R (1977). Mycotoxins of *Aspergillus oryzae* strains for use in the food industry as starters and enzyme producing molds. Ann. Nutr. Alim. 31:617-624.

Pariza MW, Johnson EA (2001). Evaluating the safety of microbial enzyme preparations used in food processing: update for a new century. Reg. Toxicol. Pharmacol. 33:173-186.

Reddy TV, Viswanathan L, Venkitasubramanian TA (1979). Factors affecting aflatoxin production by *Aspergillus parasiticus* in a chemically defined medium. J. Gen. Microbiol. 114:409-413.

Soares LMV, Rodrigues-Amaya DB (1989). Survey of aflatoxins, ochratoxins A zearalenona and sterigmatocystin in some Brazilian foods by using multitoxin thin layer chromatografic method. J. Assoc. Off. Anal. Chem. 72:22-26.

Tokuoka M, Seshime Y, Fujii I, Kitamoto K, Takahashi T, Koyama Y (2008). Identification of a novel polypeptide synthase-nonribosomal peptide synthetase (PKS-NRPS) gene required for the biosynthesis of cyclopiazonic acid in Aspergillus oryzae. Fungal Genet. Biol. 45:1608-1615.

Urano T, Trucksess MW, Matusik J (1992). Liquid chromatographic determination of cyclopiazonic acid in corn and peanuts. J. Assoc. Off. Anal. Chem. 75:319-322.

Vinokurova NG, Ivanushkina NE, Khmel'nitskaya II, Arindabarov MU (2007). Synthesis of a-cyclopiazonic acid by fungi of the genus Aspergillus. Appl. Biochem. Microb. 43:435-438.

Wiedmeier RD, Arambel MJ, Walters JL (1987). Effect of yeast culture and *Aspergillus oryzae* fermentation extract on ruminal characteristics and nutrient digestibility. J. Dairy Sci. 70:2063-2068.

27

Generation of a cDNA library reveals the presence of a *Ty-1*/copia transposon in *Phellinus rimosus* (Hymenochaetaceae, Basidiomycota)

Luis Cesar Maffei Sartini Paulillo[1], Sizenando José de Andrade Junior[2], Abelmon da Silva Gesteira[3], Marco Brotto[4,5] and Aristóteles Góes-Neto[2]

[1]Faculdade de Tecnologia e Ciências - FTC, Mestrado Profissional em Bioenergia. Av. Luis Viana Filho, 8812, Pituaçu, Salvador, CEP 41.741-490, BA - Brasil.
[2]PPG Biotecnologia UEFS/Fiocruz-BA. Av. Transnordestina s.nº, Novo Horizonte, Feira de Santana, – CEP 44.036-900, BA – Brasil.
[3]Embrapa Mandioca e Fruticultura Tropical. Rua Embrapa, s/nº. Cruz das Almas, CEP 44380-000, BA – Brasil.
[4]Muscle Biology Research Group-MUBIG, Schools of Nursing, Medicine, and Biological Sciences, University of Missouri-Kansas City, Kansas City, 64108. Missouri. USA.
[5]International Visiting Professor and Researcher, Conselho Nacional de Pesquisa-CNPq and Faculdade de Tecnologia e Ciências-FTC, Salvador, Brasil.

Lignocellulolytic fungi have significant potential as bio-convertors with a particular capacity to degrade recalcitrant natural composts, such as lignin. We generated a cDNA library of a lignocellulolytic fungus, the *Phellinus rimosus* (Hymenochaetaceae, Basidiomycota), and a typical species of polypore mushroom from the Brazilian semi-arid region. To the best of our knowledge, this is the first report ever to describe the existence of a *Ty1*-copia retrotransposon in the *P. rimosus* species. Given the taxonomic complexity in the Brazilian semi-arid region, our findings will be useful as they provide a molecular identity to this important parasite species of the semi-arid region of Brazil.

Key words: Lignocellulolytic fungi, biotechnology, *Phellinus rimosus*, *Ty-1 copia*, Brazilian semi-arid region, molecular marker.

INTRODUCTION

Most species of the order Hymenochaetales Oberwinkler (Basidiomycota) shows a pattern of tropical distribution and are able to degrade wood, causing white rot. To date, studies in Brazil involving such organisms have been limited to general taxonomic surveys, specific inventories or ecological studies (Drechsler-Santos et al., 2010; 2008; Góes-Neto and Groposo, 2005; Azevedo and Guerrero, 1993). There is a clear lack of studies aimed at functional genomics characterization of this species in Brazil, particularly in the semi-arid region.

The study of molecular diversity of native species of lignocellulolytic Hymenochaetales from the semi-arid region of Brazil, among them the *Phellinus rimousus*, can be carried out through the construction and analyses of cDNA libraries, which, in turn, will enable mapping of genes of biotechnological potential for the development of wood delignification processes, therefore, potentially posing as a strategy for pulp bleaching with a consequent reduction of damaging environmental impact (Martinez et al., 2004, Coehn et al., 2002, Mayer and Staples, 2002, Blanchette, 1991). Thus, the major aim of this study was to construct and analyze a cDNA library of the species *Phellinus rimosus*.

The cDNA library was generated and clones derived from the library were sequenced and analyzed *in silico* using bioinformatics tools. To the best of our knowledge, this is the first report ever to describe the existence of a *Ty1*-copia retrotransposon in the *P. rimosus* species.

*Corresponding author. E-mail: paulillolc@yahoo.com.br.

Given the taxonomic complexity in the Brazilian semi-arid region, our novel findings could be useful as they provide a molecular identity to this important parasite species of the semi-arid region of Brazil.

MATERIALS AND METHODS

Sample collection and processing

Phellinus rimosus basidiomata were removed from the trunk of live tree of *Cesalpinia pyramidalis* (Leguminosae) in the Brazilian semi-arid region (Ipirá, Bahia, Brazil, 12°10'49"S; 39°46'14"W) in an area of Caatinga vegetation. Basidiomata were then appropriately placed in a waxed paper bag and transported in a cooler to be immediately processed. A fresh portion of the tissue of the internal part of the pileus was used as source for extraction of total RNA. The remaining basidiomata were then dehydrated under a constant flow of warm air at 40°C in an electric dehumidifier and subsequently deposited in HUEFS (Herbarium of the *Universidade Estadual de Feira de Santana* - BA) under the number 118024 (http://herbario.uefs.br).

Taxonomic identification

The taxonomic identification of the host tree species was determined by comparison of identified vouchers in HUEFS with the aid of an expertise botanist in Leguminosae. In addition, it was based on the analysis of macro and micromorphological features as previously described by Ryvarden 1991, and host identity by Dreschler-Santos et al. Briefly, features included in our macro-analysis of dehydrated basidiomes comprised the type and longevity of basidiome colour, form, attachment, surface, and size of pileus (when present), as well as the colour and form of abhymenial margin (when present), and also the colour, fertility and size of hymenial margin, form and number of pores, form and thickness of dissepiments, the presence or the absence of black line, and stratification, colour, and size of tubes. Micro-analysis features included the type of the hyphal system, colour, wall, form, ramification, septation and diameter of context and tramal generative and skeletal hyphae; the presence or the absence of crystal hyphae, and extra-hymenial setae; the presence or the absence, colour, wall, and form of hymenial setae; presence or absence, colour, wall, and form of cystidioles; colour (in water and KOH), wall, form, size, and length/width ratio of spores. Macroscopic features were analyzed with the aid of a stereomicroscope and sections were made for the study of microscopic characteristics. Slide preparations were stained with 1% phloxine, 5% KOH, and Melzer's reagent was used to test the iodine reaction.

RNA extraction

Total RNA was extracted using the RNeasy Mini Protocol for Isolation of Total RNA from Plant Cells and Tissues, and Filamentous Fungi (Qiagen, Germantown, MD, USA). Next, we used the kit BD Creator Smart cDNA Library Construction (Clontech, Mountain View, CA, USA) to obtain the cDNA library. DNA fragments were fractioned using a HIS Trap column (FF crude, GE Healthcare, Waukesha, WI, USA). The specific profile of fractions was further verified by gel electrophoresis. Selected fragments were annealed to the vector pTZ57R/T (Figure 1) (Fermentas) following the manufacturer instructions. Transformation was performed using a BIO-RAD *Gene Pulser Cuvette* 0.2 cm electrod, employing electro-competent DH5ß cells (Invitrogen,

Carlslab, CA, USA). Plating was carried out between the cells undergoing transformation and blank colonies were pre-selected. Twelve positive clones were used as template, for proper control of fragment sizes inserted in the transformation medium through a PCR reaction using the following primers (M13-F 5' GTAAAACGACGGCCAGT 3' and M13-R 5' CAGGAAACAGCTATGAC 3'). Vector extraction was accomplished through Mini-prep and sequencing reactions were processed by using *DYEnamic* ^{TM} *ET* Dye Terminator Cycle Sequencing kit for *MegaBACE DNA* Analysis Systems (Amersham Biosciences, Sweden). Generated sequences were then converted to the FASTA file format and quality scores were simultaneously obtained with PHRED. Values equal or higher to 10 were discarded and final processing achieved with utilization of the CROSS-MATCH program. Sequences were then analyzed using Genbank/tBLASTx (NCBI, MD, USA) (Johnson et al., 2008).

Cloning procedures

First strand cDNA were synthesized using the Clontech BD SMART PCR cDNA Synthesis KIT (Mountain View, CA) as recommended by the manufacturer. 0.5 to 1 µg of total RNA was incubated at 72°C for 2 min with 1 µl 3' BD SMART CDS Primer II A (12 µM) and 1 µl BD SMART II A Oligonucleotide (12 µM) in a total volume of 5 µl. Then 2 µl 5X First-Strand Buffer, 1 µl DTT (20 mM); 1 µl dNTP Mix (10 mM of each dNTP), 1 µl BD PowerScript Reverse Transcriptase were added and the mixture was incubated at 42°C for 1 h in an air incubator. Next, 3 µl Biotin-dATP (Invitrogen), 3 µl Biotin-dCTP (Invitrogen), 1 µl 5'-NVVVVV-3' prime, 2 µl 5X First-Strand Buffer, 1 µl BD PowerScript Reverse Transcriptase were added, and the mixture was kept at 42°C for 30 min. For capture of the unfinished strand, the reaction was mixed with 600 µl of Streptavidine MagneSphere Paramagnetic Particles (Promega) and eluted as recommended by the manufacturer. For the cDNA amplification step, a 2 µl aliquot from the first strand synthesis was used in the LD PCR (Clontech). Each reaction was performed with 80 µl deionized water, 10 µl 10X BD Advantage 2 PCR Buffer, 2 µl 50X dNTP Mix (10 mM of each dNTP), 4 µl 5' PCR Primer II A (12 µM), 2 µl 50X BD Advantage 2 Polymerase Mix in a 98 µl total volume. The PCR reaction consisted of 18 to 25 PCR cycles at 95°C for 15 s, 65°C for 30 s, 68°C for 6 min, following with a final extension at 70°C for 10 min. After comparison of fragment sizes with those of model species, fragment sizes of some cDNA libraries were improved using cDNA size fractionation. These libraries were submitted to an "agarase step" after 18 cycles PCR. Double-stranded cDNA was separated on 1% low-melting agarose gel and the DNA ladder "lane" was stained and photographed. Two size fractions (< 1.2 kb and > 1.2 kb) were excised from the unstained cDNA "lane" based on the DNA ladder "lane". cDNAs were extracted from the gel slices with agarase (Fermentas). After a gelase digestion, the cDNA was precipitated with one volume of isopropanol. The pellets were dried and suspended in ribonuclease free water. Four to five additional PCR cycles were performed in order to improve the efficiency of ligation in pGEM®-T Easy Vector. One µl of the second strand product was cloned in pGEM®-T Easy Vector Systems (Promega) and transformed by electroporation in the DH10B T1 resistant strain of Escherichia coli (Invitrogen). Last, transformation products were platted on LB-ampicillin agar plates and incubated overnight at 37°C. White colonies were picked using a Qpix 2 XT biorobot (Genetix) and stored in 384 well plates at -80°C.

RESULTS AND DISCUSSION

Our first step was to clone specific inserts in to the

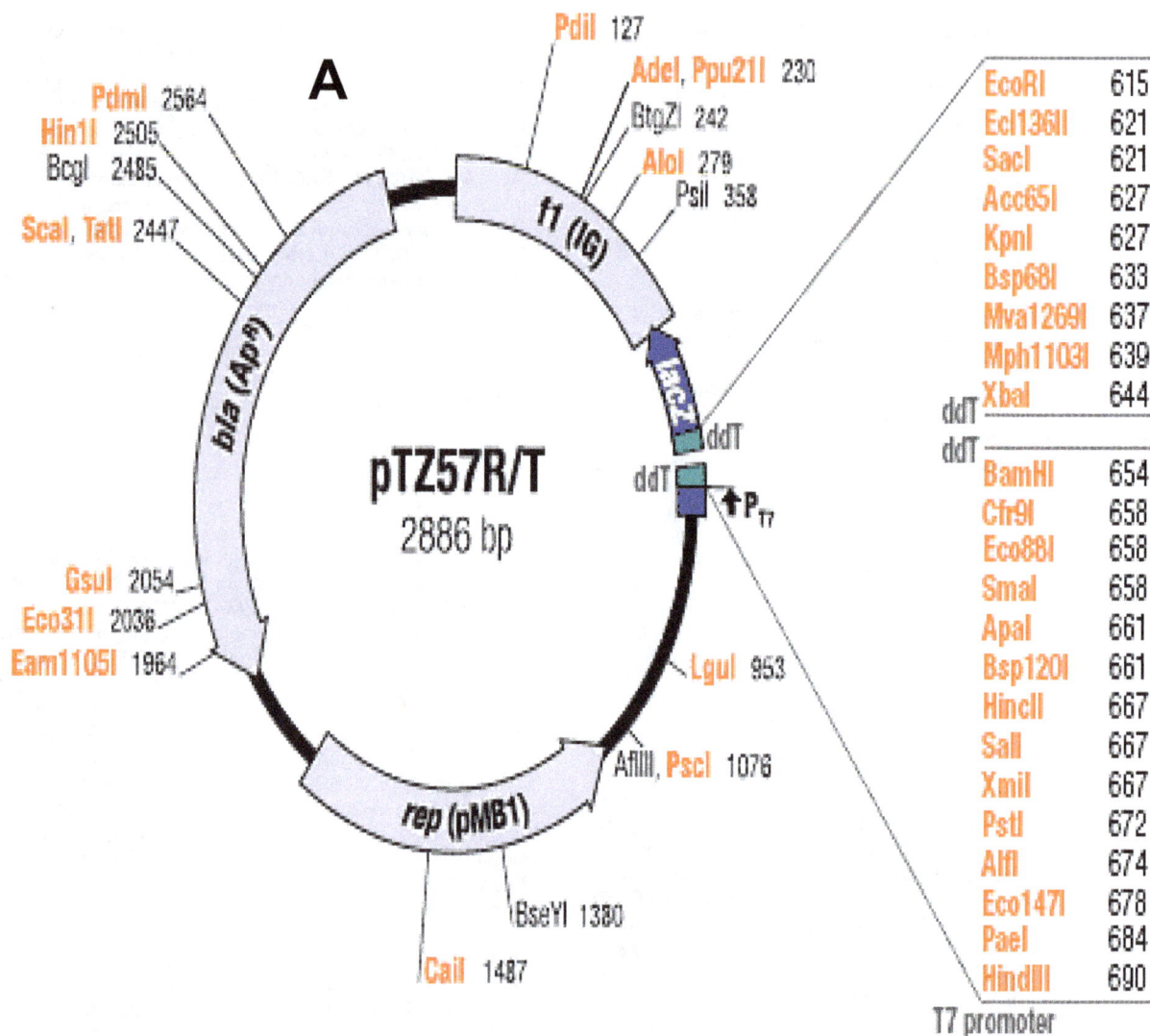

Figure 1A. Schematic diagram of the vector pTZ57R/T. This vector was used in the cloning of fragments amplified by RT-PCR.

pTZ5R/T vector. To this end, a total of 86 inserts were cloned into the pTZ57R/T vector shown in Figure 1A. They had an average size of 665.8 bp (116-1782) and a PHRED valued of ≥ 10 (Table 1). Among the analyzed sequences, most of them demonstrated significant similarity with rRNA sequences of species belonging to the genus *Phellinus* (Table 1), in agreement with other studies (Grant et al., 2006, Martinez et al., 2006). Next, as shown in Figure 1B, total RNA samples of the species *P. rimosus* were efficiently extracted. PCR of the colonies revealed that the majority of cloned fragments had a size superior to 500 bp as demonstrated in Figure 2. Importantly, these results support the robustness of methods employed in this study and also demonstrate the efficiency of our bacterial transformation.

We also found that most of the sequences had an E value near zero, indicating a high degree of similarity between the compared sequences. It also demonstrates

that alignment did not occur merely by chance. Furthermore, results shown in Table 1 demonstrate that clone F06 shares a high degree of homology (75%) with the coding gene of the small rRNA subunit of *Phellinus kawakamii,* a species that is phylogenetically related to the *Phellinus rimosus* (Larsson et al., 2006). The samples (field-collected basidiomata) were undoubtedly identified as *Phellinus rimosus,* based on morphology and comparison with well-identified vouchers in HUEFS from the same region (additional details are in materials and methods section). Furthermore, *Phellinus rimosus* exhibits host recurrence (Cesalpinia spp.) in the Caatinga vegetation of Brazilian Semi-arid region (Drescheler-Santos, 2010) and our samples were collected in this host (more specifically, Cesalpinia pyramidalis). In addition, the morphological identification of this voucher specimen was also reviewed (Drechsler-Santos et al., 2009). Surprisingly, the D05 clone showed significant

Table 1. cDNA library clones of the *Phellinus rimosus* present significant homology to sequences deposited in *GenBank*. Surprisingly, clone D05 corresponds to a RT coding gene segment of a retrotransposon providing a novel way of molecular identity for this species.

Clone	Homology	Size (Bp)	Access (Hits)	E-value
A08	*Phellinus laevigatus* small subunit ribosomal RNA gene	812	AF230363 AF465617 L47584	3e-22 9e-04 0.003
A12	*Laccaria bicolour* S238N-H82 hypothetical protein	485	XM_001880189 XM_001830397 XM_002475341	4e-14 1e-13 5e-12
B03	Uncultured soil basidiomycete ITS1, 5.8S rRNA gene and ITS2	328	FM866368 DQ873641 EF011124	0.010 0.014 0.014
C02	*Clavibacter michiganensis* subsp. Michiganensis	881	AM711867 AM849034	0.021 0.069
C07	*Aspergillus oryzae* RIB40 DNA, SC026	629	AP007159	0.68
C10	*Coprinopsis cinerea* okayama7#130 hypothetical protein	929	XM_001829195 XM_002395956 XM_770463	1e-07 1e-06 1e-04
D05	*Zingiber officinale* retrotransposon putative reverse transcriptase (RT) gene	585	Q983234	9e-33
D10	*Phellinus laevigatus* small subunit ribosomal RNA gene	962	AF230363	0.51
E11	*Phellinus laevigatus* small subunit ribosomal RNA gene	562	AF230363 EF204913 FJ591062	9e-11 6e-09 5e-07
F02	*Coprinopsis cinerea* okayama7#130 hypothetical protein	558	XM_001834637 XM_001883187	2e-17 1e-15
F06	*Phellinus kawakamii* 25S large subunit ribosomal RNA gene	536	AY059028 AM269847 AM269856	3e-25 4e-25 4e-25

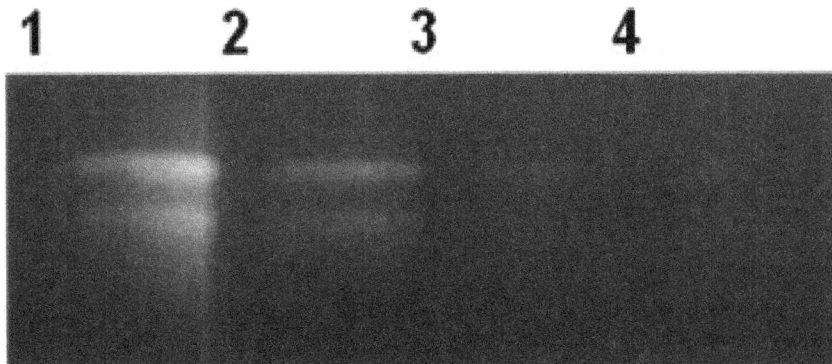

Figure 1B. Effective purification of RNA samples after treatment with DNase I. The first 2 lanes (Lanes 1 and 2) correspond to the first elution of the column, whiles Lanes 3 and 4 to the second elution column.

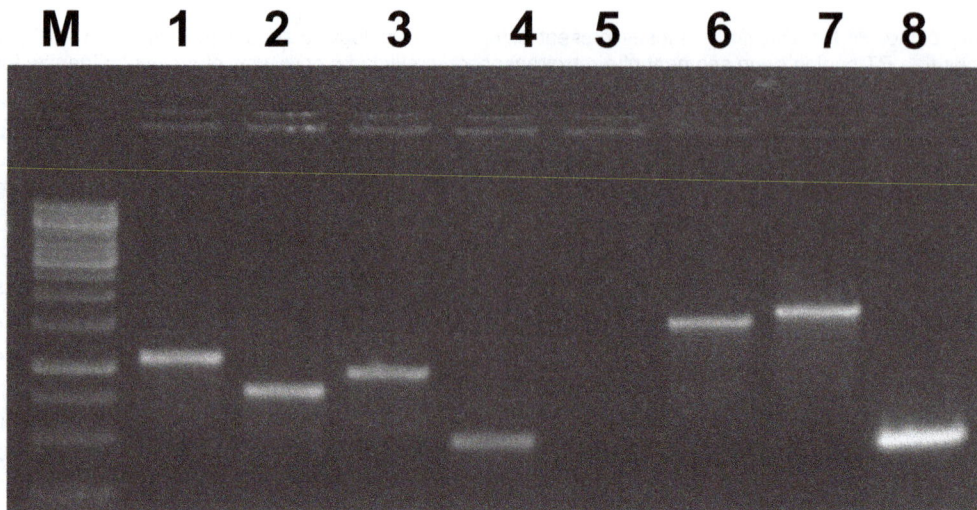

Figure 2. Robust amplification of cDNA molecules. Molecules obtained from colonies of the *P. rimosus* library were of different molecular weights, with most fragments having molecular weights equal or superior to 500 bp (Lanes 1-8= colonies; M = molecular weight marker 1 Kb).

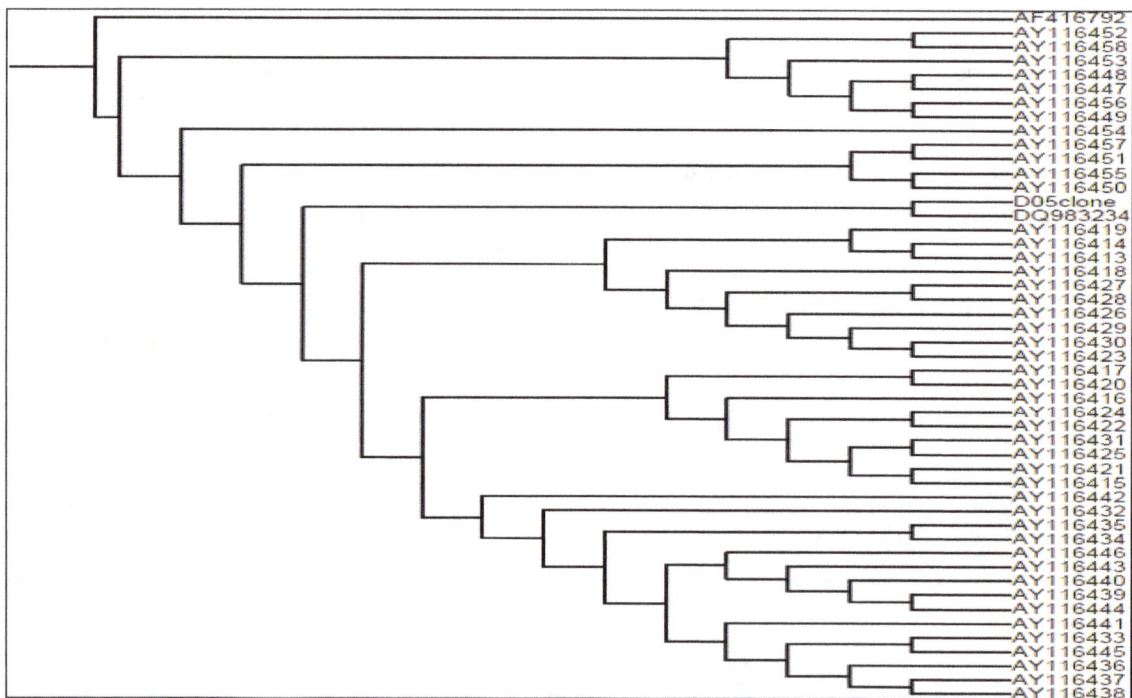

Figure 3. Phylogenetic tree based on sequences of Diez et al. 2003, D05 clone and *Zingiber officinale* DQ983234 (least squares distance matrix method, outgroup with Ty1copia RT retrotransposon of *Musa* sp. AAB Group clone 3B, AF416792).

homology with sequences of Reverse Transcriptase (RT) of plants (*Zingiber officinale* DQ983234, *Cicer arietinum* AJ535749 and *Beta nana* AJ489197). This clone corresponds to a RT coding gene segment of a retrotransposon, previously described by Díez et al. (2003), which is a new RT class (that is, *Ty1/copia*), characteristic of the Basidiomycota, and related to the RT type *Ty1/copia* of plants and not of fungi. To further

confirm the identification of this clone we conducted a phylogenetic analysis (least squares distance matrix method, outgroup with Ty1copia RT retrotransposon of *Musa* sp. AAB Group clone 3B AF416792) using the sequences in Díez et al. (2003) that led to the same information from our molecular biology studies (Figure 3 and Table 2). To the best of our knowledge, this work is the first report to demonstrate the existence of a type

Table 2. Correspondence between codes in phylogenetic tree (Figure 3) and sampled taxa.

Code	Taxon
D05 clone	*Phellinus rimosus*
AY116413	*Pisolithus microcarpus*
AY116414	*Pisolithus microcarpus*
AY116429	*Laccaria bicolour*
AY116430	*Laccaria bicolour*
AY116428	*Laccaria bicolour*
AY116423	*Pisolithus microcarpus*
AY116426	*Laccaria bicolour*
AY116427	*Laccaria bicolour*
AY116419	*Pisolithus microcarpus*
AY116418	*Pisolithus microcarpus*
AY116415	*Pisolithus microcarpus*
AY116421	*Pisolithus microcarpus*
AY116422	*Pisolithus microcarpus*
AY116424	*Pisolithus microcarpus*
AY116420	*Pisolithus microcarpus*
AY116445	*Pisolithus* sp
AY116433	*Pisolithus* sp
AY116434	*Pisolithus* sp
AY116436	Pisolithus sp
AY116438	*Pisolithus* sp
AY116437	*Pisolithus* sp
AY116416	*Pisolithus microcarpus*
AY116425	*Pisolithus microcarpus*
AY116431	*Laccaria bicolour*
AY116417	*Pisolithus microcarpus*
AY116442	*Pisolithus* sp
AY116441	*Pisolithus* sp
AY116443	*Pisolithus* sp
AY116440	*Pisolithus* sp
AY116435	*Pisolithus* sp
AY116432	*Pisolithus* sp
AY116444	*Pisolithus* sp
AY116446	*Pisolithus* sp
AY116439	*Pisolithus* sp
AY116447	*Eucalyptus globulus*
AY116449	*Eucalyptus globulus*
AY116456	*Eucalyptus globulus*
AY116448	*Eucalyptus globulus*
AY116453	*Eucalyptus globulus*
AY116458	*Eucalyptus globulus*
AY116452	*Eucalyptus globulus*
AY116450	*Eucalyptus globulus*
AY116455	*Eucalyptus globulus*
AY116451	*Eucalyptus globulus*
AY116454	*Eucalyptus globulus*
AY116457	*Eucalyptus globulus*
AF416792	*Musa* sp. AAB Group clone 3B (OUTGROUP)

Ty1-copia retrotransposon, identified through its RT characteristic in the *P. rimosus* species of the order Hymenochaetales of Basidiomycota. This is very important since the Ty1-copia is normally found in yeasts and plants and not in fungi. Thus, our novel finding allows for the application of such knowledge as a molecular marker (Díez et al., 2003) for the genetic studies of populations of this species, an important species that is a parasite of typical trees of the Brazilian semi-arid region and belonging to a taxonomic complex of very difficult delimitation and interpretation (Drechsler-Santos et al., 2010). Transposons are DNA segments characterized by their ability to move inside the genome. They can cause significant impact in the chromossomal organization, gene expression patterns, as well as induce somaclonal variations in fungi and plants (Daboussi 1997; Hirochika et al., 1996). Retrotransposons move through an intermediary RNA by utilizing RT for reverse transcription and insertion of cDNA into the genome (Daboussi, 1997). LTR retrotransposons are further characterized by two repeated terminals and two ORFs that are similar to the retroviruses genes *gag* and *pol*, which are subdivided into two main categories: *Ty1-copia e Ty3-gypsy* (Daboussi, 1997). *Ty1-copia* elements occur mainly in yeasts and mycelial Ascomycota, but apparently, are very rare in Basidiomycota, having only been described in very small number of species (Díez et al., 2003).

Therefore, our exciting findings provide a way of establishing a molecular identity for the *Phellinus rimosus*, a tree parasite species of high significance for the Brazilian semi-arid region, particularly the Northeast Caatinga. The presence of RT that is characteristic of plants, and the presence of a *Ty-1 copia* transposon in the *Phellinus rimosus* might further suggest that this species could have potential biotechnology applications as a bioconversor in the area of wood delignification. Additional studies will now be required to fully test the utilization of *Phellinus rimosus* as a tool in wood delignification for reduction of the damaging effects to the environment of such process.

ACKNOWLEDGEMENTS

We would like to thank all the people that contributed directly or indirectly to this work and, especially the Graduate Program in Biotechnology (PPGBiotec UEFS, http://www2.uefs.br/ppgbiotec) and Fundação de Amparo à Pesquisa do Estado da Bahia (FAPESB,http://www.fapesb.ba.gov). Support to Dr. Marco Brotto was provided by the Conselho Nacional de Desenvolvimento Científicoe Tecnológico, Brasil (CNPq, Grant 314605-2009-6; 12/2009-01/2011) and Fundação de Amparo à Pesquisa da Bahia-FAPESBA (02/2012-02/2013) through a Visiting Professorship and Scholar position at the Faculdade de Tecnologia e Ciências de Salvador – FTC (www.ftc.br).

REFERENCES

Azevedo CPL, Guerrero RT (1993). Estudo biossistemático de espécies do gênero Hymenochaete (Basidiomycetes) no Rio Grande do Sul. Insula, 22: 143-176.

Blanchette RA (1991). Delignification by wood-decay fungi. Annu. Rev. Phytopathol, 29: 381-403.

Coehn R, Persky L, Hadar Y (2002). Biotechnological application and potential of wood-degrading mushrooms of the genus Pleurotus. Appl. Microbiol. Biotechnol., 58: 582-594.

Daboussi MJ (1997). Fungal transposable elements and genome evolution. Genetica, 100: 253–260.

Díez J, Beguiristain T, Le Tacon F, Casacuberta JM, Tagu D (2003). Identification of Ty1-copia retrotransposons in three ectomycorrhizal basidiomycetes: evolutionary relationships and use as molecular markers. Curr. Genet., 43: 34–44.

Drechsler-Santos ER, Gibertoni TB, Góes-Neto, A, Cavalcanti MAQ (2009). A re-evaluation of the lignocellulolytic Agaricomycetes from the Brazilian semi-arid region. Mycotaxon, 108: 241-244.

Drechsler-Santos ER, Groposo C, Loguercio-Leite C (2008). New records of lignocellulytic fungi (Basyidiomycetes) from the Atlantic Rain Forest in State of Santa Catarina, Brazil. Hoehnea, 35: 57-61.

Drechsler-Santos ER, Santos PJP, Gibertoni TB, Cavalcanti MAQ (2010). Ecological aspects of Hymenochaetaceae in an area of Caatinga (semi-arid) in Northeast Brazil. Fungal Diversity, 42: 71-78.

Drechsler-Santos ER, Santos PJP, Gibertoni TB, Cavalcanti MAQ (2010). Ecological aspects of Hymenochaetaceae in an area of Caatinga (semi-arid) in Northeast Brazil. Fungal Diversity, 42: 71-78

Góes-Neto A, Groposo C. 2005. Hymenochaetales. Hymenochaetoid clade. Hymenochaetaceae. Version 10 June 2005. http://tolweb.org/Hymenochaetoid_clade/20547/2005.06.10 in The Tree of Life Web Project, http://tolweb.org/ (Visited on August 26, 2010).

Grant S, Grant WD, Cowan DA, Jones BE, Ma Y, Ventosa A, Heaphy S (2006). Identification of eukaryotic open reading frames in metagenomic cDNA libraries made from environmental samples. Appl. Environ. Microbiol., 72: 135-143.

Hirochika H, Sugimoto K, Otsuki Y, Tsugawa H, Kauda M (1996). Retrotransposons of rice involved in mutations induced by tissue culture. Proc Natl Acad. Sci. USA, 93: 7783–7788.

Johnson M, Zaretskaya, I, Raytselis Y, Merezhuk Y, McGinnis S, Madden, TL (2008). "NCBI BLAST: a better web interface" Nucleic Acids Res., 36: W5-W9.

Larsson KH, Parmasto E, Fischer M, Langer E, Nakasone KK, Redhead AS (2006). Hymenochaetales: a molecular phylogeny for the hymenochaetoid clade. Mycologia, 98: 926-936.

Martinez D, Larrondo LF, Putnam N (2004). Genome sequence of the lignocellulose degrading fungus Phanerochaete chrysosporium strain RP78. Nature Biotechnol., 22: 695-700.

Martinez RJ, Mills HJ, Story S, Sobercky PA (2006). Prokaryotic diversity and metabolically active microbial population in sediments from an active mud volcano in the Gulf of Mexico. Environ. Microbiol., 8: 1783-1796.

Mayer AM, Staples RC (2002). Laccase: new functions for an old enzyme. Phytochemistry, 60: 551-565.

Ryvarden L (1991). Genera of Polypores-Nomenclature and Taxonomy. Synopsis Fungorum 5, Fungiflora, Oslo.

28

Growth and *in vitro* phosphate solubilising ability of *Scleroderma sinnamariense*: A tropical mycorrhiza fungus isolated from *Gnetum africanum* ectomycorrhiza root tips

Eneke Esoeyang Tambe Bechem

Department of Plant and Animal Sciences, Faculty of Science, University of Buea, P. O. Box 63 Buea, Cameroon.
E. mail: tamenekeso@yahoo.co.uk.

The growth and ability of *Scleroderma sinnamariense* isolated from *Gnetum africanum* ectomycorrhiza root tips to solubilise calcium tetrahydrogen diorthophosphate, calcium phytate, hydroxyapatite and amorphous iron phosphate was assessed in axenic culture under a range of conditions. *S. sinnamariense* grew on all P sources when NH_4^+ was sole nitrogen source, but failed to grow on amorphous iron phosphate in unbuffered media with peptone as sole nitrogen source. The fungus solubilised calcium tetrahydrogen diorthophosphate and hydroxyapatite in unbuffered media with NH_4^+ as nitrogen source. The internal phosphorus status of the mycelia had no influence on the ability of the fungus to solubilise the substrates. The data confirmed that mycorrhizal fungi can access P sources which are normally unavailable and may pass on the absorbed P to the plant partner in case of symbiosis.

Key words: Phosphorus solubilisation, *Scleroderma sinnamariense,* mycorrhiza fungus.

INTRODUCTION

Phosphorus (P) is the most likely element to limit plant growth, next to nitrogen. This is because a large proportion of P in the soil is poorly soluble and most of the P is rapidly fixed in soil into fractions that are unavailable to plant roots (Sanyal and De Datta, 1991). Phosphorus is present in soil in many different physical and chemical forms, which include fluorapatite, hydroxyapatite, oxyapatite, chlorapatite, wavellite, monetite and inositol phosphates (Lapeyrie et al., 1991). In most acid soils like the type found in Cameroon, most P is deposited as various forms of amorphous and crystalline iron and aluminium phosphates (Sanyal and De Datta, 1991). Plants have developed differing physiological mechanisms and strategies for acquiring P from soil, with mycorrhiza formation being one of the most widespread ways of overcoming P deficiency by land plants. There is increasing evidence that mycorrhizal plants may be able to use soil P fractions held in specific physico-chemical forms that are otherwise unavailable to uncolonised plants (Koide and Kabir, 2000; Smith et al., 2000; Tibbett and Sanders, 2002; Smith and Read,

2008). This may be achieved by mycelia extending beyond the depletion zone, thereby increasing surface area for absorption. It may also be possible via the production of acids, which solubilize the phosphate making it available for absorption.

In Cameroon *Gnetum* spp. grow principally in tropical rain forests in mixed communities supported by acid oxisols and ultisols characterised by low soluble inorganic P. This leads to competition for limited nutrients such as P. Transformation of P derived from organic matter would be paramount in the P nutrition of such plants, as well as in nutrient cycling in such ecosystems (Johnson et al., 2003). Some researchers have suggested that the release of organic acids from EM fungi may solubilise recalcitrant mineral phosphates (Leyval and Berthelin, 1986; Lapeyrie et al., 1991). Others believe that before insoluble organic P can be broken down by fungal phosphatases and the resulting phosphate absorbed by the roots, they would firstly have to be solubilised (Lapeyrie et al., 1991). There is therefore a necessity to evaluate the solubilisation abilities of fungi *in vitro*.

The main objective of the work reported here was to evaluate the ability of *Scleroderma sinnamariense* to solubilise P and grow in the presence of different sparingly soluble phosphorus sources with either peptone or ammonium as the N source. The effect of buffering on growth and solubilisation was also assessed. *Pisolithus tinctorius* was included for comparison.

MATERIALS AND METHODS

The method used here was reported by Lapeyrie et al. (1991) and by Van Leerdam et al. (2001).

Fungal culture

S. sinnamariense was isolated from *Gnetum* roots and identified using molecular techniques as described in Bechem (2004). One set of starter cultures for this experiment had been grown for 30 days on modified Melin-Nokran (MMN) agar with ammonium-N; another set was grown on MMN agar with peptone-N, whilst the last set was grown on MMN agar with ammonium-N in absence of a phosphorus source. This was to check for the influence of internal mycelia P status and N source on solubilisation.

Basal medium

The basal medium consisted of $MgSO_4.7H_2O$ 140 mg/L, $CaCl_2$ 50 mg/L, $ZnSO_4$ 3 mg/L, FeEDTA mg/L, thiamine HCl 10 µg/L and agar 10 g/L, as modified in this study from Marx (1969).

Nitrogen sources

The nitrogen sources were $(NH_4)_2SO_4$ and peptone. When $(NH_4)_2SO_4$ was added at 0.284 g/L, glucose was 3.004 g/L. Peptone was added at 0.380 g/L with a corresponding glucose amount of 2.100 g/L. This gave a C:N ratio of approximately 20:1 in both cases.

P sources

The P sources used were calcium tetra hydrogen di-orthophosphate $(CaH_4(PO_4)_2)$ (BDH®), calcium phytate $(C_6H_6Ca_6O_{24}P_6)$ (Sigma®), hydroxyapatite $(Ca_5(PO_4)_3OH)$ (Sigma®) and amorphous iron phosphate. Hereafter, these P sources would be referred to as $CaPO_4$, phytate, apatite and $FePO_4$ respectively. These P sources were chosen because most inorganic P in soils occurred in these forms (Richardson, 2001) with iron and aluminium phosphates being predominant in tropical acid soils (Sanyal and De Datta, 1991; Bolan, 1991). The preparation of the amorphous iron phosphate was based on the method first reported by Carter and Hartshorne (1923) and subsequently used by Cate et al. (1959) and Bolan et al. (1987) for the preparation of amorphous and eventually crystalline iron phosphate.

In the preparation process, 22.5 g of carbonyl iron powder was dissolved in reagent-grade orthophosphoric acid (270 ml of 50% solution). The solution was diluted to 600 ml, filtered and the iron content oxidised by addition of 30% hydrogen peroxide. Diluting the solution to 25 L precipitated amorphous ferric phosphate. The precipitate was washed several times using deionised water, followed by decantation and vacuum filtration between successive washings. The appearance of faint brownish-yellow coloration on the precipitate was assumed to indicate that the phosphorus-iron ratio had attained unity (Carter and Hartshorne, 1923). The precipitate was then allowed to dry at room temperature before storage in a dessicator. The resulting powder was used in the assay as amorphous iron phosphate.

Preparation of P sources, culture media and inoculation

Distilled water was added to 5 g of each P source and autoclaved for 15 min at 121 °C. Following cooling, contents were filtered through a 0.45 µm Millipore filter followed by several washes in sterile distilled water. This was done under sterile conditions. For each N and P source, one batch was buffered with 2- (N-morpholino) ethane-sulphonic acid (MES) at 50 mM (Van Leerdam et al., 2001), whilst the other was unbuffered. This was to check for the effect of pH on solubilisation. The pH of the medium was adjusted to 5.5 prior to autoclaving at 121 °C for 15 min. Peptone was filter sterilised and added to an already autoclaved medium. Basal medium was then poured into a 9 cm diam. Petri dish to give a thin layer. Each previously autoclaved P sources was then added to a portion of basal medium resulting in a suspension, which was poured into the plates on top of the first layer of agar and allowed to solidify. Each dish was inoculated with a 5 mm diam fungal plug cut from the edge of an actively growing mycelial mat. Each phosphate, nitrogen and buffer treatment was done in three replicates for each inorganic phosphorus (Pi) starved or Pi sufficient isolate. Incubation was at 30 °C in the dark. The mycelial, media, P source and buffer combinations used in the assay are shown in Table 1.

Plates were observed visually every 10 days for zones of clearing around the mycelial mat. Plates, which showed no zones of clearing, were observed using a dissecting and compound microscope to look for zones of clearing which might be smaller than the fungal colony. The diameter of fungal colony and that of zone of clearing was measured by taking the longest axis of fungal colony and any cleared ring at the end of the experiment. The experiment ran for 23 days; however, the plates were checked again after 60 days of incubation to see if there was any change in solubilisation with time.

Statistical analysis

Data were analysed using the General Linear Model of Minitab 13. Three-way ANOVA was used to test for the effect of P sources ($CaPO_4$, $FePO_4$, Apatite), internal mycelia P status and buffer on colony diameter (growth). A two-way ANOVA to evaluate the effect of inoculum P status and P sources ($CaPO_4$, $FePO_4$, Apatite, phytate) on colony diameter in absence of buffer was also run.

RESULTS

Growth of fungi

Generally, colony diameter ranged from 5 mm to a maximum of 71 mm. Growth was considered to have taken place when the colony diameter at the end of the experiment was greater than 5 mm.

Growth of fungi on peptone-N

P sufficient *Sclerodema* showed growth on all P sources except on $FePO_4$ in unbuffered media (Figure 1a). The

Table 1. Inoculum status, assay medium, P source and buffer combinations used in the solubilisation experiment.

Internal mycelial N/P status	Media N/P status	P source	Presence/absence of buffer
Peptone + P	Peptone + P	$CaPO_4$	MES
Peptone + P	Peptone + P	$CaPO_4$	No MES
Peptone + P	Peptone + P	Phytate	MES
Peptone + P	Peptone + P	Apatite	MES
Peptone + P	Peptone + P	Apatite	No MES
Peptone + P	Peptone + P	$FePO_4$	MES
Peptone + P	Peptone + P	$FePO_4$	No MES
Ammonium + P	Ammonium + P	$CaPO_4$	MES
Ammonium + P	Ammonium + P	$CaPO_4$	No MES
Ammonium + P	Ammonium + P	Phytate	MES
Ammonium + P	Ammonium + P	Apatite	MES
Ammonium + P	Ammonium + P	Apatite	No MES
Ammonium + P	Ammonium + P	$FePO_4$	MES
Ammonium + P	Ammonium + P	$FePO_4$	No MES
Ammonium - P	Ammonium + P	$CaPO_4$	MES
Ammonium - P	Ammonium + P	$CaPO_4$	No MES
Ammonium - P	Ammonium + P	Phytate	MES
Ammonium - P	Ammonium + P	Apatite	MES
Ammonium - P	Ammonium + P	Apatite	No MES
Ammonium - P	Ammonium + P	$FePO_4$	MES
Ammonium - P	Ammonium + P	$FePO_4$	No MES

source of medium P, the presence of buffer and the interaction of these two factors had significant effects ($P<0.001$) on colony diameter (Table 2). Maximum growth of 40 mm was observed on $FePO_4$ in buffered conditions. Growth of fungus in buffered medium was superior to that observed in unbuffered medium. The difference was greatest for $FePO_4$ (Figure 1a). Growth of fungus on phytate in unbuffered medium was comparable to that on $CaPO_4$ and apatite but better than that on $FePO_4$ (Figure 1a). P sufficient *Pisolithus* also showed growth on all P sources in both buffered and unbuffered media (Figure 1b) except when grown on Ca phytate in unbuffered medium.

The source of P had a significant effect ($P<0.001$) on growth. Maximum growth of 62 mm was observed following growth of fungus on $CaPO_4$ in buffered medium. Growth of the fungus on $CaPO_4$ and $FePO_4$ in buffered medium was superior to that in unbuffered medium on these same P sources. However, growth on apatite in unbuffered medium was superior to that observed in buffered medium (Figure 1b). The effect on growth of the interaction between the buffer and P source was also significant at $P<0.05$. Growth of the fungus on $CaPO_4$ was comparable to that in apatite in unbuffered medium but superior to that observed on $FePO_4$ and phytate (Figure 1b). A summary of the statistical analysis of data from this experiment is presented on Table 2.

Growth of fungi on ammonium-N

Scleroderma

The fungus grew on all P sources but growth on $CaPO_4$ and apatite was better than growth on $FePO_4$ (Figure 2a). P sufficient inoculum grew better in comparison to P starved inoculum on $CaPO_4$ and apatite but not on $FePO_4$. Growth on buffered medium by the fungus was better than growth on medium devoid of buffer and the difference was greatest for $FePO_4$ (Figure 2a). Growth on phytate in unbuffered medium was comparable to that on apatite and $CaPO_4$ and better than that on $FePO_4$ (Figure 2b).

Pisolithus

The fungus grew on all P sources except when P sufficient fungus was assayed on $FePO_4$. P sufficient inoculum showed better growth in comparison to P starved inoculum but the difference was not significant (Table 3). Growth of the fungus on apatite in unbuffered medium was superior to growth in buffered medium whereas on $CaPO_4$, growth in buffered medium was greater than that in unbuffered medium (Figure 3a). The fungus was unable to grow on phytate in unbuffered medium, an observation, which was similar to that of P

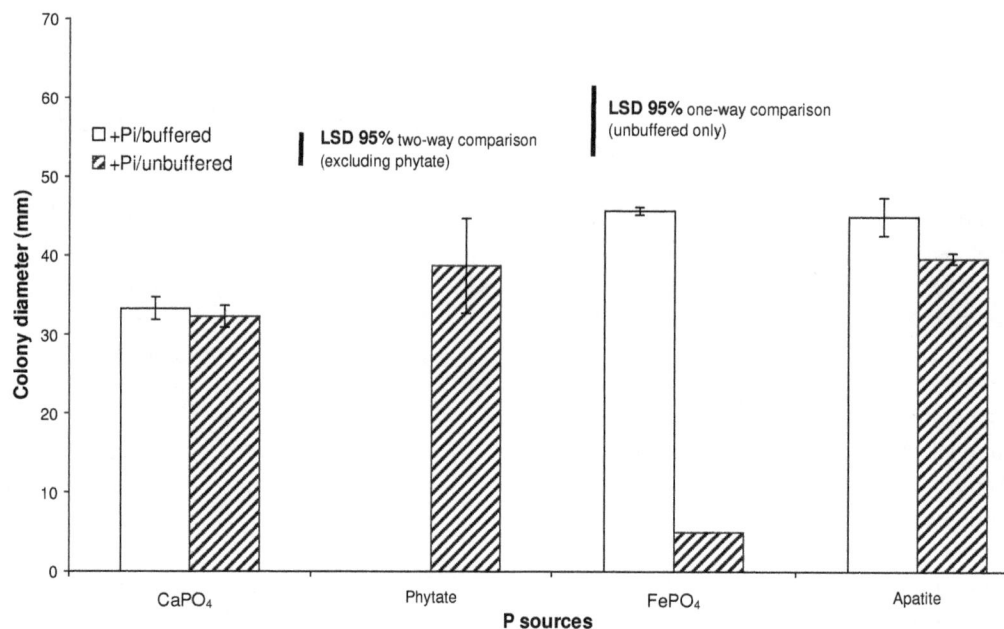

a

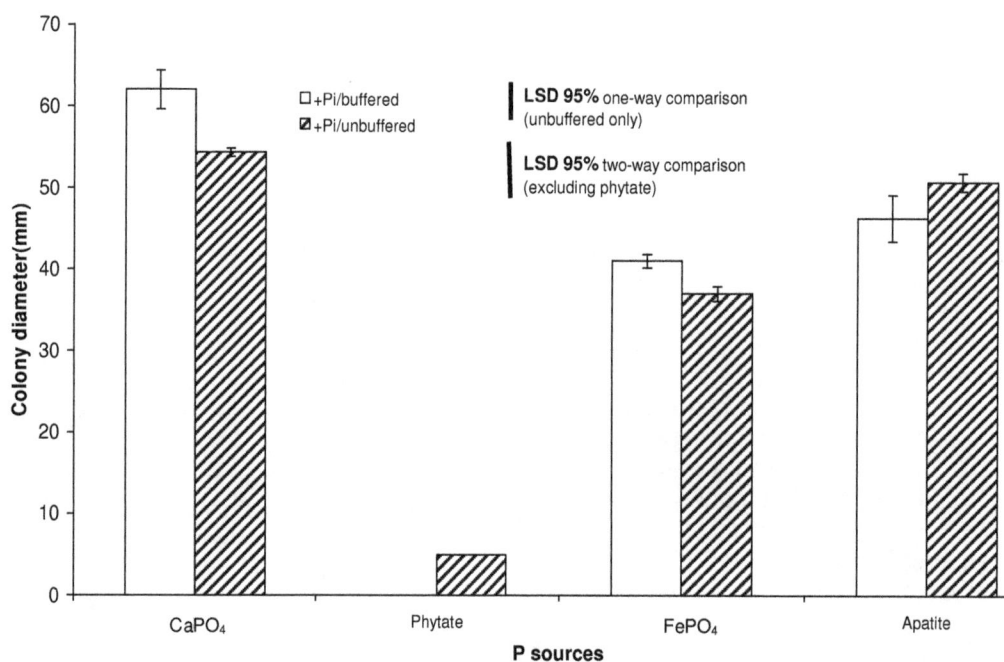

b

Figure 1. Growth (mm diameter) variation of *S. sinnamariense* (a) and *P. tinctorius* (b) on different sparingly soluble P sources with peptone-N in the presence or absence of a buffer. The + Pi in the legend represents the internal Pi status of mycelia inoculum. Each value is a mean of three replicates. Vertical bars represent standard error of the mean. CaPO₄, Phytate, FePO₄ and Apatite on the X-axis represent Calcium tetrahydrogen diorthophosphate, calcium phytate, amorphous iron phosphate and hydroxyapatite respectively.

sufficient fungus on FePO₄ (Figure 3b). P starved fungus showed a growth pattern on FePO₄ which was unusual (Figure 4). A summary of the statistical analysis of data are shown in Tables 3 and 4.

Solubilisation of sparingly soluble P sources

The results of solubilisation by both *Scleroderma* and *Pisolithus* of the different P sources and in the presence

Table 2. Summary of two-way analysis of variance of the effects of buffer and P source on growth of Scleroderma and Pisolithus on MMN containing peptone-N. Starter inoculum was P sufficient.

Species	Source of variation	DF	F	P
Scleroderma	Buffer	1	80.28	* * *
	P sources	2	52.97	* * *
	Buffer* P source	2	113.40	* * *
Pisolithus	Buffer	1	1.88	ns
	P sources	2	38.45	* * *
	Buffer* P source	2	3.96	*

n = 3, *P<0.05, ** P<0.01, *** P<0.001. ns: not significant. P sources are $CaPO_4$, $FePO_4$, Apatite.

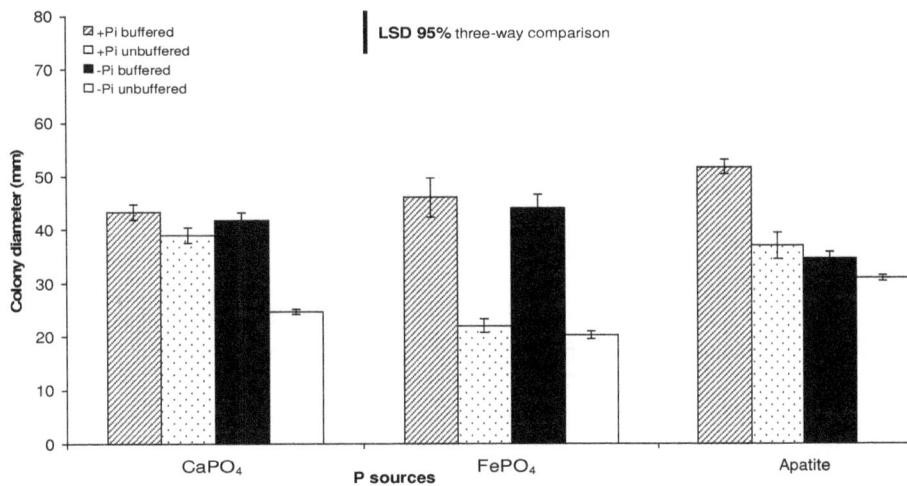

a

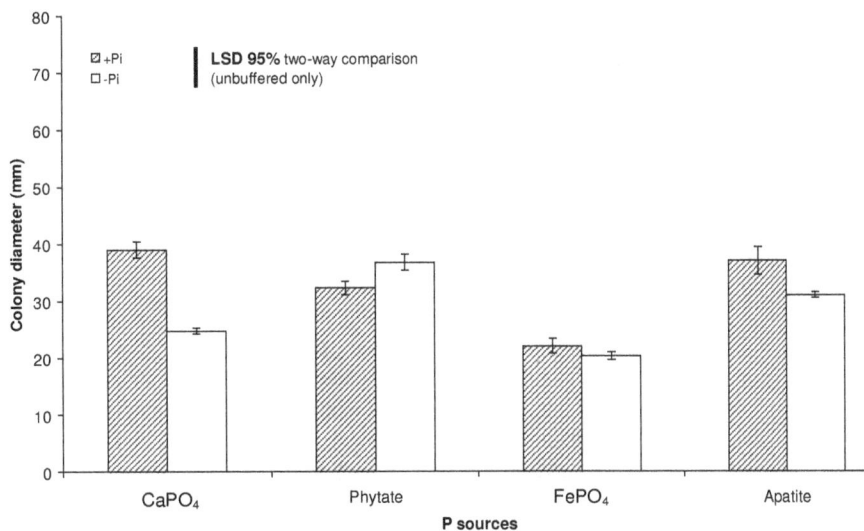

b

Figure 2 Growth of Scleroderma on different sparingly soluble P sources in MMN containing ammonium-N. (a) Growth on buffered and unbuffered medium. (b) Growth on unbuffered medium only. Each value is a mean of three replicates. Vertical bars represent standard error of mean. P in legend represents the internal P status of inoculum. CaP04, Phytate, FePO4 and Apatite on the X-axis represent Calcium tetrahydrogen diorthophosphate, calcium phytate, amorphous iron phosphate and hydroxyapatite respectively.

Table 3. Summary of three-way analysis of variance of the effects of inoculum P status, presence or absence of buffer in growth medium and source of inorganic P, on colony diameter following growth of *Scleroderma* and *Pisolithus* for 23 days on MMN agar medium containing sparingly soluble P sources.

Species	Source of variation	DF	F	P
	P source	2	7.29	0.003**
	Inoculum	1	33.92	0.000***
	Buffer	1	142.12	0.000***
Scleroderma	Inoculum*Buffer	1	0.03	0.857[ns]
	Inoculum*P source	2	5.36	0.012*
	Buffer*P source	2	14.56	0.000***
	Inoculum*Buffer*Psource	2	7.85	0.002**
	P source	2	69.16	0.000***
	Inoculum	1	1.89	0.182[ns]
	Buffer	1	102.83	0.000***
Pisolithus	Inoculum*Buffer	1	0.00	1.000ns
	Inoculum*P source	2	9.68	0.001***
	Buffer*P source	2	81.25	0.000***
	Inoculum*Buffer*Psource	2	4.75	0.018*

P sources are $CaPO_4$, $FePO_4$, Apatite.

of NH_4^+-N are shown in Table 5. Cleared zones were observed in two cases for *Scleroderma* (Figure 5) but none was observed with *Pisolithus*. The mean diameter of solubilisation rings are shown in Figure 6. There was solubilisation of $CaPO_4$ and hydroxyapatite by both P starved and P sufficient *Scleroderma*. Solubilisation was only possible when N source of medium was ammonium and in the absence of a buffer. Maximum solubilisation ring diameter of 38 mm was recorded on hydroxyapatite, by the P starved fungus. P starved fungus demonstrated an ability to solubilise apatite and $CaPO_4$ which was superior to that shown by P sufficient fungus. This difference was greatest for apatite. Solubilisation of $CaPO_4$ by P sufficient fungus was comparable to solubilisation of apatite. When expressed as a percentage of fungal colony diameter, solubilisation by P starved fungus of hydroxyapatite was 119 and 128 for $CaPO_4$, as calculated from Figures 2 and 6 using the following equation:

Ring diameter $\div$ colony diameter $\times$ 100% -----------------(1)

Pisolithus on the other hand was unable to solubilise any of the P sources in both buffered and unbuffered conditions.

DISCUSSION

Our observations in the solubilisation experiment are similar to those in investigations using ericoid and other EM fungi species. In the study c arried out by Van

Leerdam et al. (2001) the *Hymenoscyphus* isolates only solubilised hydroxyapatite in the presence of ammonium-N and in unbuffered medium. This observation, quite similar to that of *Scleroderma* isolate in this study, is probably due to proton expulsion. The resulting pH change in the medium might then lead to the observed solubilisation. This phenomenon is confirmed by studies showing that during ammonium transport by the hyphae, there is H^+ secretion into the medium (Jennings, 1997). Some fungi like *Paxillus involutus* have been shown to produce oxalic acid during growth on nitrate-N (Lapeyrie et al., 1987). It is thought that such acid helps in P solubilisation by acidification (Jones et al., 1980). So far, no study had looked at solubilisation on N sources other than ammonium and nitrate. In an assay on growth of these *Scleroderma* and *Pisolithus* isolates on various N sources including peptone, it was observed that growth on peptone was followed by a fall in pH of medium (Bechem, unpublished). But the resulting decrease in pH was not significant enough to bring about solubilisation by *Scleroderma* of hydroxyapatite and $CaPO_4$ in peptone-N as observed on ammonium-N. Nonetheless, the fact that both fungi did grow on these P sources means there was some solubilisation to permit growth. It is probable that an unknown proportion of the insoluble P provided must have dissolved, thus supporting growth without any apparent indication of solubilisation. It is unfortunate that other similar studies such as Van Leerdam et al. (2001) did not indicate whether growth was recorded in the absence of solubilisation or not.

There might be several reasons for the inability of the *Pisolithus* isolate to grow on Ca phytate in this study. One

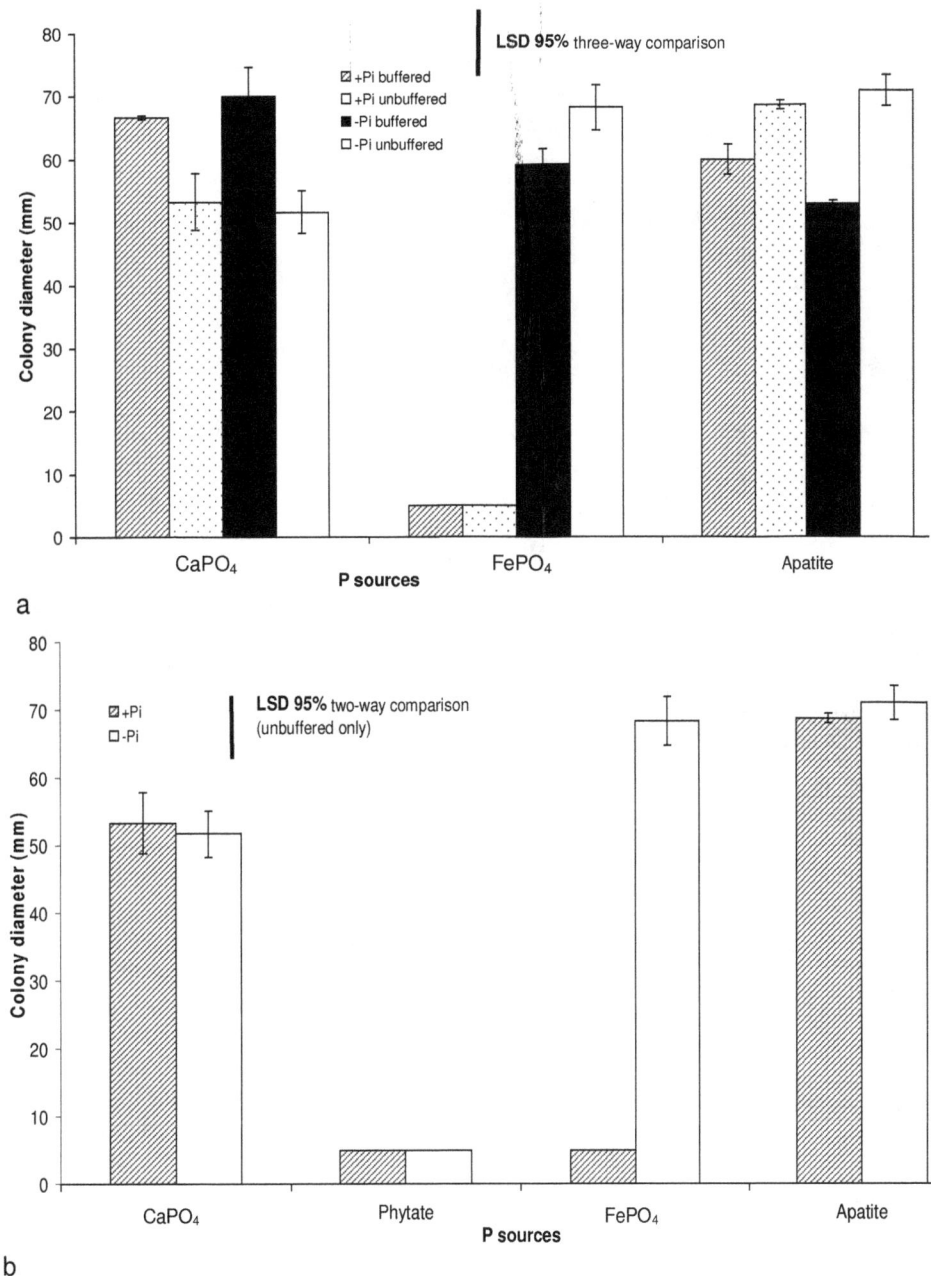

Figure 3. Growth of *Pisolithus* on different sparingly soluble P sources in MMN containing ammonium-N. (a) Growth on buffered and unbuffered medium. (b) Growth on unbuffered medium only. Each value is a mean of three replicates. Vertical bars represent standard error of mean. P in legend represents the internal P status of inoculum. $CaPO_4$, Phytate, $FePO_4$ and Apatite on the X-axis represent Calcium tetrahydrogen diorthophosphate, calcium phytate, amorphous iron phosphate and hydroxyapatite respectively.

possibility is that the amount of calcium phytate used in this experiment may have been toxic to the fungus. The *Pisolithus* isolate used in this study might have failed to produce phosphatase under the experimental conditions and therefore could not utilise the available P source. Both *Pisolithus* isolates used by Lapeyrie et al. (1991) grew on Ca phytate although only Isolate II could

solubilise this P source when ammonium-N was provided. In the present study, the absence of intracellular P might have induced changes in the type of compounds exuded by the *Pisolithus* isolate. This alteration may be the reason for the observed strange growth pattern of the P starved fungus on $FePO_4$. Such an observation suggested specific selective synthesis

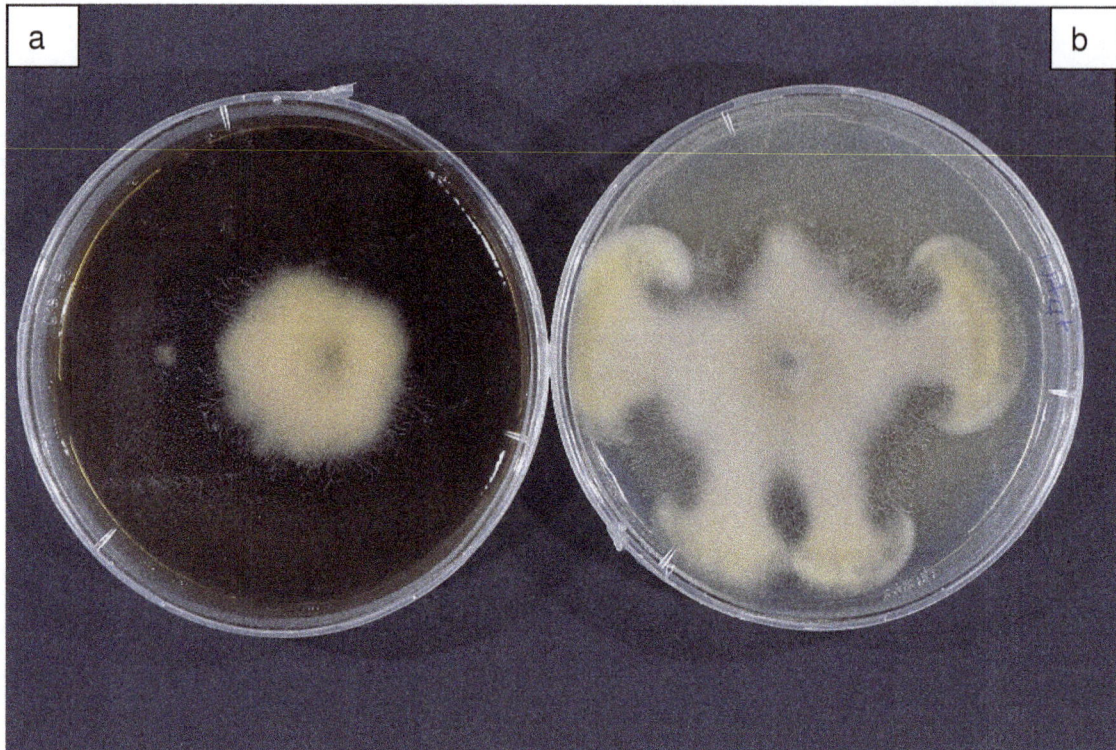

Figure 4. Growth of *Pisolithus* on nutrient agar. (a) Amorphous FePO$_4$ as opposed to the commonly observed radial growth pattern on MMN nutrient agar and (b) shows the unusual pattern observed with growth of fungus on MMN agar (minus potassium dihydrogen orthophosphate) containing ammonium-N.

Table 4. Summary of two-way analysis of variance of the effects of inoculum P status and source of inorganic P, on colony diameter following growth of Scleroderma and Pisolithus for 23 days on MMN agar medium containing sparingly soluble P sources in absence of buffer.

Species	Source of variation	DF	F	P
Scleroderma	P source	3	34.48	0.000***
	Inoculum	1	17.23	0.000***
	Inoculum*P source	3	13.62	0.000***
Pisolithus	P source	3	156.23	0.000***
	Inoculum	1	52.67	0.000***
	Inoculum*P source	3	51.35	0.000***

P sources are CaPO$_4$, FePO$_4$, Apatite, Phytate; ns: not significant; *** $P<0.001$, ** $P<0.01$, * $P<0.05$.

and exudation of some compounds enabling the fungus to grow on FePO$_4$. The inability of *Scleroderma* and *Pisolithus* to solubilise some of the P sources assayed does not necessarily imply that they could not solubilise these P sources under natural conditions, with the help of some bacteria. Villegas and Fortin (2002) showed that the extramatrical mycelium of *Glomus intraradices* could solubilise sparingly soluble calcium phosphate *in vitro* only when grown in association with *Pseudomonas*

aeroginosa and *Pseudomonas putida*. These bacteria were also unable to solubilise this P source without the help of the fungus. There is therefore a possibility that in association with some mycorrhiza helping bacteria, many more P sources could be accessed *in situ*.

It was also interesting to note that solubilisation of a P source did not necessarily mean better growth on that P source. In this study, it was noted within the same P source that, in situations where solubilisation occurred, the fungal growth diameter was inferior to that observed in the absence of solubilisation. A probable reason for such an observation may be that the fall in pH which is thought to have assisted in the solubilisation process demonstrated adverse effects on the fungal culture, causing a reduction in growth in comparison to the situation where solubilisation did not occur. Alternatively, the observed solubilisation of the P sources led to an increase in the amount of available P in the growth medium. The attained levels may eventually become toxic to the culture. Lower pH values are known to cause hydroxyapatite to dissolve and the protons react with the hydroxyl ion, as well as the phosphate groups in the lattice according to the mass balance as illustrated in Van Leerdam et al. (2001):

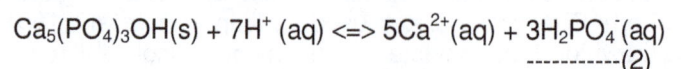

$$Ca_5(PO_4)_3OH(s) + 7H^+ (aq) \iff 5Ca^{2+}(aq) + 3H_2PO_4^-(aq)$$
$$----------(2)$$

Table 5. Abilities of *Scleroderma* and *Pisolithus* to solubilise some organic and inorganic P sources in buffered and unbuffered media with ammonium as sole nitrogen source, after 23 days of growth.

Assay media		Inoculum pretreatment			
		Scleroderma sinnamariense		*Pisolithus tinctorius*	
		NH$_4^+$-N			
		+ Pi	- Pi	+ Pi	- Pi
NH$_4^+$-N CaH$_4$(PO$_4$)$_2$	Buffered	-	-	-	-
	Unbuffered	+	+	-	-
NH$_4^+$-N Ca phytate	Buffered	-	-	-	-
NH$_4^+$-N FePO$_4$	Buffered	-	-	-	-
	Unbuffered	-	-	-	-
NH$_4^+$-N hydroxyapatite	Buffered	-	-	-	-
	Unbuffered	+	+	-	-

+ Indicates zone of clearing around colony, - indicates no zone of clearing around colony. Consistent results were obtained in three replicate plates for each isolate.

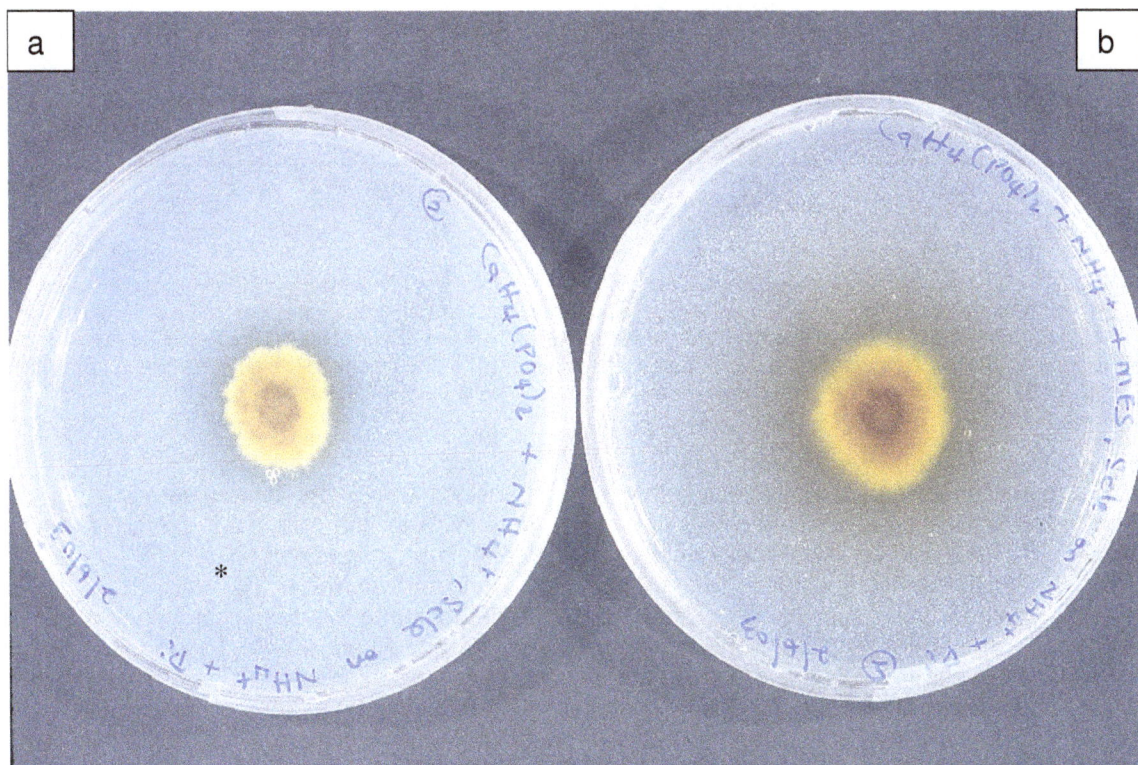

Figure 5. Growth and solubilisation of CaPO$_4$ in MMN agar medium by *Scleroderma*. (a). Solubilisation in unbuffered media. Asterik shows zone of clearing. (b). Inability of fungus to solubilise the same P source in presence of buffer.

Organic acids produced during growth of fungi may chelate the associated cations such as Ca^{2+} with a resulting enhancement in solubilisation (Attiwill and Leeper, 1987). Fungi have been shown to excrete a range of organic acids, including oxalic, citric and gluconic acids, during growth and the ability of fungi to solubilise sparingly soluble phosphates in the presence of NO$_3^-$ nitrogen has been attributed to the acidifying and chelating abilities of these compounds (Lapeyrie et al., 1991).

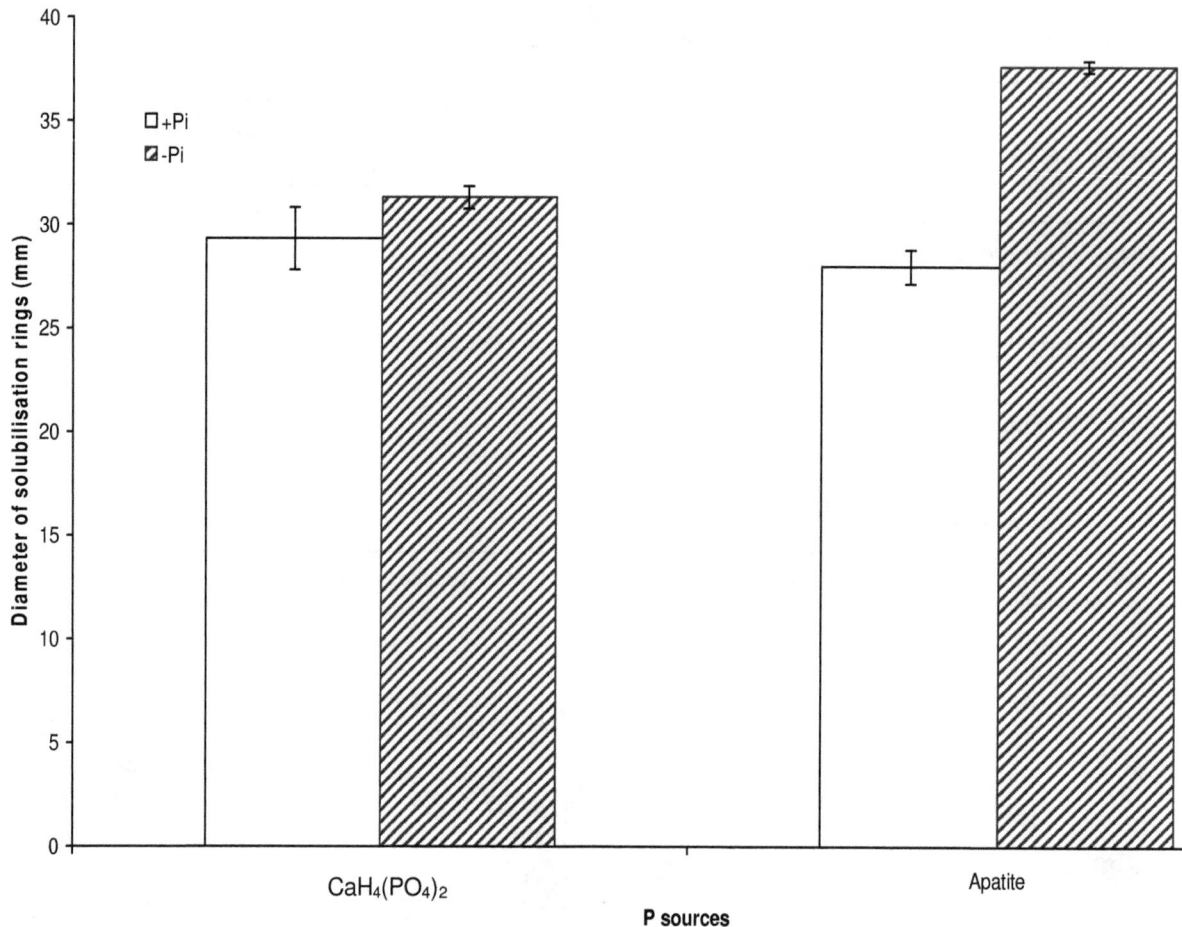

Figure 6. Variation in mean diameter (mm) of solubilisation rings following 23 days of growth of *Scleroderma* on MMN agar with insoluble hydroxyapatite and calcium hydrogen phosphate as only P source. Each value is a mean of three replicates. Vertical bars represent standard error of the mean. $CaH_4(PO_4)_2$ and Apatite on the X-axis represent calcium tetrahydrogen diorthophosphate and hydroxyapatite respectively. $\pm$ P in the legend represents the internal phosphorus status of mycelia inoculum, +P means phosphorus sufficient, -P means phosphorus starved.

Nonetheless, the resulting chelated complexes may have an adverse effect on the growth of the fungi in pure culture. Cairney and Smith (1992) found that inoculum P status does influence hyphal absorption of soluble phosphate. This defined the scope of this study to investigate the extent to which this might influence the abilities of *Scleroderma* and *Pisolithus* to solubilise the P sources tested. The internal P status of fungus inoculum had an effect on the solubilisation of hydroxyapatite but not of $CaPO_4$. Van Leerdam et al. (2001) did not observe any effect of P status of inoculum on solubilisation in their study. Although they concluded that solubilisation activities were not dependent on internal phosphorus concentrations, we believe from our assay that this may vary with P source and with the fungal species. The fact that *Scleroderma* could solubilise hydroxyapatite and calcium tetrahydrogen di-orthophosphate *in vitro* does not necessarily imply that its plant partner can utilise these P sources *in situ*. It would be necessary to demonstrate that solubilisation of a P source leads to uptake and eventual translocation to the host plant. Such a demonstration confirms that the host plant can utilise these P sources via their fungal partners.

During mycorrhization, both the plant and fungus undergo some physiological and biochemical changes (Smith and Read, 2008). These might affect their ability to access phosphorus in different forms. Although *in vitro* experiments give an idea of the different potentials of the fungi, we still cannot conclude as to what happens in their natural environment. To be able to do this, it would be necessary to look at the ability of the fungus-plant association to access different P sources in conditions quite similar to those of their natural environment.

ACKNOWLEDGEMENTS

The author acknowledges Professor Ian Alexander for his advice and Commonwealth Scholarship Scheme UK for her financial assistance.

REFERENCES

Attiwill PM, Leeper GW (1987). Forest soils and nutrient cycles. (Melbourne University Press: Melbourne).

Bechem EE (2004). Mycorrhizal Status of Gnetum spp. in Cameroon: evaluating diversity with a view to ameliorating domestication efforts. PhD thesis, University of Aberdeen, Scotland, UK.

Bolan NS (1991). A critical review on the role of mycorrhiza fungi in the uptake of phosphorus by plants. Plant and Soil, 134: 189-207.

Bolan NS, Robson AD, Barrow NJ (1987). Effects of vesicular-arbuscular mycorrhiza on the availability of iron phosphates to plants. Plant and Soil, 99: 401-410.

Cairney JWG, Smith SE (1992). Influence of intracellular phosphorus concentration on phosphate absorption by the ectomycorrhizal basidiomycete Pisolithus tinctorius. Mycol. Res., 96(8): 673-676.

Carter SR, Hartshorne NH (1923). The system of ferric oxide-phosphoric acid-water. A new phosphate. J. Chem. Soc., 123: 2223-2233.

Cate WE, Huffman EO, Deming ME (1959). Preparation of crystalline ferric phosphates. Soil Sci., 88: 130-132.

Jennings DH (1997). The physiology of fungal nutrition. Cambridge University Press, Cambridge.

Johnson AH, Frizano J, Vann DR (2003). Biogeochemical implications of labile phosphorus in forest soils determined by the Hedley fractionation procedure. Oecologia, 135: 487-499.

Jones D, Wilson MJ, Tait JM (1980). Weathering of a basalt by Pertusaria corallina. Lichenologist (Oxford), 12: 2770-289.

Koide RT, Kabir Z (2000). Extra radical hyphae of the mycorrhizal fungus Glomus intraradices can hydrolyse organic phosphate. The New Phytologist, 148: 511-517.

Lapeyrie F, Ranger J, Vairelles D (1991). Phosphate-solubilising activity of ectomycorrhiza fungi in vitro. Can. J. Bot., 69: 342-346.

Lapeyrie F, Chilvers GA, Behm CA (1987). Oxalic acid synthesis by the mycorrhizal fungus Paxillus involutus. New Phytologist, 106: 139-146.

Leyval C, Berthelin J (1986). Comparison between the utilization of phosphorus from insoluble mineral phosphates by ectomycorrhizal fungi and rhizobacteria. In: Physiological and Genetical Aspects of Mycorrhizae (ed. V.Gianinazzi-Pearson & S. Gianinazzi), pp. 340-345. INRA: Paris, France.

Marx DH (1969). The influence of ectotrophic mycorrhizal fungi on resistance of pine roots to pathogenic infections. I. Antagonism of mycorrhizal fungi to root pathogenic fungi and soil bacteria. Phytopathology, 59: 153-163.

Richardson AE (2001). Prospects for using soil microorganisms to improve the acquisition of phosphorus by plants. Aust. J. Plant Physiol., 28: 897-906.

Sanyal SK, De Datta SK (1991). Chemistry of phosphorus transformations in soil. Adv. Soil Sci., 16: 1-20.

Smith FA, Jakobsen I, Smith SE (2000). Spatial differences in acquisition of soil phosphate between two arbuscular mycorrhizal fungi in symbiosis with Medicago trunculata. New Phytologist, **147**: 357-366.

Smith SE, Read DJ (2008). Mycorrhiza Symbiosis, 3rd. ed. London , Academic Press.

Tibbett M, Sanders FE (2002). Ectomycorrhizal symbiosis can enhance plant nutrition through improved access to discrete organic nutrient patches of high resource quality. Ann. Bot., 89: 783-789.

Van Leerdam DM, Williams PA, Cairney JWG (2001). Phosphate – solubilising abilities of ericoid mycorrhizal endophytes of Woollsia pungens (Epacridaceae). Aust. J. Bot., 49: 75-80.

Villegas J, Fortin JA (2002). Phosphorus solubilisation and pH changes as a result of the interactions between soil bacteria and arbuscular mycorrhizal fungi on a medium containing NO_3^- as nitrogen source. Can. J. Bot., 80(5): 571-576.

29

Effect of carbon, nitrogen and trace elements on growth and sporulation of the *Termitomyces striatus* (Beeli) Heim

Simerjit Kaur

Department of Applied Sciences, Rayat-Bahra Institute of Engineering and Biotechnology, Mohali Campus (Punjab), India. E-mail: dr.simer07@yahoo.com.

Nutritional studies namely carbon, nitrogen and trace element requirements of *Termitomyces striatus* have been carried out. Amongst all the carbon compounds used, the most favourable in order of effectiveness are D (+) glucose, D (+) sucrose, maltose and D (+) raffinose. The fungus showed poor growth with lactose. Sodium nitrite served as the best inorganic nitrogen source for the growth of this fungus. Ammonium acetate, ammonium phosphate, ammonium oxalate, potassium nitrate and sodium nitrate supported fairly good growth of the fungus. Among different amino acids tested, maximum average mycelial dry weight was obtained with L-arginine followed by glycine and DL-tryptophan. The fungus showed poor growth with L-α-amino-n-butyric acid, L-cystine, L-cysteine HCl and DL-serine. The selected concentrations of trace elements also affected the mycelial growth of this fungus to a significant level. There is a gradual increase in growth (average mycelial dry weight) from control to optimum concentration of required trace elements beyond which it decreases. None of the trace elements required for growth are found to be completely fungistatic for the growth of this fungi. The fungus formed asexual conidia similar to that formed in the sporodochial stage in the termite garden of termites of the subfamily *Macrotermitinae*.

Key words: *Termitomyces,* edible mushrooms, cultural studies, C, N nutrition, mycelial growth.

INTRODUCTION

Edible mushrooms are very rich source of various active substances. Nutritional requirements for mushroom mycelium are relatively simple but various nutrient supplements play an essential role in metabolism as their coenzymes. Therefore, mycelial growth is a very much important factor that is responsible for fruiting in mushrooms (Pokhrel and Ohga, 2007; Pokhrel et al., 2009). The species of the paleotropical genus *Termitomyces* Heim are edible agarics, obligately symbiotic with termites belonging to the subfamily, Macrotermitinae. These are usually characterized by the termite association, prominent perforatorium on the pileus and the subterranean pseudorhiza connected to the comb in the termite nest. The sporodochia in the termitarium give rise to basidiome on the onset of rainy season. Review of literature reveals that a very little work has been done on nutritional requirements of this fungus.

Though this fungus is edible in nature and has been used as a food in most part of the world but it cannot be grown under laboratory conditions. This fungus grows only under natural conditions. Despite various *in vitro* investigations, only anamorphs formation has been recorded on some species on routine synthetic media (Botha and Eicker, 1991). For successful commercial cultivation is to determine the nutritional factors that are very much essential for mycelial growth and fruiting. Fruit body formation is influenced by different factors, including genetic make-up of the strain, environmental parameters and the nutrition of the medium (Pascoal et al., 2011). Therefore, studies were initiated to find out the optimum physical factors for the growth and sporulation of *Termitomyces striatus* (Beeli) Heim, to be employed later on for commercial exploitation of this fungus. Further investigations are in progress regarding this.

MATERIALS AND METHODS

Methods of morphological studies

For detailed microscopic examinations, the dried specimen of the

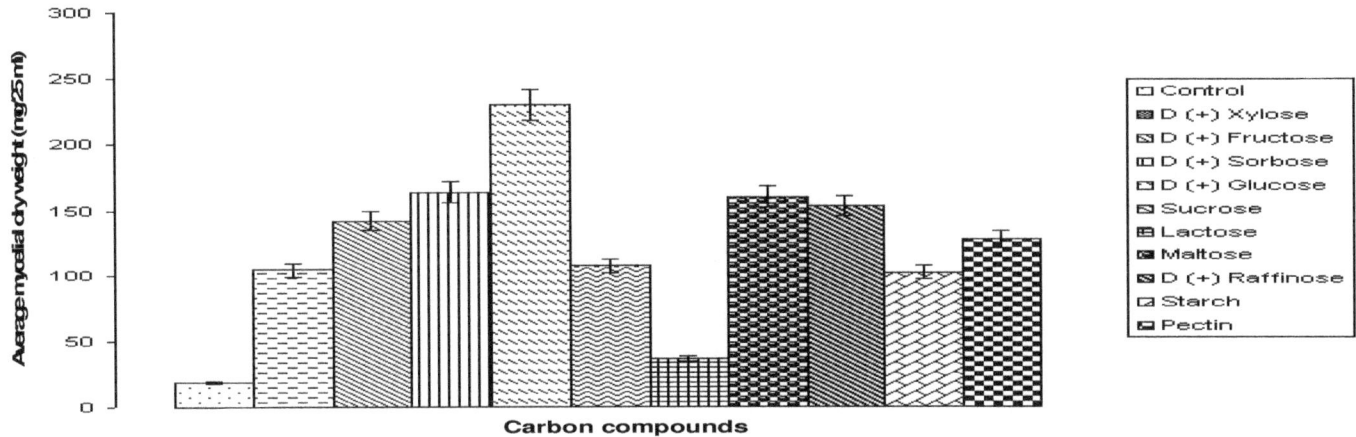

Figure 1. Growth (average mycelial dry wt.-mg/25ml) of *Termitomyces striatus* with different carbon compounds at optimum temperature (32 °C), pH (7.0) and days (12) of incubation. S.E.- 1.7490737, S.Ed.- 2.58732716, C.D (at 5% level)-5.11746543.

fungus was revived in 2% KOH for 10 to 15 min. The specimen was microscopically studied from macerations and free hand sections for various structures like basidiospore, basidia and cystidia. For determining the presence or absence of clamps and septation of hyphae, 1% aqueous solutions of Congo red in 5% ammonia and 2% aqueous solutions of phloxine were used. Cotton blue (lactic acid 50% 30 ml, cotton blue 0.05 g) was used to study the cyanophilous nature of the spores or hyphae after Le Gal (1947). To note the amyloidity of the spores, Melzer's reagent (Chloral-hydrate 22.0 g, potassium iodide 1.5 g, iodine 0.5 g, distilled water 20 ml) was used after Singer (1962). For studying the cystidia, sulphobenzaldehyde (water 1.5 ml, pure H_2SO_4, 5.0 ml and benzaldehyde 4.5 ml) was used after Slysh (1960).

Methods of physiological studies

The fungus was isolated from the pileus of the basidiome by single hyphal tip isolate technique and maintained on PDA (Potatoes 200 g, Dextrose 20 g, Agar 15 g and distilled water to make 1 L). Glucose–peptone medium (glucose 10 g, peptone 2 g, KH_2PO_4 1.5 g, $MgSo_4.7H_2O$ 0.5g/L), adjusted to pH 7.0 was used in all the experiments. To study the effect of carbon sources on mycelial growth, the quantity of carbon sources calculated on the basis of carbon present in 10 g of glucose was added to the basal medium but in the experiments on nitrogen sources utilization, the nitrogen sources were added in amount equivalent to that of peptone present in the basal medium.

The solution of each carbon and nitrogen sources was sterilized separately after adjusting to pH 7.0 (neutral) in order to avoid the possibility of their breakdown during autoclaving. Each of the solution was then added aseptically and proportionately to get the normal strength of the basal medium. A medium without carbon and nitrogen sources was used as control in the respective experiments. In the experiments to study the effect of trace elements on growth and sporulation of *T. striatus*, the peptone was replaced by sodium nitrite. Eight trace elements (Fe, Zn, Mn, Ca, Cu, Co, Mo and B) with selected concentrations (0.000001 to 400 ppm) were tested in the experiments on trace element requirements. A medium devoid of trace elements was used as control. The "materials and methods" are same as described by Prasher and Rawla (1988).

An aliquot of 25 ml of the medium was dispensed in each 250 Erlenmeyer flask, plugged and sterilized at 15 lbs psi steam pressure for 15 min. Each flask was seeded with 1 ml of standardized mycelial suspension having 2.5 mg dry weight/ml and incubated at optimum temperature (32 °C) for optimum days (13). At the termination of the experiment, the data were recorded on average mycelial dry weight (mg/25 ml), microscopic deficiency symptoms and biochemical deficiency symptoms (in terms of total sugars, soluble proteins and free amino acids) of each treatment.

Statistical analysis

The data on growth of all the experiments were analyzed statistically with respect to dry weight of the mycelium of individual replicate with variables by applying one-way ANOVA (SPSS 11 for windows) in terms of significance and non-significance of the data. The significance is denoted by statistical error (S.E.), statistical error of difference (S.Ed.) and critical difference (C.D.) at 5% level.

RESULTS

The study of Figure 1 reveals that out of 12 carbon sources tested for *T. striatus,* growth response of the fungus was different with various carbon sources. D (+) glucose was found to be best carbon source followed by D (+) sucrose, maltose, D (+) raffinose, D (-) fructose, pectin, sucrose, D (+) xylose and starch. The fungus showed poor growth with lactose. The study of Figures 2 and 3 reveals that inorganic sources of nitrogen had shown better growth of this fungus as compared to organic sources. Among inorganic nitrogen sources, sodium nitrite served as the best nitrogen source for the growth of this fungus. Ammonium acetate, ammonium phosphate, ammonium oxalate, potassium nitrate and sodium nitrate supplemented fairly good growth. L-arginine was the best nitrogen source followed by glycine and Dl-tryptophan. The fungus showed poor growth with L-α-amino-n-butyric acid, L-cystiene HCl, L-Cystine and Dl-serine.

The observation recorded in Figure 4 reveals that all the selected eight trace elements were required for the growth of this fungus. There was gradual increase in growth (average mycelial dry weight) from control to the optimum concentrations of the required elements, beyond

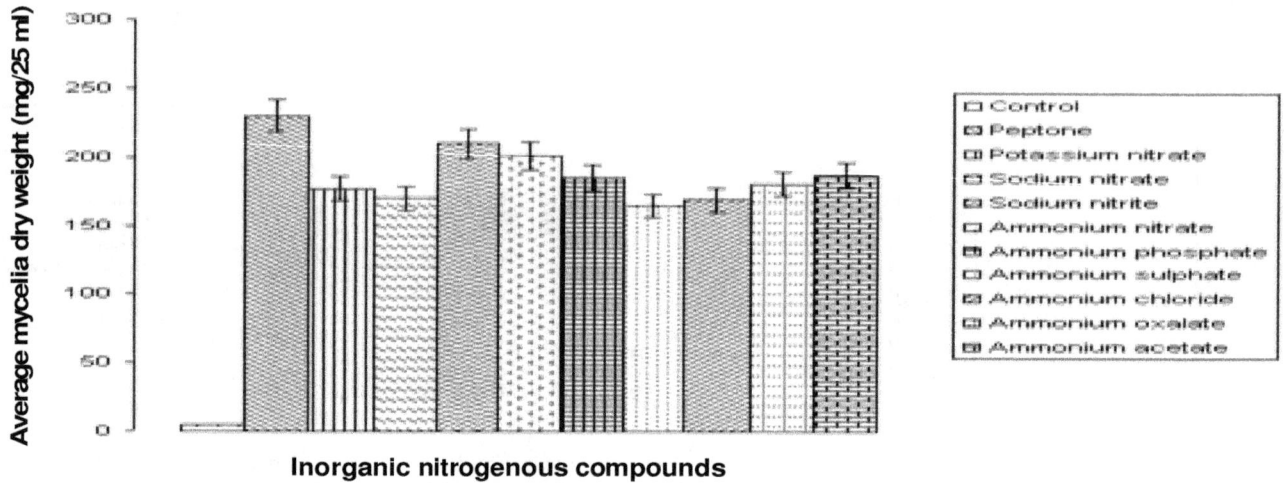

Figure 2. Growth (average mycelial dry weight-mg/25 ml) of *Termitomyces striatus* with different inorganic compounds at optimum temperature (32°C), pH (7.0) and days (12) of incubation. S.E.-1.76876524, S.Ed.-2.32379100, C. D. (at 5% level) -3.50479463.

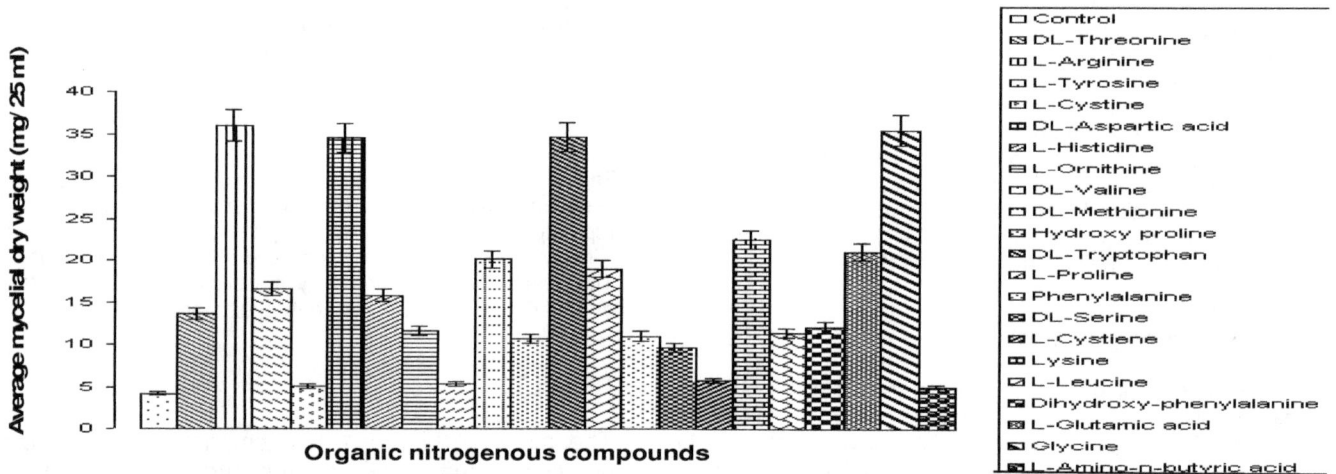

Figure 3. Growth (average mycelial dry weight-mg/25 ml) of *Termitomyces striatus* with different organic compounds at optimum temperature (32°C), pH (7.0) and days (12) of incubation. S.E.-1.4981263, S.Ed.-3.0301891, C.D. (at 5% level)-4.7484701.

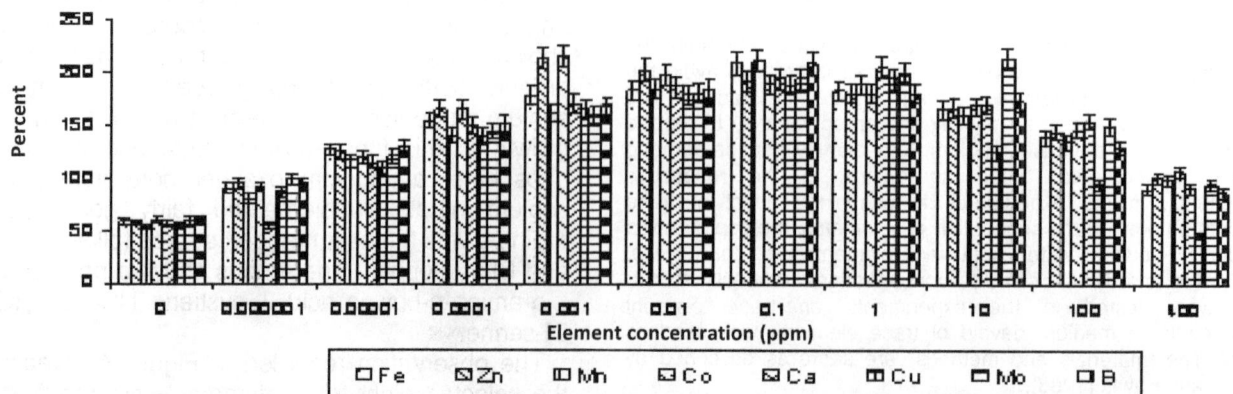

Figure 4. Growth (average mycelial dry wt.-mg/25ml) of *Termitomyces striatus* with selected concentrations of trace elements at its optimum temperature (32°C), pH (7.0) and days (12) of incubation.

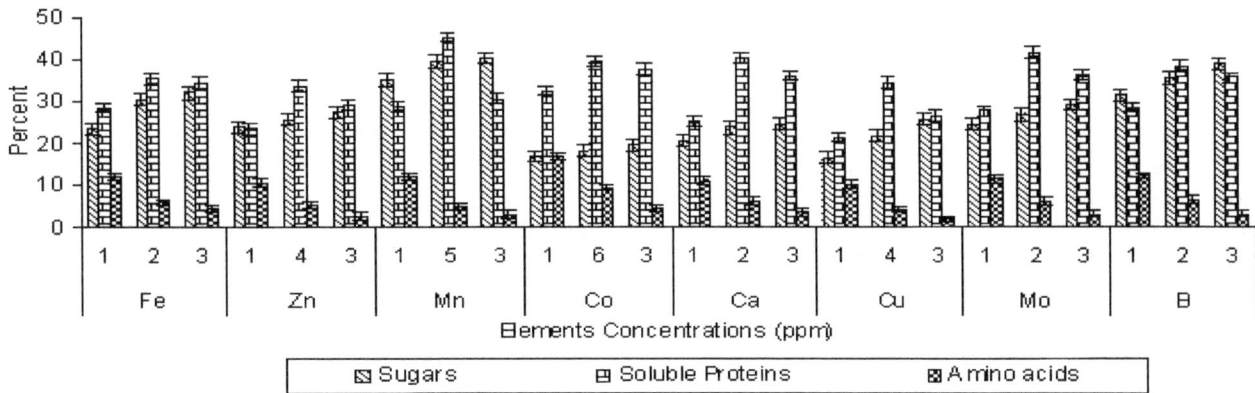

Figure 5. Relationship between the amount of total sugars, soluble proteins, free amino acids and the essential trace elements at their nil, optimum and inhibitory concentrations in the mycelium of *Termitomyces striatus*.

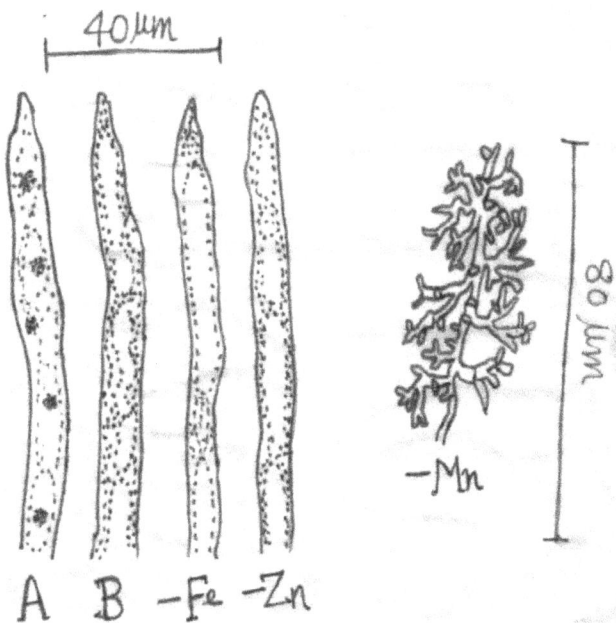

Figure 6. A) Basal medium supplemented with optimum concentration of all trace elements.B) Basal medium not supplemented with optimum concentration of all trace elements.

which it started decreasing. Zn, Co 0.001 ppm; Ca, Cu 1 and Mo 10; Mn, Fe and B 0.1 ppm supported optimum growth of the fungus. None of the trace elements required for the growth was found to be completely fungistatic for this fungus. The study of Figure 5 reveals that the amount of sugars decreased in *T. striatus* in the deficiency of required trace elements. It increased beyond optimum concentrations of the required trace elements. Likewise, amount of total soluble proteins also decreased in deficiency of trace elements. It however, decreased beyond optimum concentrations of trace elements required for growth. The amount of total free amino acids increased in the deficiency of trace elements

individually, as well as in control and decreased otherwise. The dry weight of the replicates did not vary significantly in all the experiments. The fungus showed trace element deficiency symptoms in the deficiency of Fe, Zn and Mn (Figure 6). No deficiency symptoms were shown by the fungus at the microscopic level in the deficiency of Ca, Mo, B, Cu and Co. The deficiency of Fe and Zn made the hyphae vacuolated whereas in the deficiency of Mn there was excessive hyphal branching near apex in the fungus.

DISCUSSION

In utilizing glucose, D (+) raffinose, D (-) fructose *T. striatus* resembles *Morchella hybrida* (Sharma, 2003) and *Ustilago esculenta* (Chung and Tzeng, 2004). In utilizing nitrite source of nitrogen, it resembles with edible fungi like *Morchella esculenta* and many other fungi (Morton and MacMillan, 1954; Dutt and Bedi, 1974). Nitrite is found to be toxic for most of the fungi and the poor growth of many fungi in nitrite was attributed to the toxic effect exerted by the pyruvic acid accumulated in the mycelium (Nord and Mull, 1945). But *T. striatus* has shown maximum growth on sodium nitrite. Detailed investigations with certain fungi have previously indicated that hyphal output was maximum at alkaline pH and the fungi attained good growth on nitrite nitrogen if initial pH was adjusted to 6.5-7.5. More growth on alkaline nitrite medium suggests that it is the free unionized acid rather than nitrite ion which is toxic to fungi (Bilgrami and Verma, 1978). Varied utilization of amino acids tested during these investigations is in accordance with many other workers (Bilgrami and Verma, 1978; Sharma, 2003; Jonathan and Fasidi, 2001; Chung and Tzeng, 2004).

In requiring Fe, Zn and Mn, this fungus resembles many other fungi (Prasher and Rawla, 1988; Carlile et al., 2001; Sharma et al., 2005). The fungus shows interesting behaviour in requiring Co becausee no member of basidiomycota, so far, has been reported to require Co for its growth, except only a few fungi belonging to other groups have been reported to require Co (Madan and Thind, 1979; Singh, 1979; Prasher and Rawla, 1988). The excessive branching in the deficiency of Mn in this fungus indirectly reflects the occurrence of factor/factors promoting, excessive branching which most probably might be concerned with softening of hyphal wall. Vacuolization of the hyphae with age or under stress is a natural phenomenon among fungi. Biochemical changes include a decrease in total sugars, soluble proteins and increase in free amino acids. The decreased sugars and soluble proteins suggest the marked effect on their synthesis and the enzymes concerned in their metabolism. This is further evidenced by the increase in total free amino acids in the trace element deficient cultures.

T. striatus reproduced asexually producing conidia with carbon sources, inorganic nitrogen sources and all the selected trace elements but it has shown nil sporulation with amino acids. Fe, Zn, Mn and Ca are essential for conidial production whereas Boron is promotary for conidial production. The role of Fe, Zn, Mn and Ca in conidial production is in agreement with earlier findings of Prasher and Rawla (1988). The development of anamorph in cultures of *T. striatus* under optimum physical conditions, which have been obtained from basidiome of the fungus in the present study and similar findings of development of anamorph in cultures obtained from basidiome context (Botha and Eicker, 1991) are indicative of the fact that factor/s other than studied during investigations may be required for the stimulation of the basidiome context cultures to revert to the perfect phase. Studies are in progress in this direction.

ACKNOWLEDGEMENTS

The author is thankful to the Department of Botany, Panjab University, Chandigarh, India, for providing the laboratory facilities and valuable technical assistance to carry out the present work.

REFERENCES

Bilgrami KS, Verma RN (1978). *Physiology of Fungi.* Vikas Publishing house Pvt. N. Delhi, p. 507.

Botha WJ, Eicker A (1991). Cultutral studies on the genus *Termitomyces* in South Africa. 1. Macro and Microscopic characters of basidiome & context culture. Mycol. Res., 95: 435-443.

Carlile MJ, Watkinson SC, Gooday GW (2001). *The Fungi.* Academic Press, New York, p. 598.

Chung KR, Tzeng DD (2004). Nutritional requirements of the edible gall-producing fungus *Ustilago esculenta.* J. Biol. Sci., 4(2): 246-252.

Dutt S, Bedi PS (1974). Effect of carbon and nitrogen nutrition on the growth and sporulation of *Helminthosporium speciferum.* Indian J. Mycol. Plant Pathol., 4: 190-193.

Jonathan SG, Fasidi IO (2001). Effect of carbon, nitrogen and mineral sources on growth of *Psathyerella atroumbonata* Pegler, a Nigerian edible mushroom. Food Chem., 72: 479-483.

Le Gal (1947). Researches Sur Les ornamentation sporales des discmycetes opercules. Ann. Sci. Nat. Bot., 11: 118-193.

Madan M, Thind KS (1979). Role of trace elements on the growth and sporulation of *Alternaria chartarum* and *A. solani.* Proc. Indian Nat. Sci. Acad. (B), 45: 628-632.

Morton AG, MacMillan A (1954). The assimilation of nitrogen from ammonium salts and nitrate by fungi. J. Exp. Bot., 5: 232-252. fide. Bilgrami and Verma (1978). Physiology of Fungi.

Nord FF, Mull RP (1945). Recent progress in the biochemistry of *Fusaria.* In: Advances in Enzymology ed by Nord, F.F. and Werkmen, C.H., 5: 165-205.

Pascoal JGJ, Marcia MT, Rosane FS, Danny LR, Eustaquio SD (2011). Nutritional requirements for growth of *Agaricus brasiliensis.* Maringa, 33: 93-97.

Pokhrel CP, Ohga S (2007). Submerged culture conditions for mycelia yield and polysaccharides production by *Lyophyllum decastes.* Food Chem., 105: 641-646.

Pokhrel CP, Yadav RKP, Ohga S (2009). Fourier transform infrared spectroscopic study on glycolkaloid concentration in varieties of *solanum tuberosum.* J. Ecobiotech., 1(1): 46-49.

Prasher IB, Rawla GS (1988). Trace element requirements of some members of Saprolegniaceae. In Rawla (Ed.) Advances in Mycology. Panjab University, pp. 224-236.

Sharma M (2003). Nutritional studies on *Morchella esculenta* (Sow) Pers. M.Phil thesis. Panjab University, p. 24.

Sharma VP, Sharma SR, Kumar S (2005). Nutritional requirements for mycelial growth and cultivation of *Flamulina velutipes.* Mushroom Res., 14(1): 13-18.

Singh S (1979). Studies on the nutrition of some members of Ustilaginales. Ph.D. Thesis, Panjab University, Chandigarh, India.

Singer R (1962). Mushrooms and Truffles. Interscience Publishers Inc. New York.

Slysh AR (1960). The genus *peniophora* in New York State and adjacent regions. St. Univ. College of Forestry Tech. Publ., 83-95 pp.

Permissions

All chapters in this book were first published in JYFR, by Academic Journals; hereby published with permission under the Creative Commons Attribution License or equivalent. Every chapter published in this book has been scrutinized by our experts. Their significance has been extensively debated. The topics covered herein carry significant findings which will fuel the growth of the discipline. They may even be implemented as practical applications or may be referred to as a beginning point for another development.

The contributors of this book come from diverse backgrounds, making this book a truly international effort. This book will bring forth new frontiers with its revolutionizing research information and detailed analysis of the nascent developments around the world.

We would like to thank all the contributing authors for lending their expertise to make the book truly unique. They have played a crucial role in the development of this book. Without their invaluable contributions this book wouldn't have been possible. They have made vital efforts to compile up to date information on the varied aspects of this subject to make this book a valuable addition to the collection of many professionals and students.

This book was conceptualized with the vision of imparting up-to-date information and advanced data in this field. To ensure the same, a matchless editorial board was set up. Every individual on the board went through rigorous rounds of assessment to prove their worth. After which they invested a large part of their time researching and compiling the most relevant data for our readers.

The editorial board has been involved in producing this book since its inception. They have spent rigorous hours researching and exploring the diverse topics which have resulted in the successful publishing of this book. They have passed on their knowledge of decades through this book. To expedite this challenging task, the publisher supported the team at every step. A small team of assistant editors was also appointed to further simplify the editing procedure and attain best results for the readers.

Apart from the editorial board, the designing team has also invested a significant amount of their time in understanding the subject and creating the most relevant covers. They scrutinized every image to scout for the most suitable representation of the subject and create an appropriate cover for the book.

The publishing team has been an ardent support to the editorial, designing and production team. Their endless efforts to recruit the best for this project, has resulted in the accomplishment of this book. They are a veteran in the field of academics and their pool of knowledge is as vast as their experience in printing. Their expertise and guidance has proved useful at every step. Their uncompromising quality standards have made this book an exceptional effort. Their encouragement from time to time has been an inspiration for everyone.

The publisher and the editorial board hope that this book will prove to be a valuable piece of knowledge for researchers, students, practitioners and scholars across the globe.

List of Contributors

Juan Carlos González-Hernández
Laboratorio de Bioquímica del Departamento de Ingeniería Bioquímica, Instituto Tecnológico de Morelia; Avenida Tecnológico 1500. C. P. 58120. Morelia, Michoacán, México

Thomas S. Davis
Northern Arizona University, School of Forestry, P. O. BOX 15018, Flagstaff, Arizona, USA, 86011

Richard W. Hofstetter
Northern Arizona University, School of Forestry, P. O. BOX 15018, Flagstaff, Arizona, USA, 86011

Kier D. Klepzig
USDA Forest Service, Southern Research Station, Pineville, Louisiana, USA

Jeffrey T. Foster
Center for Microbial Genetics and Genomics, Northern Arizona University, Flagstaff, Arizona, USA

Paul Keim
Center for Microbial Genetics and Genomics, Northern Arizona University, Flagstaff, Arizona, USA

Mukaram Shikara
Biotechnology Division, Applied Sciences Department, University of Technology, Baghdad, Iraq

Avijgan Majid
Iranian Traditional Medicine research Center, Isfahan University of Medical Sciences, Isfahan, Iran

Mahboubi Mohaddesse
Barij Essence pharmaceutical Company, Isfahan, Kashan, Iran

Darabi Mahdi
Barij Essence pharmaceutical Company, Isfahan, Kashan, Iran

Saadat Mahdi
Shahr-e-kord University of Medical Sciences, Shahr-e-kord, Iran

Sarikhani Sanaz
Iranian Traditional Medicine research Center, Isfahan University of Medical Sciences, Isfahan, Iran

Nazilla Kassaiyan
Iranian Traditional Medicine research Center, Isfahan University of Medical Sciences, Isfahan, Iran

A. H. M. El-Said
Department of Botany, Faculty of Science, South Valley University, Qena, Egypt

T. H. Sohair
Department of Botany, Faculty of Science, South Valley University, Qena, Egypt

A. G. El-Hadi
Department of Botany, Faculty of Science, Az-Zawiyah University, Libya

Ravinder Nagpal
Dairy Microbiology Division, National Dairy Research Institute, Karnal 132001, (Haryana), India

Anil Kumar Puniya
Dairy Microbiology Division, National Dairy Research Institute, Karnal 132001, (Haryana), India

Jatinder Paul Sehgal
Dairy Cattle Nutrition Division, National Dairy Research Institute, Karnal 132 001 (Haryana), India

Kishan Singh
Dairy Microbiology Division, National Dairy Research Institute, Karnal 132001, (Haryana), India

Prashant Kumar Lathar
Department of Biotechnology, Jay Pee University of Information Technology, Waknaghat 173234, Solan (H. P), India

Arti Sharma
Department of Biotechnology, Jay Pee University of Information Technology, Waknaghat 173234, Solan (H. P), India

Isha Thakur
Department of Biotechnology, Jay Pee University of Information Technology, Waknaghat 173234, Solan (H. P), India

E. I. Eziashi
Nigerian Institute for Oil Palm Research (NIFOR), Plant Pathology Division, P. M. B. 1030, Benin City, Edo State, Nigeria

I. B. Omamor
Nigerian Institute for Oil Palm Research (NIFOR), Plant Pathology Division, P. M. B. 1030, Benin City, Edo State, Nigeria

C. E. Airede
Nigerian Institute for Oil Palm Research (NIFOR), Plant Pathology Division, P. M. B. 1030, Benin City, Edo State, Nigeria

C. V. Udozen
Nigerian Institute for Oil Palm Research (NIFOR), Plant Pathology Division, P. M. B. 1030, Benin City, Edo State, Nigeria

N. Chidi
Nigerian Institute for Oil Palm Research (NIFOR), Plant Pathology Division, P. M. B. 1030, Benin City, Edo State, Nigeria

Donatha Damian Tibuhwa
Department of Molecular Biology and Biotechnology, University of Dar es Salaam, P.O. Box 35179, Dar es Salaam, Tanzania

Emira Noumi
Laboratoire d'Analyse, Traitement et Valorisation des Polluants de l'Environnement et des Produits, Département de Microbiologie, Faculté de Pharmacie, Monastir, Tunisia
Departamento de Microbiología y Ecología, Facultad de Farmacia, Universidad de Valencia, Burjassot, Valencia, Spain

Mejdi Snoussi
Laboratoire d'Analyse, Traitement et Valorisation des Polluants de l'Environnement et des Produits, Département de Microbiologie, Faculté de Pharmacie, Monastir, Tunisia
Laboratoire de Traitement et de Recyclage des Eaux. Centre de Recherches et des Technologies des eaux, Technopôle de Borj-Cédria, BP 901, 2050 Hammam-Lif, Tunisia

Maria del Pilar Vercher
Departamento de Microbiología y Ecología, Facultad de Farmacia, Universidad de Valencia, Burjassot, Valencia, Spain

Eulogio Valentin
Departamento de Microbiología y Ecología, Facultad de Farmacia, Universidad de Valencia, Burjassot, Valencia, Spain

Lucas DelCastillo
Departamento de Microbiología y Ecología, Facultad de Farmacia, Universidad de Valencia, Burjassot, Valencia, Spain

Amina Bakhrouf
Laboratoire d'Analyse, Traitement et Valorisation des Polluants de l'Environnement et des Produits, Département de Microbiologie, Faculté de Pharmacie, Monastir, Tunisia

Rohit Sharma
Department of Biological Sciences, Mycological Research Laboratory, R. D. University, Jabalpur- 482 001, Madhya Pradesh, India
Microbial Culture Collection, Affiliated to National Centre for Cell Science, University of Pune, Ganeshkhind, Pune- 411 007, Maharashtra, India

Ram. C. Rajak
Department of Biological Sciences, Mycological Research Laboratory, R. D. University, Jabalpur- 482 001, Madhya Pradesh, India

Akhilesh. K. Pandey
Department of Biological Sciences, Mycological Research Laboratory, R. D. University, Jabalpur- 482 001, Madhya Pradesh, India

Aiping Zheng
Rice Research Institute, Sichuan Agricultural University, Wenjiang 611130, China

Yanran Wang
Rice Research Institute, Sichuan Agricultural University, Wenjiang 611130, China

Neveen S. M. Soliman
Department of Food Hygiene, Faculty of Veterinary Medicine, Cairo University, Egypt

Salwa A. Aly
Department of Food Hygiene, Faculty of Veterinary Medicine, Cairo University, Egypt

Vagner G. Cortez
Universidade Federal do Paraná, Rua Pioneiro 2153, Jardim Dallas, CEP 85950-000, Palotina, PR, Brazil

Iuri G. Baseia
Departamento de Botânica, Universidade Federal do Rio Grande do Sul, Av. Bento Gonçalves 9500, CEP 91501-900, Porto Alegre, RS, Brazil

Rosa Mara B. Silveira
Departamento de Botânica, Ecologia, Universidade Federal do Rio Grande do Norte, e Zoologia, CEP 59072-970, Natal, RN, Brazil

V. B. Hosagoudar
Tropical Botanic Garden and Research Institute, Palode - 695 562, Thiruvananthapuram, Kerala, India

J. Thomas
Tropical Botanic Garden and Research Institute, Palode - 695 562, Thiruvananthapuram, Kerala, India

D. K. Agarwal
Plant pathology Division, IARI, New Delhi 110 012, India

T. Panda
Department of Botany, S. N. College, Rajkanika 754 220, Kendrapara, Orissa, India

Karim Dagno
Unit of Plant Pathology, University of Liege, Gembloux Agro Bio Tech, Passage des Déportés 2, B-5030 Belgium

Julien Crovadore
Plants and Pathogens Group, Institute Earth Nature and Landscape, University of Applied Sciences of Western Switzerland, 150 route de Presinge, Jussy, Geneva 1254, Switzerland

François Lefort
Plants and Pathogens Group, Institute Earth Nature and Landscape, University of Applied Sciences of Western Switzerland, 150 route de Presinge, Jussy, Geneva 1254, Switzerland

Rachid Lahlali
Agriculture and Agri-Food Canada, Saskatoon Research Centre, 107 Science Place, S7N0X2, Canada

Ludivine Lassois
Unit of Plant Pathology, University of Liege, Gembloux Agro Bio Tech, Passage des Déportés 2, B-5030 Belgium

M. Haïssam Jijakli
Unit of Plant Pathology, University of Liege, Gembloux Agro Bio Tech, Passage des Déportés 2, B-5030 Belgium

Suhaib A. Bandh
Centre of Research for Development, University of Kashmir Srinagar-190006, India

Azra N. Kamili
Centre of Research for Development, University of Kashmir Srinagar-190006, India

Bashir A. Gana
Department of Biochemistry, University of Kashmir Srinagar-190006, India

Samira Saleem
Centre of Research for Development, University of Kashmir Srinagar-190006, India

Bashir A. Lone
Centre of Research for Development, University of Kashmir Srinagar-190006, India

Humera Nissa
Centre of Research for Development, University of Kashmir Srinagar-190006, India

Valerie Gonsalves
Department of Microbiology, Goa University, Taleigao Plateau, Goa-403206, India

Shweta Nayak
Department of Microbiology, Goa University, Taleigao Plateau, Goa-403206, India

Sarita Nazareth
Department of Microbiology, Goa University, Taleigao Plateau, Goa-403206, India

Radha Pyati
Department of Chemistry, University of North Florida, Jacksonville, FL 32224, USA

Lindsay L. Elvir
Department of Chemistry, University of North Florida, Jacksonville, FL 32224, USA

Erica C. Charles
Department of Chemistry, University of North Florida, Jacksonville, FL 32224, USA

Umawattee Seenat
Department of Chemistry, University of North Florida, Jacksonville, FL 32224, USA

Tom D. Wolkow
Department of Biology, University of Colorado at Colorado Springs, Colorado Springs, CO 80933, USA

Meriem Amina Rezki
Laboratoire de Biotechnologie des Rhizobia et Amélioration des Plantes (LBRAP), Département de Biotechnologie, Faculté des Sciences, Université d'Es-Senia, BP 1524 El Mnaouar 31 000 Oran, Algérie
Université de Toulouse; INSA, UPS, Laboratoire d'Ingénierie des Systèmes Biologiques et des Procédés, INRA- UMR792 and CNRS-UMR5504; 135 Avenue de Rangueil; F-31400 Toulouse, France

Laurent Benbadis
Université de Toulouse; INSA, UPS, Laboratoire d'Ingénierie des Systèmes Biologiques et des Procédés, INRA- UMR792 and CNRS-UMR5504; 135 Avenue de Rangueil; F-31400 Toulouse, France

Gustavo DeBillerbeck
Université de Toulouse; INSA, UPS, Laboratoire d'Ingénierie des Systèmes Biologiques et des Procédés, INRA- UMR792 and CNRS-UMR5504; 135 Avenue de Rangueil; F-31400 Toulouse, France
INP-ENSAT, Avenue de l'Agrobiopole, F-31326 Castanet-Tolosan Cedex, France

Zoubida Benbayer
Laboratoire de Biotechnologie des Rhizobia et Amélioration des Plantes (LBRAP), Département de Biotechnologie, Faculté des Sciences, Université d'Es-Senia, BP 1524 El Mnaouar 31 000 Oran, Algérie

Jean Marie François
Université de Toulouse; INSA, UPS, Laboratoire d'Ingénierie des Systèmes Biologiques et des Procédés, INRA- UMR792 and CNRS-UMR5504; 135 Avenue de Rangueil; F-31400 Toulouse, France

Kathiresan Kandasamy
Centre of Advanced Study in Marine Biology, Annamalai University, Parangipettai: 608502, Tamil Nadu, India

Nabeel M. Alikunhi
Centre of Advanced Study in Marine Biology, Annamalai University, Parangipettai: 608502, Tamil Nadu, India

Manivannan Subramanian
Centre of Advanced Study in Marine Biology, Annamalai University, Parangipettai: 608502, Tamil Nadu, India

N.S. AFSHAN
Centre for Undergraduate Studies, University of the Punjab Quaid-e-Azam Campus, Lahore, 54590, Pakistan

A. N. KHALID
Department of Botany, University of the Punjab Quaid-e-Azam Campus, Lahore, 54590, Pakistan

A. R. NIAZI
Department of Botany, University of the Punjab Quaid-e-Azam Campus, Lahore, 54590, Pakistan

Shumin Yao
College of Life Science, Qufu Normal University, Qufu 273165, People's Republic of China

Jing Huang
College of Life Science, Qufu Normal University, Qufu 273165, People's Republic of China

Jingli Tan
College of Life Science, Qufu Normal University, Qufu 273165, People's Republic of China

Fekadu Alemu
Department of Biology, College of Natural and Computational Sciences, Dilla University, P.O.Box., 419, Dilla, Ethiopia

A. Sosa
Instituto de Ciencia Animal, Apartado Postal 24, San José de las Lajas, La Habana, Cuba

E. Kobashigawa
Departamento de Engenharia de Alimentos, Faculdade de Zootecnia e Engenharia de Alimentos, Universidade de São Paulo

J. Galindo
Instituto de Ciencia Animal, Apartado Postal 24, San José de las Lajas, La Habana, Cuba

R. Bocourt
Instituto de Ciencia Animal, Apartado Postal 24, San José de las Lajas, La Habana, Cuba

C. H. Corassin
Departamento de Engenharia de Alimentos, Faculdade de Zootecnia e Engenharia de Alimentos, Universidade de São Paulo

C. A. F. Oliveira
Departamento de Engenharia de Alimentos, Faculdade de Zootecnia e Engenharia de Alimentos, Universidade de São Paulo

Luis Cesar Maffei Sartini Paulillo
Faculdade de Tecnologia e Ciências - FTC, Mestrado Profissional em Bioenergia. Av. Luis Viana Filho, 8812, Pituaçu, Salvador, CEP 41.741-490, BA - Brasil

Sizenando José de Andrade Junior
PPG Biotecnologia UEFS/Fiocruz-BA. Av. Transnordestina s.nº, Novo Horizonte, Feira de Santana, – CEP 44.036-900, BA – Brasil

Abelmon da Silva Gesteira
Embrapa Mandioca e Fruticultura Tropical. Rua Embrapa, s/nº. Cruz das Almas, CEP 44380-000, BA – Brasil

Marco Brotto
Muscle Biology Research Group-MUBIG, Schools of Nursing, Medicine, and Biological Sciences, University of Missouri-Kansas City, Kansas City, 64108. Missouri. USA International Visiting Professor and Researcher, Conselho Nacional de Pesquisa-CNPq and Faculdade de Tecnologia e Ciências-FTC, Salvador, Brasil

Aristóteles Góes-Neto
PPG Biotecnologia UEFS/Fiocruz-BA. Av. Transnordestina s.nº, Novo Horizonte, Feira de Santana, – CEP 44.036-900, BA – Brasil

Eneke Esoeyang Tambe Bechem
Department of Plant and Animal Sciences, Faculty of Science, University of Buea, P. O. Box 63 Buea, Cameroon

Simerjit Kaur
Department of Applied Sciences, Rayat-Bahra Institute of Engineering and Biotechnology, Mohali Campus (Punjab), India